NASA OFFICE OF SPACE SCIENCE EDUCATION
AND
PUBLIC OUTREACH CONFERENCE

COVER ILLUSTRATION:

The Sagittarius Star Cloud.

Credit: Hubble Heritage Team (AURA/STScI/NASA).

ASTRONOMICAL SOCIETY OF THE PACIFIC CONFERENCE SERIES

A SERIES OF BOOKS ON RECENT DEVELOPMENTS IN
ASTRONOMY AND ASTROPHYSICS

Volume 319

EDITORIAL STAFF

Managing Editor: J. W. Moody
Publication Manager: Enid L. Livingston
Computer Tech/Trouble Shooter: Andrew W. Johnson
Electronic Publication Specialist/e-books: Lisa B. Roper

PO Box 24463, Room 211 - KMB, Brigham Young University, Provo, Utah, 84602-4463
Phone: (801) 422-2111   Fax: (801) 422-0624   E-mail: aspcs@byu.edu

LaTeX Consultant: T. J. Mahoney (Spain) – tjm@ll.iac.es

PUBLICATION COMMITTEE:

Joss Bland-Hawthorn
George Jacoby
James B. Kaler
J. Davy Kirkpatrick

A listing of all other ASP Conference Series and IAU volumes published by the ASP may be found at the back of this volume.

ASTRONOMICAL SOCIETY OF THE PACIFIC
CONFERENCE SERIES

Volume 319

NASA OFFICE OF SPACE SCIENCE EDUCATION AND PUBLIC OUTREACH CONFERENCE

Proceedings of a meeting held at the
Union League Club of Chicago, Chicago, Illinois, USA
12-14 June, 2002

Edited by

Carolyn Narasimhan
DePaul University, Chicago, Illinois, USA

Bernhard Beck-Winchatz
DePaul University, Chicago, Illinois, USA

Isabel Hawkins
University of California Berkeley, Berkeley, California, USA

and

Cassandra Runyon
College of Charleston, Charleston, South Carolina, USA

SAN FRANCISCO

ASTRONOMICAL SOCIETY OF THE PACIFIC

390 Ashton Avenue
San Francisco, California, 94112-1722, USA
Phone: 415-337-1100
Fax:415-337-5205
E-mail: service@astrosociety.org
Web Site: www.astrosociety.org

All Rights Reserved
© 2004 by Astronomical Society of the Pacific.
ASP Conference Series - First Edition

No part of the material protected by this copyright notice may be reproduced or utilized in any form or by any means – graphic, electronic, or mechanical including photocopying, taping, recording or by any information storage and retrieval system, without written permission from the Astronomical Society of the Pacific .

ISBN: 1-58381-181-8

Library of Congress Cataloging in Publication Data
Main entry under title
Card Number: 2004113062

Printed in United States of America by Sheridan Books, Ann Arbor, Michigan

For Jeffrey Rosendhal

Contents

Dedication	v
Contents	vii
Preface	xvii
Letter from the Editors	xix
Participants	xxi

Opening Remarks — 1

Welcome Message	3
J. D. Rosendhal	
NASA Space Science – Our Commitment to E/PO	5
E. J. Weiler	
Knock, Knock ... Who's There? Looking into Our Schools	17
M. A. Lopez Freeman	

Session on Science Education Research — 23

Context for the Panel on Science Education Research	25
The Evaluation of Adult Science Learning	26
J. D. Miller	
Educational Research: the Example of Light and Color	35
P. M. Sadler	
Is the Teacher of Science a Scientist?	44
S. A. Lee	

Sessions on Issues in Formal and Informal Science Education — 49

Context for the Panels on Formal and Informal Science Education	51
Issues in Formal Science Education	53
Keynote Address	55
G. D. Nelson	
Summary of Formal Education Discussions	73
T. J. Teays	
Professional Development For SOFIA's E/PO Program	77
E. K. DeVore and M. Bennett	

The Committee for Action Program Services (CAPS)   85
    L. Davis
The NASA Student Involvement Program   87
    F. Mahootian
Systemic Change in Delaware and Its Use of NASA-Generated
    Materials   88
    H. Shipman
Building a Technology Culture in 29 Chicago Schools   94
    D. G. York
Achieving Coherence, Effectiveness and Widespread Usage in NASA-
    Related Standards-Based Curriculum Development   95
    J. Barber
New Opportunities Through Minority Initiatives in Space Science .   102
    A. J. Kawakami and R. Crowe
Guide to Inquiry-Based Learning Using NASA Resources   107
    D. Barstow
Yerkes Observatory Professional Development Programs   109
    V. Hoette
Astronomy in the Ice: Bringing Neutrino Astronomy to the Secondary
    Schools   111
    J. Madsen, R. Atkins, M. Briggs, J. Cooley, B. Dingus,
    F. Halzen, S. Millar, T. Millar, K. Rawlins, and D. Steele
Research-Based Public Outreach and Education   114
    G. Bothun
NASA/JPL Solar System Educators Program   115
    E. Brunsell
Empowering Teachers to Address Space Science Content Standards in
    the Classroom   120
    T. G. Guzik, R. McNeil, and E. Babin
A Research Based Approach to Developing Engaging Classroom Ready
    Curriculum   125
    E. E. Prather and T. F. Slater
Exploring the Solar System: A Science Enrichment Course for Gifted
    Elementary School Students   130
    W. S. Kiefer, R. R. Herrick, A. H. Treiman, and P. B. Thompson
Issues in Formal Education as Systemic Programs   133
    G. Pollock

Issues in Informal Science Education   137
Keynote Address   139
    P. Knappenberger
Summary of Informal Education Discussions   147
    I. Hawkins
Creating Full-Dome Experiences in the New Digital Planetarium .   155
    C. Sumners and P. Reiff
Coyote, NASA, and other Informal Educators   160
    L. Moroney

Creating Authentic and Compelling Web-Based Science Experiences for the Public: Examples from *Live @ The Exploratorium* Series   163
R. Semper
Understanding Planetariums as Media   165
J. Sweitzer
The Big Picture: Planetariums, Education and Space Science . . .   169
R. Wyatt
Planetariums in K-12 Science Education   174
S. Laatsch
An Urban Partnership: Inservice and Science Enrichment Programs for a Hispanic Serving Charter School   175
P. A. Morris, O. G. Garza, M. Lindstrom, J. Allen, J. Wooten, and V. D. Obot
Windows to the Universe – A Web Resource Spanning Formal and Informal Science Education   178
R. M. Johnson, C. J. Alexander, J. J. Bergman, C. R. Deardorff, L. Gardiner, J. Genyuk, S. Henderson, M. LaGrave, D. Mastie, and R. Russell
The International Planetarium Society   182
M. Ratcliffe
The Accessible Universe: Making Space Science Accessible to People with Special Needs   185
N. Grice
Collaborative Support for Solar Eclipse 2001 Activities   187
V. L. Thomas, G. R. Carruthers, and E. Takamura
Reaching the Spanish-speaking Community through the Adler Course Program   191
J. F. Salgado
Spacelink: Providing Updates in Space Science and Astronomy . .   194
F. J. Mendez and J. P. O'Leary
The Small Planetarium: Its Nature, Mission, and Needs   196
J. E. Bishop
Journey through the Universe – Taking Underserved Communities to the Frontier   200
J. Goldstein, M. Bobrowsky, T. Livengood, S. Smith, and B. Riddle
StarDate/*Universo* Astronomy for the Masses   202
S. Preston and M. K. Hemenway

Session on Scientists' Participation in Education and Public Outreach   203

Context for the Panel on Scientists' Participation in Education and Public Putreach   205
Space Science Education: The Emergence of a Professional Community   207
R. E. Lopez

Scientists' Participation in Education and Outreach 214
  A. Fraknoi
Impressions from the OSS Conference on Education and Outreach 220
  C. H. McGruder
The Importance of Involving Research Scientists in Education and
  Outreach 221
  L. F. Fortson

Poster Session 229
Summary of the Poster Session 231
  B. Beck-Winchatz and C. Runyon
NASA/MSFC/NSSTC Science Communication Roundtable 235
  M. Adams, D. L. Gallagher, and R. Koczor
Issues in Informal Education: Event-Based Science Communication
  Involving Planetaria and the Internet 237
  M. Adams, D. L. Gallagher, and A. Whitt
Student Participation in Prototype Mars Rover Field Tests 239
  R. E. Arvidson, C. D. Bowman, D. M. Sherman, and
  S. W. Squyres
Discovery Missions: Unique Approaches to Education and Public
  Outreach 241
  S. E. Asplund
Bringing the Mysteries of Space Down to Earth and Into Your
  Classroom 244
  M. Baguio, M. W. Fischer, and W. T. Fowler
The Role of Amateur Astronomers in Informal Education and
  Outreach 246
  M. Bennett
Working with Informal Education: A Partnership with the Girl Scouts
  of the USA (GSUSA) on a National Level 249
  R. Betrue
NASA Robotics Education Project 252
  C. D. Bowman, F. Boyer, J. Hering, and M. J. León
The International Space Station Amateur Telescope 254
  O. H. Brettman and B. Beaman
Space Science Education and Outreach Activities in the Washington,
  DC Metropolitan Area 257
  G. R. Carruthers and V. L. Thomas
The Educator Ambassador Program for NASA's GLAST Mission . 260
  L. R. Cominsky and P. Plait
Our Sun – the Star of Classroom Activities and Public Outreach Events 262
  N. Craig, A. Miller-Bagwell, and M. B. Larson
Seven Years of Growing Space Science Education 263
  N. Craig, A. Miller-Bagwell, and I. Hawkins
Effective Use of Space Science Data and Research in the Classroom 264
  S. K. Croft

NOMISS: Integrating Space Science and Culture Through Summer
   Programs and Professional Development Experiences   267
   R. Crowe and A. J. Kawakami
The National Organization for the Professional Advancement of Black
   Chemists and Chemical Engineers   270
   D. L. Davis
Developing a Simulation-Based Curriculum: Challenges and
   Opportunities   272
   I. Doxas
Making the Connection Between Formal and Informal Learning . .   275
   P. B. Dusenbery and C. A. Morrow
Developing Exhibitions through Public/Private Partnerships: A Case
   Study of the Space Weather Center Exhibit   278
   P. B. Dusenbery and L. Mayo
IDEAS GRANT: An Opportunity for Creative Collaboration . . .   281
   B. Eisenhamer and H. Bradbury
The American Geological Institute's National Earth Science Week   284
   L. Ellis, C. Callahan, C. Martinez, and M. Smith
The Educational Activities of the Astronomical Society of the Pacific   286
   A. Fraknoi, S. Chippindale, and M. Bennett
Project ASTRO: A National Network Helping Teachers & Families
   with Astronomy Education   289
   A. Fraknoi, E. Howson, S. Chippindale, and D. Schatz
The Silicon Valley Astronomy Lectures: An Example of Institutional
   Cooperation for Public Outreach   292
   A. Fraknoi, K. Burton, and D. Morse
Designing Collaborative Learning into the Curriculum   295
   K. Garvin-Doxas
Building STARBASE: Robotic Telescopes for Hands-On Science
   Education   298
   R. Gelderman
Touch The Universe: A NASA Braille Book of Astronomy   301
   N. Grice, B. Beck-Winchatz, B. Wentworth, and M. R. Winchatz
Elements of Successful Websites for Educators   305
   A. Hammon
Society for the Advancement of Chicanos and Native Americans in
   Science (SACNAS)   307
   L. Haro and L. Hundt
The SOFIA EXES Teacher Associate Program   309
   M. K. Hemenway, J. H. Lacy, D. T. Jaffe, and M. J. Richter
Engaging Minority Undergraduate Students in Space Science . . .   312
   L. P. Johnson, S. Austin, and E. Zirbel
The Invisible Universe Online: A Distance Learning Course on Astro-
   nomical Origins for Teachers   315
   J. Keller, E. E. Prather, and T. F. Slater
Image Processing Experiments for the Classroom   318
   W. S. Kiefer and K. Leung

Pre-College Students Contribute to the Cassini-Jupiter Millennium
Flyby   321
   M. J. Klein and J. P. Roller

Involving Students in Active Planetary Research Using 2001 Mars
Odyssey THEMIS Camera Data: The Mars Student Imaging
Project   323
   S. L Klug, P. R. Christensen, K. Watt, and P. Valderrama

The National Education Standards "Quilts": A Display Method Aiding
Teachers to Link NASA Educational Materials to National
Education Standards   326
   R. Knudsen and A. Hammon

NASA Educational CD-ROMs – Research and Evaluation   329
   R. Knudsen

SNOOPY: A Novel Payload Integrated Education and Public Outreach
Project   333
   K. R. Kuhlman, M. H. Hecht, D. E. Brinza, J. E. Feldman,
   S. D. Fuerstenau, T. P. Meloy, L. E. Möller, K. Trowbridge,
   J. Sherman, L. Friedman, L. Kelly, J. Oslick, K. Polk, C. Lewis,
   C. Gyulai, G. Powell, A. M. Waldron, C. A. Batt, and
   M. C. Towner

NASA Planetary Data: Applying Planetary Satellite Remote Sensing
Data in the Classroom   337
   P. Liggett, E. Dobinson, D. Hughes, M. Martin, D. Martin,
   and B. Sword

The Wisconsin Idea: Bringing Knowledge to the Community Beyond
the Campus   341
   S. S. Limaye and R. A. Pertzborn

The Imagine the Universe! E/PO Program   345
   J. C. Lochner

Customer Interviews to Improve NASA Office of Space Science Education and Public Outreach Leveraging Success   348
   L. L. Lowes and G. Jew

Using Scientific Results from the Hubble Space Telescope to Create
Curriculum Support Tools   352
   D. McCallister, B. Eisenhamer, J. Eisenhamer, and D. A. Smith

Working with Scientists in the A.A.S. Division for Planetary Sciences   355
   E. D. Miner

An Investigation Into Student Understanding of Astrobiology
Concepts   359
   E. G. Offerdahl, Edward E. Prather, and Tim F. Slater

ASTRO-VENTURE: Using Astrobiology Missions and Interactive Technology to Engage Students in the Learning of Standards-Based
Concepts   361
   C. O'Guinn

Project FIRST and Eye on the Sky: Space Science in the Early Grades   364
   R. Paglierani and S. Feldman

Scientists Actively Involved in Education and Public Outreach: A
   Successful Experience at the MIT Center for Space Research   365
   I. L. Porro
The Search for Life in the Universe Online: A Distance Learning Course
   on Astrobiology for Teachers   368
   E. E. Prather and T. F. Slater
MarsQuest Online   370
   C. Randall, J. B. Harold, and P. Andres
Balancing Relevancy and Accuracy: Presenting Gravity Probe B's
   Science Meaningfully   372
   S. K. Range
Space Update: A Fun Way to Teach Space Science   374
   P. Reiff and C. Sumners
Issues in Formal Education: Building and Delivering Standards-Based
   E/PO Products and Activities   377
   J. Ristvey and J. Behne
NASA Space Science Diversity Initiatives   380
   P. J. Sakimoto
Why Things Fall: What College Students Don't Understand About
   Gravity and How To Teach It Better   384
   H. L. Shipman, N. W. Brickhouse, Z. R. Dagher, and W. J. Letts
Interdisciplinary Approaches to Teaching Astronomy: Special Topics
   and Team-Taught Courses   385
   H. L. Shipman
A Systemic Approach to Improving K-12 Astronomy Education Using
   the Internet   386
   T. F. Slater, E. E. Prather, and G. T. Tuthil
Partnering Scientists and Educators: A Model for Professional and
   Product Development   388
   D. A. Smith, B. Eisenhamer, T. J. Teays, and D. McCallister
The Art and Science of Storytelling in Presenting Complex Information
   to the Public, or, Give 'em More than Just the Facts   390
   A. M. Sohus and A. S. Wessen
Using Sloan Digital Sky Survey Data in the Classroom   394
   R. Sparks, C. Stoughton, and M. J. Raddick
Integrating Radio Physics Projects into Education: INSPIRE and
   Radio JOVE   397
   W.W. L. Taylor and J.R. Thieman
How Does an Educator Find NASA Space Science Resources? . . .   400
   T. J. Teays, C. Rest, D. A. Smith, and B. Eisenhamer
Schmidt Crater: Making Data from the Mars Global Surveyor
   Accessible to Introductory Astronomy Students   402
   F. J. Thomas
A Successful Formula for Teacher Retention and Renewal: The Teacher
   Leaders in Research-Based Science Education Program   405
   C. E. Walker and S. M. Pompea

NOAO: Supporting Astronomy Education Across the Spectrum .. 406
    C. E. Walker and S. M. Pompea
PROJECT ASTRO-TUCSON: The Art Of Learning About The Cosmos Around Us 407
    C. E. Walker and S. M. Pompea
HOKU: An Online Astronomy Newsletter for Educators and Parents 408
    E. L. Wells and L. Bryson
Astronomy Education Review: A New Journal/Magazine for
    Astronomy and Space Science Education 410
    S. Wolff and A. Fraknoi

Closing Remarks 413

Appendices 419
Appendix A – The NASA Office of Space Science Education and Public Outreach Program 421
The NASA Office of Space Science Education and Public Outreach
    Program 423
    J. D. Rosendhal, P. J. Sakimoto, R. A. Pertzborn, and L. Cooper
Appendix B – The Office of Space Science E/PO Support Network 439
The Origins Education Forum 441
    I. Griffin, D. A. Smith, and C. Rest
NASA's Solar System Exploration Education and Public Outreach
    Forum 443
    E. D. Miner, L. L. Lowes, and S. E. Asplund
Universe Education Forum: NASA - SAO Forum on the Structure and
    Evolution of the Universe 445
    R. Gould
The Sun-Earth Connection Education Forum 446
    J. Thieman and I. Hawkins
DePaul University Space Science Center for Education and Outreach 448
    C. Narasimhan, J. Sweitzer, and B. Beck-Winchatz
Mid-Atlantic Region Space Science Broker 450
    N. Naik
NESSIE: New England Space Science Initiative in Education ... 453
    C. Sneider, C. Clemens, and W. Waller
SCORE – South Central Organization for Researchers and Educators 455
    S. Shipp
Southeast Regional Clearinghouse (SERCH): a NASA Office of Space
    Science Broker/Facilitator 457
    C. Runyon

The NASA Office of Space Education and Public Outreach Broker/
   Facilitator Program at the Space Science Institute of Boulder,
   CO 459
   C. A. Morrow, J. B. Harold, and C. L. Edwards
Space Science Network Northwest 460
   J. Lutz
**Appendix C – Implementing the Office of Space Science
Education/Public Outreach Strategy: A Critical
Evaluation at the Six-Year Mark** 461
   SScAC E/PO Task Force
Author Index 473
Acronym Index 476

Preface

The NASA Office of Space Science Conference on Education and Public Outreach took place June 12-14, 2002 at the Union League Club of Chicago. There were 278 participants who identified themselves in a variety of ways: 50 scientists, 74 formal science educators, 82 informal science educators, 37 science education researchers, 59 NASA Office of Space Science (OSS) or Education and Public Outreach (E/PO) staff, 49 members of the OSS E/PO Support Network, 6 NASA Education staff, and 60 others. Indeed, one characteristic of the participants was that most of them have multiple roles in education and public outreach.

The goals and outcomes of this first NASA OSS-sponsored conference focusing on E/PO are summarized in a Letter to the Reader on page xviii. In this space, I would like to recall special circumstances related to the conference and acknowledge and thank the many people who contributed to its success.

The special circumstances relate to September 11, 2001. The OSS conference was originally scheduled for September 12-14 of that year. Twenty-five people had arrived in Chicago by the time flights were cancelled. The Union League Club graciously stayed open on the 11th, housed those who could not get back home for up to five days, and provided whatever was needed as we dealt individually and collectively with the shock. As the group talked about how to proceed, we very nearly came to the decision to cancel the conference altogether. That the conference eventually took place in June of 2002 was due to the reactions and expressions of support on the part of the E/PO community, and to the clear realization that in the wake of the tragic events, it was all the more important to hold the conference that we had planned.

In my acknowledgements, therefore, I would like to thank first of all the participants, those members of the space science E/PO community who by their early support for the idea of the conference and their willingness to contribute their best thoughts and engage openly and honestly with the issues, were truly responsible for making the conference a success. At the same time, I must thank Jeffrey Rosendhal, Education and Public Outreach Director of OSS in June 2002, who was responsible for making the conference happen. His willingness to listen to and question half-formed ideas throughout the period of development strengthened that process, and his full support for the ultimate plan was invaluable.

The Organizing Committee consisted of OSS Support Network members Bernhard Beck-Winchatz, Troy Cline, Isabel Hawkins, Kathleen Johnson, Rebecca Knudsen, Elaine Lewis, Leslie Lowes, Louis Mayo, Cherilynn Morrow, Carolyn Narasimhan (chair), Greg Schultz, Anita Sohus, and Terry Teays. Lucy Fortson, Andrew Fraknoi, Jesus Pando, and Gwen Pollock served as advisors to the Organizing Committee. Special thanks are due to Victoria Simek and Karen Cullen from DePaul University for making the conference run so smoothly.

The Adler Planetarium & Astronomy Museum was a partner in this project from the early planning stages to the popular evening event they hosted on the second day of the conference. I thank Paul Knappenberger, President of the Adler, for his support and for agreeing to give one of the plenary addresses.

The other plenary speakers were Edward Weiler, Maria Alicia Lopez Freeman, and George Nelson; members of the opening panel on Science Education Research, Jon Miller, Moderator, Shelley Lee, Philip Sadler, and Michael Zelik; and members of the closing panel on Scientists' Participation in Education and Public Outreach, Carl Pilcher, Moderator, Lucy Fortson, Andrew Fraknoi, Charles McGruder, and Ramon Lopez. These participants contributed significantly to providing both context for and reflections on the discussions that took place over the three days. I thank them also for their contributions to these proceedings.

During the second day of the conference, thirty-two participants gave panel presentations on exemplary E/PO programs and initiatives in formal and informal science education, and many of them have contributed articles to these proceedings. Thanks to the moderators of these panels: Larry Bilbrough, Lynn Cominsky, Geoffrey Haines-Styles, Mary Kay Hemenway, Parvin Kassaie, Larry Lebofsky, Dennis Schatz, and Jim Sweitzer.

The hosts of the conference were the members of the OSS Support Network. In addition to those who served on the Organizing Committee listed above, many of them acted as facilitators and recorders for the panel sessions. I would like to acknowledge the contribution of Katherine Mitchell, Principal Analyst for Organizational Consulting, University of California, Berkeley, who helped us develop a plan for facilitation of the breakout sessions. And special thanks go to Roy Gould, Isabel Hawkins, and Terry Teays who both hosted and summarized the parallel panel sessions.

Finally, I would like to thank those who worked on these proceedings: Karin Hauck at UC Berkeley, and Victoria Simek and Lara Petropoulos at DePaul University, who provided technical support, and my three co-editors, Bernhard Beck-Winchatz, DePaul University, Isabel Hawkins, UC Berkeley, and Cassandra Runyon, College of Charleston, who put in many hours communicating with authors, reading, editing, and sharing in decisions on the organization of the volume. Isabel and Cass join me in acknowledging the incredible effort that Bernhard made in preparing the volume for publication by the Astronomical Society of the Pacific. My co-editors turned this project, too long delayed, into a collaborative effort that has reminded us of the excitement of the conference and the future we all talked about during that event. I hope that the proceedings will do the same for you.

Carolyn Narasimhan
DePaul University
May 1, 2004

Dear Reader,

This Conference Proceedings is the result of a gathering in Chicago in June 2002 of space scientists and both formal and informal educators. As the first such conference sponsored by NASA's Office of Space Science (OSS), the Conference goals were to strengthen and deepen the education and public outreach (E/PO) efforts of OSS, and to enhance the ability of the space science community to contribute to these efforts. To achieve these goals, the Conference brought together about 280 scientists and educators with an interest in space science education and outreach, tapping expertise within and outside of NASA to provide learning and networking opportunities. Through a mixture of presentations and discussions, participants focused on the following areas:

- Facilitating scientist participation in education and outreach.

- Issues and challenges faced by educators in formal and informal education settings to incorporate results from space science research in exhibits, the classroom, and other venues.

- Ways in which science education research can help inform space science education and public outreach efforts.

- Best practices in space science education and public outreach as exemplified by a broad range of initiatives, including those that demonstrated effective participation of scientists, outreach to underutilized/underserved communities, and use of technology.

During plenary talks, panel presentations, breakout and poster sessions, and unstructured time, participants had the opportunity to hear about current issues in space science education and to engage in discussions about future endeavors.

Several major conclusions emerged from the conference:

- There is an emerging space science E/PO community and members of this community need professional development opportunities in all areas of education and public outreach;

- There needs to be more coherence in the development of materials, activities, and products;

- Creating and sustaining partnerships is key to the success of the OSS effort;

- Assessment and evaluation are acknowledged as important, but the community needs help in doing this.

This Proceedings volume follows the arrangement of talks and panels as they were presented during the Conference. We were fortunate to hear from keynote and panel speakers who are at the cutting edge of their field, and participants benefited from interactions in the form of structured opportunities and informal discussions.

Several of the keynote speakers chose to make their contributions to these Proceedings in a conversational style, to preserve the flavor and enthusiasm of their presentations during the Conference. When appropriate, we have included references to other talks or resources throughout the volume. We hope you find this volume informative and useful in your current and future space science education efforts.

We would like to thank all the presenters, the members of the OSS Education and Public Outreach Support Network, the members of the Conference organizing committee, NASA Office of Space Science, the Adler Planetarium, and especially the conference participants for making this successful gathering successful educational, collaborative, and fun.

Sincerely,

Carolyn Narasimhan (DePaul University),
Bernhard Beck-Winchatz (DePaul University),
Isabel Hawkins (UC Berkeley), and
Cassandra Runyon (College of Charleston)

Participants

MITZI ADAMS, NASA/MSFC/NSSTC, NSSTC/320 Sparkman Drive, Huntsville, AL 35805 ⟨mitzi.adams@nasa.gov⟩

DAVID ALEXANDER, Rice University, Department of Physics and Astronomy, MS 108, P.O. Box 1892, Houston, TX 77251-1892 ⟨dalex@rice.edu⟩

JACLYN ALLEN, Lockheed Martin/ JSC, NASA Johnson Space Center C23, Houston, TX 77058 ⟨jaclyn.allen1@jsc.nasa.gov⟩

RICHARD ALVIDREZ, Jet Propulsion Laboratory, 4800 Oak Grove Drive, Pasadena, CA 91109 ⟨alvidrez@jpl.nasa.gov⟩

ANDREA ANGRUM, Jet Propulsion Laboratory, Voyager/Ulysses Project, 4800 Oak Grove Dr., M/S 264-112, Pasadena, CA 91109 ⟨andrea.angrum@jpl.nasa.gov⟩

CRAIG ANTHONY, SERCH – NASA OSS Broker/Facilitator, 66 George Street, Charleston, SC 29424 ⟨anthonym@cofc.edu⟩

SHARI ASPLUND, Jet Propulsion Laboratory, 4800 Oak Grove Dr. 156-230, Pasadena, CA 91109 ⟨shari.e.asplund@jpl.nasa.gov⟩

LISA AUSTGEN, Challenger Learning Center of Northwest Indiana, c/o Purdue University Calumet, 2300 173rd Street, Hammond, IN 46323 ⟨austgen@clcnwi.com⟩

SHERMANE AUSTIN, Medgar Evers College/MUSPIN, 1150 Carroll Street, Brooklyn, NY 11225 ⟨shermane@bellatlantic.net⟩

MARGARET BAGUIO, Texas Space Grant Consortium, 3925 W. Braker Lane, Suite 200, Austin, TX 78759 ⟨baguio@tsgc.utexas.edu⟩

JACQUELINE BARBER, Lawrence Hall of Science, University of California at Berkeley, Berkeley, CA 94720 ⟨jbarber@uclink4.berkeley.edu⟩

LOUIS BARBIER, NASA/Goddard Space Flight Center, Code 661, Greenbelt, MD 20771 ⟨lmb@cosmicra.gsfc.nasa.gov⟩

DANIEL BARSTOW, TERC, 2067 Massachuetts Avenue, Cambridge, MA 02140 ⟨dan_barstow@terc.edu⟩

LINDSAY BARTOLONE, MAP EPO Coordinator, 1300 S. Lake Shore Drive, Chicago, IL 60605 ⟨clark@astro.princeton.edu⟩

ALICIA BATURONI, NASA Ames Research Center, NASA Ames/Planners Collaborative, Mailstop 226-4, Moffett Field, CA 94035 ⟨alicia@spaceed.org⟩

BARRY BEAMAN, Astronomical League – ISS-AT, Praesepe Observatory / 6804 Alvina Road, Rockford, IL 61101 ⟨praesepe44@aol.com⟩

BERNHARD BECK-WINCHATZ, NASA Space Science Center at DePaul University, 990 W. Fullerton Ave, Suite 4400, Chicago, IL 60614 ⟨bbeck@condor.depaul.edu⟩

SANDRA BEGAY-CAMPBELL, Sandia National Laboratories, PO Box 5800, MS 0753, Albuquerque, NM 87185-0753 ⟨skbegay@sandia.gov⟩

JACINTA BEHNE, McREL, 2550 S. Parker Rd., Suite 500, Aurora, CO 80014 ⟨jbehne@mcrel.org⟩

KERRI BEISSER, JHU/APL, 11100 Johns Hopkins Rd., 4-262, Laurel, MD 20723 ⟨kerri.beisser@jhuapl.edu⟩

REBECCA BELEI, Challenger Learning Center of Northwest Indiana, c/o Purdue University Calumet, 2300 173rd Street, Hammond, IN 46323 ⟨belei@calumet.purdue.edu⟩

JEANNETTE BENAVIDES, NASA, Goddard Space Flight Center, Greenbelt, MD 20771 ⟨jbenavid@pop300.gsfc.nasa.gov⟩

MIKE BENNETT, Astronomical Society of the Pacific, 390 Ashton Ave., San Francisco, CA 94112 ⟨mbennett@astrosociety.org⟩

JOHN BERRYMAN, PBCC, 2414 NW 31st Street, Boca Raton, FL 33431 ⟨profjohn@profjohn.com⟩

ROSALIE BETRUE, JPL – SSE E/PO Forum, 4800 Oak Grove Drive, MS 180-109, Pasadena, CA 91109 ⟨Rosalie.Betrue@jpl.nasa.gov⟩

LARRY BILBROUGH, NASA Headquarters, Education Division, Washington, DC 20546 ⟨larry.bilbrough@hq.nasa.gov⟩

JEANNE BISHOP, Westlake Schools Planetarium, 24525 Hilliard Road, Westlake, OH 44145 ⟨jeanbishop@aol.com⟩

JERRY BLACKMON, California University of PA, 250 University Ave., California, PA 15419 ⟨blackmon@cup.edu⟩

DIANE BOLLEN, Cornell University, 426 Space Sciences Building, Cornell University, Ithaca, NY 14853 ⟨diane@astro.cornell.edu⟩

GREG BOTHUN, University of Oregon, Dept. of Physics, Eugene, OR 97403 ⟨nuts@bigmoo.uoregon.edu⟩

ARTHUR BOWMAN, Hampton University Science Center, Hampton Univ. / Tyler St., Hampton, VA 23668 ⟨arthur.bowman@hamptonu.edu⟩

CATHERINE BOWMAN, Raytheon ITSS/NASA Ames Research Center, MS 269-3, Moffett Field, CA 94035-1000 ⟨cbowman@mail.arc.nasa.gov⟩

HEATHER BRADBURY, Space Telescope Science Institute, 3700 San Martin Drive, Baltimore, MD 21218 ⟨hbradbur@stsci.edu⟩

ROBERT BRAZZLE, IMSA and SSEP, 1500 W. Sullivan Road, Aurora, IL 60506 ⟨brazzle@imsa.edu⟩

ORVILLE BRETTMAN, Astronomical League, 13915 Hemmingsen RD, Huntley, IL 60142 ⟨rivendell.astro@worldnet.att.net⟩

MATTHEW BRIGGS, U. Wisconsin-Madison, 1150 University Ave, Madison, WI 53711 ⟨meb@physics.wisc.edu⟩

Participants

GINA BRISSENDEN, CDES, 303-E Eagle Heights Dr., Madison, WI 53705 ⟨gina@fauxhead.astro.wisc.edu⟩

ERIC BRUNSELL, Space Education Initiative, 2830 Ramada Way, Suite 203, Green Bay, WI 54304 ⟨eric@spaceed.org⟩

GEORGE CARRUTHERS, Naval Research Laboratory, Code 7645, Washington, DC 20375-5320 ⟨george.carruthers@nrl.navy.mil⟩

DIANA CHALLIS, Adler Planetarium & Astronomy Museum, 1300 S . Lakeshore Dr., Chicago, IL 60605 ⟨challis@adlernet.org⟩

EVERETT CHAVEZ, AISES, 2305 Renard Pl, SE, Albuquerque, NM 87106 ⟨everett@aises.org⟩

TROY CLINE, NASA/Goddard Space Flight Center, Code 633, Building 26, Room 201, Greenbelt, MD 20771 ⟨cline@mail630.gsfc.nasa.gov⟩

SUSAN COHEN, Program Evaluation and Research Group, 29 Everett Street, Cambridge, MA 02138 ⟨suecohen@mail.lesley.edu⟩

LYNN COMINSKY, Sonoma State University, 1801 East Cotati Avenue, Rohnert Park, CA 94928 ⟨lynnc@charmian.sonoma.edu⟩

RACHEL CONNOLLY, American Museum of Natural History, Rose Center for Earth and Space, Central Park West at 79th St., New York, NY 10024 ⟨connolly@amnh.org⟩

SARAH CONNOLLY, Program Evaluation and Research Group, Lesley University 29 Everett Street, Cambridge, MA 02138 ⟨sconnoll@mail.lesley.edu⟩

DENISE COOK-CLAMPERT, Ball Aerospace & Technologies Corp., 1600 Commerce Street, Boulder, CO 80301 ⟨dcookcla@ball.com⟩

LARRY COOPER, NASA Headquarters, 300 E Street, SW Washington, DC 20546 ⟨Larry.P.Cooper@nasa.gov⟩

PETER COPPIN, Principal Investigator, EventScope, STUDIO for Creative Inquiry, Carnegie Mellon University, 5000 Forbes Avenue, Pittsburgh, PA 15213-3890 ⟨coppin@cmu.edu⟩

NAHIDE CRAIG, The Center for Science Education, Space Sciences Laboratory, 7 Gauss Way, MC 7450, Berkeley, CA 94720-7450 ⟨ncraig@ssl.berkeley.edu⟩

STEVEN CROFT, Center for Educational Technologies, CET, Wheeling Jesuit University, Weeling, WV 26003 ⟨scroft@cet.edu⟩

RICHARD CROWE, University of Hawaii Hilo / NOMISS, 200 West Kawili Street, Hilo, HI 96720 ⟨rcrowe@hubble.uhh.hawaii.edu⟩

KAREN CULLEN, NASA Space Science Center at DePaul University, 990 W. Fullerton Ave, Suite 4400, Chicago, IL 60614 ⟨kcullen@depaul.edu⟩

SANDRA DALY, Harvard-Smithsonian Center for Astrophysics, 60 Garden St, Cambridge, MA 02138 ⟨sdaly@cfa.harvard.edu⟩

ROLF DANNER, Jet Propulsion Laboratory, 4800 Oak Grove Drive, Pasadena, CA 91109-8099 ⟨rolf.danner@jpl.nasa.gov⟩

DORIS DAOU, California Institute of Technology, Infrared Processing & Analysis Center/Spitzer Science Center, 770 South Wilson Ave., Suite #224, M/S 100-22, Pasadena, CA 91125 ⟨daou@ipac.caltech.edu⟩

DENNIS DAVIDSON, Center for Cartographic Design, 104 8th Avenue, #5FS, New York, NY 10011 ⟨davidson3d@earthlink.net⟩

ANITA DAVIS, Earth Science Enterprise Informal Education Lead, Goddard Space Flight Center MS 900.2, Greenbelt, MD 20771 ⟨adavis@pop100.gsfc.nasa.gov⟩

DARRELL DAVIS, NOBCChE, 910 Bentle Branch Lane, Cedar Hill, TX 75104 ⟨darrelldavis@worldnet.att.net⟩

LINDA L. DAVIS, CAPS, 910 Bentle Branch Lane, Cedar Hill, TX 75104 ⟨lindadavis@worldnet.att.net⟩

MIKE DAVIS, Harold Washington College, 30 E. Lake St., Chicago, IL 60601 ⟨mdavis@ccc.edu⟩

PHIL DAVIS, Jet Propulsion Laboratory, OSS Solar System Forum, MS 180-109, 4800 Oak Grove Dr., Pasadena, CA 90807 ⟨phillips.w.davis@jpl.nasa.gov⟩

ANGELA DAVRE, Medical College of Wisconsin, 8701 Watertown Plank Road, Milwaukee, WI 53226 ⟨angda@mcw.edu⟩

DENNIS DAWSON, Western Connecticut State U., 181 White St, Danbury, CT 06810 ⟨astroDen@netscape.net⟩

RICHARD DELZENERO, Northeastern Illinois University, 5500 N. St. Louis, Chicago, IL 60625 ⟨R-Delzenero@neiu.edu⟩

WANDA DEMAGGIO, National Aeronautics and Space Administration, Code FA00, Building 1100, Stennis Space Center, MS 39529-6000 ⟨wanda.demaggio@ssc.nasa.gov⟩

EDNA DEVORE, SETI Institute & SOFIA E/PO, 2035 Landings Dr., Mountain View, CA 94043 ⟨edevore@seti.org⟩

KAREN DODSON, NASA Astrobiology Institute, Ames Research Center, MS 240-1, Moffett Field, CA 94035 ⟨kdodson@mail.arc.nasa.gov⟩

PAUL DUSENBERY, Space Science Institute, 4750 Walnut St., Suite 205, Boulder, CO 80301 ⟨dusenbery@spacescience.org⟩

MARY DUSSAULT, Harvard-Smithsonian Center for Astrophysics, 60 Garden St, Cambridge, MA 02138 ⟨mdussault@cfa.harvard.edu⟩

CHRISTY EDWARDS, Space Science Institute, 4750 Walnut St., Suite 205, Boulder, CO 80301 ⟨edwardcl@spacescience.org⟩

JEFF EHMEN, NASA Marshall Space Flight Center, Mail Code CD60, Huntsville, AL 35812 ⟨jeff.ehmen@nasa.gov⟩

BONNIE EISENHAMER, Space Telescope Science Institute, 3700 San Martin Drive, Baltimore, MD 21218 ⟨bonnie@stsci.edu⟩

SUE ELLIS, NASA – AESP, 1024 Silver Lake Blvd., Frankfort, KY 40601 ⟨sue@aesp.nasa.okstate.edu⟩

HEATHER ENOS, University of Arizona, 1629 E. University Blvd, Tucson, AZ 85721 ⟨heather@gamma1.lpl.arizona.edu⟩

JACQUELINE FAHERTY, American Museum of Natural History – Moveable Museum, AMNH Education Dept. Central Park West @ 79th Street, New York, NY 10024 ⟨jfaherty@amnh.org⟩

SALLY FELDMAN, Washington Elementary School, 565 Wine Street, Richmond, CA 94801 ⟨feldmom@aol.com⟩

KAY FERRARI, Jet Propulsion Laboratory Solar System Ambassadors Program, 4800 Oak Grove Drive, MS 264-788, Pasadena, CA 91109-8099 ⟨Kay.A.Ferrari@jpl.nasa.gov⟩

LUCY FORTSON, Adler Planetarium/Univ. of Chicago, 1300 S. Lakeshore Dr., Chicago, IL 60605 ⟨lfortson@adlernet.org⟩

ANDREW FRAKNOI, A.S.P. & Foothill College, 390 Ashton Ave., San Francisco, CA 94112 ⟨fraknoi@fhda.edu⟩

HELENE GABELNICK, Harold Washington College, 30 E. Lake St., Chicago, IL 60601 ⟨hgabelnick@ccc.edu⟩

KATHY GARVIN-DOXAS, University of Colorado, Alliance for Technology, Learning and Society (ATLAS) Institute, Evaluation and Research Group, CB-040UCB, Boulder, CO 80309 ⟨garvindo@colorado.edu⟩

RICHARD GELDERMAN, Western Kentucky University, Dept. of Physics and Astronomy, Bowling Green, KY 42101 ⟨richard.gelderman@wku.edu⟩

NICOLE GILLESPIE, UC Berkeley, 1492 Olympus Ave, Berkeley, CA 94708 ⟨ngillesp@uclink4.berkeley.edu⟩

JEFF GOLDSTEIN, Challenger Center, 1250 North Pitt Street, Alexandria, VA 22314 ⟨jgoldstein@challenger.org⟩

EDDIE GONZALES, Jet Propulsion Laboratory, 4800 Oak Grove Drive, Pasadena, CA 91109 ⟨Eddie.Gonzales@jpl.nasa.gov⟩

ROY GOULD, Harvard-Smithsonian Center for Astrophysics, 60 Garden St, Cambridge, MA 02138 ⟨rgould@cfa.harvard.edu⟩

WILLIAM GREENE, Jet Propulsion Laboratory, 4800 Oak Grove Drive, Pasadena, CA 91109 ⟨William.M.Greene@jpl.nasa.gov⟩

NOREEN GRICE, Boston Museum of Science, Science Park, Boston, MA 02114-1099 ⟨ngrice@mos.org⟩

JENNIFER GRIER, Harvard-Smithsonian Center for Astrophysics, 60 Garden St, MS 71, Cambridge, MA 02138 ⟨jagrier@cfa.harvard.edu⟩

ARTHUR GRIFFIN, Best Practice High School & HOU(LBL), 2040 W. Adams Street, Chicago, Il 60612 ⟨agriffin62@yahoo.com⟩

IAN GRIFFIN, Space Telescope Science Institute, 3700 San Martin Drive, Baltimore, MD 21210 ⟨griffin@stsci.edu⟩

JOE GRIFFITH, PERG, Lesley University, 111 Orchard Lane, Durango, CO 81301 ⟨joegriffith@frontier.net⟩

KATHRYN GUIMOND, NASA OSS Broker/Facilitator – Southeast Regional Clearinghouse (SERCH), College of Charleston – 66 George Street, Charleston, SC 29424 ⟨serch@cofc.edu⟩

HERBERT GURSKY, Naval Research Laboratory, 4555 Overlook Ave., SW, Code 7600, Washington, DC 20375-5352 ⟨herbert.gursky@nrl.navy.mil⟩

JENNIFER GUTBEZAHL, Program Evaluation and Research Group, 29 Everett St, Cambridge, MA 02138 ⟨jgutebza@mail.lesley.edu⟩

T. GREGORY GUZIK, Louisiana State University, Department of Physics and Astronomy, Louisiana State University, Baton Rouge, LA 70803 ⟨guzik@phunds.phys.lsu.edu⟩

GEOFFREY HAINES-STILES, PASSPORT TO KNOWLEDGE / LIVE FROM..., 27 Washington Valley Road, Morristown, NJ 07960 ⟨ghs@passporttoknowledge.com⟩

HEIDI HAMMEL, Space Science Institute, 72 Sarah Bishop Road, Ridgefield, CT 06877 ⟨hbh@alum.mit.edu⟩

VIRGIL HAMMON, Jet Propulsion Laboratory E/PO, 4800 Oak Grove Dr. 180-109, Pasadena, CA 91109 ⟨ahammon@jpl.nasa.gov⟩

LUIS HARO, SACNAS, Biology, University of Texas at San Antonio, 6900 N, Loop 1604 W., San Antonio, TX 78249 ⟨lharo@utsa.edu⟩

JAMES HAROLD, Space Science Institute, 4750 Walnut St., Suite 205, Boulder, CO 80301 ⟨harold@spacescience.org⟩

ILANA HARRUS, USRA/NASA/GSFC, Building 2, code 662, Greenbelt, MD 20771 ⟨imh@lheapop.gsfc.nasa.gov⟩

ISABEL HAWKINS, UC Berkeley / Sun-Earth Connection Education Forum, UC Berkeley, 7 Gauss Way, Space Sciences Laboratory, MC 7450, Berkeley, CA 94720 ⟨isabelh@ssl.berkeley.edu⟩

MARY KAY HEMENWAY, The University of Texas at Austin, Department of Astronomy, 1 University Station, C1400, Austin, TX 78712-0259 ⟨marykay@astro.as.utexas.edu⟩

RHONDA HINES-JONES, Jet Propulsion Laboratory, 4800 Oak Grove, M/S 301-486, Pasadena, CA 91109 ⟨rhonda.r.jones@jpl.nasa.gov⟩

SUSAN HOBAN, AISRP & LTP, Mail Code 103, NASA/GSFC, Greenbelt, MD 20771 ⟨susan.hoban@gsfc.nasa.gov⟩

APRIEL K.HODARI, National Society of Black Physicists / The CNA Corporation, 4825 Mark Center Drive, Alexandria, VA 22311 ⟨hodaria@cna.org⟩

VIVIAN HOETTE, SOFIA / HAWC EPO, 400 Smythe Drive, Williams Bay, WI 53191 ⟨vhoette@hale.yerkes.uchicago.edu⟩

RUTH HUNTER, Brownsville Alliance for Science Education UTB/TSC, 80 Fort Brown, Brownsville, Texas 78520 ⟨rihunter@utb.edu⟩

FRANK IRETON, SSAI/GSFC/NASA, SSAI, 10210 Greenbelt Road, Suite 400, Lanham, MD 20706 ⟨frank_ireton@sesda.com⟩

S. Beth Jacob, NASA GSFC / SPS, Code 661, Greenbelt, MD 20771
⟨beth@milkyway.gsfc.nasa.gov⟩

Diane Jeffers, Aerospace Illinois Space Grant Consortium, 306 Talbot Lab, 104 S. Wright St., Urbana, IL 61801 ⟨dejeffer@uiuc.edu⟩

Leon Johnson, Medgar Evers College/NYC-SSRA, 1150 Carroll Street, Brooklyn, NY 11225 ⟨leon.johnson@verizon.net⟩

Roberta Johnson, University Corporation for Atmospheric Research, UCAR / 3300 Mitchell Lane, Suite 2138, Boulder, CO 80301 ⟨rmjohnsn@ucar.edu⟩

Parvin Kassaie, Jet Propulsion Laboratory, 4800 Oak Grove Drive, MS 180-109, Pasadena, CA 91109-8099 ⟨pkassaie@mail2.jpl.nasa.gov⟩

Alice Kawakami, University of Hawaii – Manoa, College of Education, 1776 University Ave., Honolulu, HI 96822 ⟨alicek@hawaii.edu⟩

Oved Kedem, Davidson Institute for Science Education at the Weizmann Institute of Science, Israel, P.O.Box 26, Rehovot, Israel, 76100 ⟨oved.kedem@weizmann.ac.il⟩

Maureen Kenney, Raytheon ITSS, 299 N. Euclid Ave Ste 500, Pasadena, CA 91101 ⟨maureen_kenney@raytheon.com⟩

Walter Kiefer, Lunar and Planetary Institute, 3600 Bay Area Blvd, Houston, TX 77058 ⟨kiefer@lpi.usra.edu⟩

Michael Klein, NASA-JPL, Pasadena, CA 91109 ⟨mike.klein@jpl.nasa.gov⟩

Sheri Klug, ASU Mars Education Program, P.O. Box 876305, Tempe, AZ 85287-6305 ⟨sklug@asu.edu⟩

Paul Knappenberger, Adler Planetarium & Astronomy Museum, 1300 S. Lake Shore Drive, Chicago, IL 60605 ⟨paul@adlernet.org⟩

Rebecca Knudsen, NASA SSE Forum, JPL, 4800 Oak Grove Dr, MS 180-109, Pasadena, CA 91109 ⟨Rebecca.Knudsen@jpl.nasa.gov⟩

Kimberly Kuhlman, Jet Propulsion Laboratory, 4800 Oak Grove Dr., M/S: 302-231, Pasadena, CA 91109 ⟨kkuhlman@jpl.nasa.gov⟩

Stephen Kulczycki, Jet Propulsion Laboratory, 4800 Oak Grove Drive, Pasadena, CA 91109 ⟨Stephen.E.Kulczycki@jpl.nasa.gov⟩

Shawn Laatsch, East Carolina University, Department of Math and Science Education, 323B Austin Hall, Greenville, NC 27858 ⟨102424.1032@compuserve.com⟩

Larry Lebofsky, University of Arizona, Lunar Lab, 1629 East University, Tucson, AZ 85721 ⟨lebofsky@lpl.arizona.edu⟩

Shelley Lee, Wisconsin Department of Public Instruction, P.O. Box 7841, Madison, WI 53707-7841 ⟨shelley.lee@dpi.state.wi.us⟩

Nancy Leon, NASA Space Place, 4800 Oak Grove Drive, Pasadena, CA 91109 ⟨nancy.j.leon@jpl.nasa.gov⟩

Thomas Levenson, Thomas Levenson Productions, 26 Lloyd Road, Watertown, MA 02472 ⟨tlevenson@post.harvard.edu⟩

ELAINE LEWIS, SECEF-Goddard Space Flight Center, Code 630 Greenbelt Rd., Greenbelt, MD 20771 ⟨lewis@mail630.gsfc.nasa.gov⟩

STEPHENIE LIEVENSE, SSE Forum/JPL, 4800 Oak Grove Drive/MS 186-114, Pasadena, CA 91109 ⟨stephenie.h.lievense@jpl.nasa.gov⟩

PATRICIA LIGGETT, Jet Propulsion Laboratory, 4800 Oak Grove Dr M/S 202-233, Pasadena, CA 91101 ⟨pat.liggett@jpl.nasa.gov⟩

SANJAY LIMAYE, Office of Space Science Education, University of Wisconsin-Madison, 1225 West Dayton Street, Madison, WI 53706 ⟨SanjayL@ssec.wisc.edu⟩

MARILYN LINDSTROM, NASA Johnson Space Center, SR NASA Johnson Space Center, Houston, TX 77058 ⟨marilyn.lindstrom-1@nasa.gov⟩

PATRICIA LINK, NASA Langley Research Center, 3B East Taylor St., Hampton, VA 23681 ⟨patricia.a.link@nasa.gov⟩

JAMES LOCHNER, NASA/GSFC & USRA, Code 662.0, Greenbelt, MD 20771 ⟨lochner@lheapop.gsfc.nasa.gov⟩

ALLISON LOONEY, Draper Laboratory, 555 Technology Square, MS 75, Cambridge, MA 02139-3563 ⟨alooney@draper.com⟩

RAMON LOPEZ, UTEP, Dept. of Physics, El Paso, TX 79968 ⟨relopez@utep.edu⟩

MARIALOPEZ FREEMAN, California Science Project, 3806 Geology Building, UCLA, Los Angeles, CA 90095-1567 ⟨mafreema@ucla.edu⟩

LESLIE LOWES, NASA Solar System Exploration E/PO Forum, Jet Propulsion Laboratory/MS 180-109/4800 Oak Grove Drive, Pasadena, CA 91109 ⟨Leslie.L.Lowes@jpl.nasa.gov⟩

PAMELA LUCAS, Head, Science Education Program, Princeton Plasma Physics Laboratory, PO Box 451, Princeton, NJ 08543 ⟨plucas@pppl.gov⟩

JULIE LUTZ, Washington Space Grant, Univ. of WA, Box 351310, Seattle, WA 98195-1310 ⟨nasaerc@u.washington.edu⟩

MOLLY MACAULEY, Resources for the Future, 1616 P Street NW, Washington, DC 20036 ⟨macauley@rff.org⟩

FARZAD MAHOOTIAN, TERC, 2067 MASS AVE, CAMBRIDGE, MA 02148 ⟨farzad_mahootian@terc.edu⟩

ATTRICE MALONE, Lunar and Planetary Institute, 3600 Bay Area Boulevard, Houston, TX 77058 ⟨malone@lpi.usra.edu⟩

BENJAMIN MALPHRUS, Morehead State University, Astrophysics Laboratory, Morehead, KY 40351 ⟨b.malphrus@morehead-st.edu⟩

JAMES MANNING, Museum of the Rockies, 600 W. Kagy Blvd., Bozeman, MT 59717 ⟨manning@montana.edu⟩

JASON MARCKS, Space Education Initiatives, 2830 Ramada Way Suite 203, Green Bay, WI 54304 ⟨jmarcks@spaceed.org⟩

LOU MAYO, GSFC/SECEF, Goddard Space Flight Center, Code 630, Greenbelt, MD 20771 ⟨lmayo@pop600.gsfc.nasa.gov⟩

JAMES DANIEL MCCALLISTER, Space Telescope Science Institute, 3700 San Martin Drive, Baltimore, MD 21218 ⟨mccallis@stsci.edu⟩

RON MCCLOSKEY, Education Manager, EventScope, STUDIO for Creative Inquiry, Carnegie Mellon University, 5000 Forbes Avenue, Pittsburgh, PA 15213-3890 ⟨ronmc@andrew.cmu.edu⟩

SHANNON MCCONNELL, Jet Propulsion Laboratory – QSS, 320 N. Halstead St. #260, Pasadena, CA 91107 ⟨shannon.l.mcconnell@jpl.nasa.gov⟩

MATTHEW MCCUTCHEON, The Latin School of Chicago, 59 W. North Blvd., Chicago, IL 60610 ⟨mmccutcheon@latinschool.org⟩

CHARLES MCGRUDER, Western Kentucky University, 1 Big Red Way, Bowling Green, KY 42101 ⟨mcgruder@wku.edu⟩

FLAVIO MENDEZ, Maryland Science Center, 601 Light Street, Baltimore, MD 21230 ⟨mendez@mdsci.org⟩

GLENN MILLER, Jet Propulsion Laboratory, 3325 Primera Ave. #7, Los Angeles, CA 90068-1571 ⟨astronomer@pacbell.net⟩

JON MILLER, Northwestern University, 303 E. Chicago Ave., Rm. 18-142, Chicago, IL 60611 ⟨j-miller8@northwestern.edu⟩

ELLIS D. MINER, SSE Forum, Jet Propulsion Laboratory, m/s 183-301, 4800 Oak Grove Drive, Pasadena, CA 91109-8099 ⟨ellis.d.miner@jpl.nasa.gov⟩

RONEN MIR, SciTech Hands On Museum, 18 W. Benton St., Aurora, IL 60506 ⟨ronen@scitech.mus.il.us⟩

WENDELL MOHLING, National Science Teachers Association, (H) 2353 Hunter Mill Rd., Vienna, VA 22181 ⟨wmohling@nsta.org⟩

LYNN MORONEY, 1944 NW 20th St., Oklahoma City, OK 73106, ⟨skyteller@aol.com⟩

PENNY MORRIS, University of Houston Downtown, Dept. Natural Science, University of Houston Downtown, 1 Main St., Houston, TX 77002 ⟨pmorris@ems.jsc.nasa.gov⟩

CHERILYNN MORROW, Space Science Institute, 4750 Walnut St., Suite 205, Boulder, CO 80301 ⟨camorrow@spacescience.org⟩

JENNIFER MULLINS, Gravity Probe B, Stanford University, HEPL Labs MC 4085, Stanford, CA 94305-4085 ⟨mullins@relgyro.stanford.edu⟩

CAROLYN NARASIMHAN, NASA Space Science Center at DePaul University, 990 W. Fullerton Ave, Suite 4400, Chicago, IL 60614 ⟨cnarasim@depaul.edu⟩

GEORGE NELSON, Western Washington University, Science, Mathematics, and Technology Education; 516 High St., Bellingham, WA 98225-9155 ⟨george.nelson@wwu.edu⟩

CAROLYN NG, NASA Sun-Earth Connection Education Forum, Code 633, NASA GSFC, Greenbelt, MD 20771 ⟨carolyn.ng@gsfc.nasa.gov⟩

MARY NOEL-BLACK, Lunar and Planetary Institute, 3600 Bay Area Blvd, Houston, TX 77058 ⟨noel@lpi.usra.edu⟩

LAWRENCE NORRIS, NSBP, ⟨norris@gwmail.usna.edu⟩

CHRISTINA O'GUINN, NASA Ames Educational Technology Team, NASA Ames Research Center, MS 226-4 Moffett Field, CA 94035 ⟨christina.m.oguinn@nasa.gov⟩

ALI OMAR, Center for Atmospheric Sciences, Hampton University, Hampton, VA 23668 ⟨ali.omar@hamptonu.edu⟩

KEVIN ORANGERS, American Museum of Natural History – Moveable Museum, New York, NY 10024 ⟨kevock@amnh.org⟩

CARLOS PABON-ORTIZ, Physics Dept., Univ of Puerto Rico at Mayaguez, PO Box 9016 Mayaguez, PR zc 00681 ⟨c_pabon@rumac.uprm.edu⟩

RUTH PAGLIERANI, UC Berkeley, Space Sciences Lab, Grizzly Peak Blvd., Berkeley, CA 94720 ⟨ruthp@ssl.berkeley.edu⟩

ZVI PALTIEL, The Weizmann Institute of Science, Rehovot, Israel 76100 ⟨zvi.paltiel@weizmann.ac.il⟩

JESUS PANDO, DePaul Univeristy, Department of Physics, Chicago, IL 60614 ⟨jpando@depaul.edu⟩

BRUCE PARTRIDGE, Haverford College, Haverford, PA 19041 ⟨bpartrid@haverford.edu⟩

JAMES PAYNE, SC State University, PO Box 7507, SCSU, Orangeburg, SC 29117 ⟨jpayne@scsu.edu⟩

LINDA PAYNE, SC State University, PO Box 7277, SCSU, Orangeburg, SC 29117 ⟨lpayne@scsu.edu⟩

ROSALYN PERTZBORN, Univ. of Wisconsin, Office of Space Science Education, 1225 W. Dayton Street, Madison, WI 35706 ⟨rosep@ssec.wisc.edu⟩

ANGELA PHELPS, Pennsylvania Space Grant Consortium, 2217 Earth-Engineering Sciences Building, University Park, PA 16802-6813 ⟨axp41@psu.edu⟩

CARL PILCHER, NASA HQ, Office of Space Science, Code SZ, Washington, DC 20546 ⟨cpilcher@hq.nasa.gov⟩

GWEN POLLOCK, Illinois State Board of Education, 100 N. First Street, C-277, Springfield, IL 62777 ⟨gpollock@isbe.net⟩

STEPHEN POMPEA, National Optical Astronomy Observatory, 950 N. Cherry Ave, Tucson, AZ 85719 ⟨spompea@noao.edu⟩

IRENE PORRO, Center for Space Research – MIT, 77 Massachusetts Ave, NE80-6079, Cambridge, MA 02139 ⟨iporro@space.mit.edu⟩

ED PRATHER, University of Arizona – Steward Observatory, 933 N Cherry Ave., Tucson, Az 85721 ⟨eprather@as.arizona.edu⟩

WAYNE PRYOR, Hampton University, Center for Atmospheric Sciences, Hampton University, Hampton, VA 23668 ⟨wayne.pryor@hamptonu.edu⟩

MICHAEL RADDICK, Johns Hopkins University, Department of Physics and Astronomy, 3400 North Charles St., Baltimore, MD 21218-2686 ⟨raddick@pha.jhu.edu⟩

CHRIS RANDALL, TERC & MarsQuest Online, 2067 Massachusetts Ave., Cambridge, MA 02140 ⟨chris_randall@terc.edu⟩

SHANNON RANGE, Gravity Probe B at Stanford University, 501 Jersey Street, No.2, San Francisco, CA 94114 ⟨kdoah@stanford.edu⟩

MARTIN RATCLIFFE, International Planetarium Society, Exploration Place, 300 N McLean Blvd, Wichita, KS 67203 ⟨mratcliffe@exploration.org⟩

GREG RAWLS, McREL, 2550 S. Parker Rd., Suite 500, Aurora, CO 80014 ⟨grawls@mcrel.org⟩

PATRICIA REIFF, Rice Space Institute, 6100 Main St. MS 108, Houston, TX 77005 ⟨reiff@rice.edu⟩

JOHN RISTVEY, McREL, McREL 2550 S. Parker Road Suite 500, Aurora, CO 80014-1678 ⟨jristvey@mcrel.org⟩

DIANNE ROBINSON, Hampton University, P.O. Box 6142, Hampton, VA 23668 ⟨dianne.robinson@hamptonu.edu⟩

ADRIENNE RODRIGUEZ, Brownsville Alliance for Science Education, UTB/TSC 80 Fort Brown, Brownsville, TX 78520 ⟨arodriguez@utb.edu⟩

JAMES ROLLER, Lewis Center for Educational Research, NASA/JPL, 17500 Mana Road, Apple Valley, CA 92307 ⟨jim@avstc.org⟩

JEFFREY ROSENDHAL, Office of Space Science, NASA HQ, Code S, Washington, DC 20546 ⟨jeffrey.rosendhal@hq.nasa.gov⟩

LAURIE RUBERG, Center for Educational Technologies, Wheeling Jesuit University, 316 Washington Avenue, Wheeling, WV 26003 ⟨lruberg@cet.edu⟩

CASSANDRA RUNYON, SERCH – NASA OSS Broker/Facilitator, College of Charleston – 66 George Street, Charleston, SC 29424 ⟨runyonc@cofc.edu⟩

PHILLIP SADLER, Harvard-Smithsonian Center for Astrophysics, 60 Garden St., MS 71, Cambridge, MA 02138 ⟨psadler@cfa.harvard.edu⟩

PHILIP SAKIMOTO, NASA Headquarters, Code SB, Washington, DC 20540 ⟨phil.sakimoto@hq.nasa.gov⟩

JOSE SALGADO, Adler Planetarium & Astronomy Museum, 1300 S Lake Shore Drive, Chicago, IL 60605-2403 ⟨salgado@adlernet.org⟩

PERRY SAMSON, University of Michigan, Atmospheric, Oceanic & Space Science, Ann Arbor, MI 48109-2143 ⟨samson@umich.edu⟩

DENNIS SCHATZ, Pacific Science Center, 200 Second Avenue No., Seattle, WA 98109 ⟨schatz@pacsci.org⟩

NANCY SCHLACK, NASA/UNCF Project: ESci Dept, NEIU, Northeastern Illinois University / Earth Science Department/ 5500 North St. Louis Ave., Chicago, IL 60625 ⟨N-Schlack@neiu.edu⟩

JACK SCHLEIN, York College of CUNY, York College, 94-20 Guy Brewer Blvd., Jamaica, NY 11451 ⟨schlein@york.cuny.edu⟩

GREG SCHULTZ, UC Berkeley, Space Sciences Lab, MC 7450, Berkeley, CA 94720-7450 ⟨schultz@ssl.berkeley.edu⟩

NICOLE SCOTT, National Association of Black Geologists and Geophysicists, 15215 BLUE ASH DR #904, HOUSTON, TX 77090 ⟨netha@worldnet.att.net⟩

ROBERT SEMPER, Exploratorium, 3601 Lyon Street, San Francisco, CA 94123 ⟨robs@exploratorium.edu⟩

JOHN SEPIKAS, Jet Propulsion Laboratory, Pasadena City College, 1570 E. Colorado, Pasadena, CA 91106 ⟨jpsepikas@paccd.cc.ca.us⟩

CHRISTOPER SHINOHARA, University of Arizona, ⟨chriss@lpl.arizona.edu⟩

HARRY SHIPMAN, University of Delaware, Sharp Laboratory, Newark, DE 19716 ⟨harrys@udel.edu⟩

VICTORIA SIMEK, NASA Space Science Center at DePaul University, 990 W. Fullerton Ave, Suite 4400, Chicago, IL 60614 ⟨vsimek@depaul.edu⟩

DENISE SMITH, Space Telescope Science Institute, 3700 San Martin Drive, Baltimore, MD 21218 ⟨dsmith@stsci.edu⟩

MICHAEL SMITH, American Geological Institute, 4220 King Street, Alexandria, VA 22302 ⟨msmith@agiweb.org⟩

CARY SNEIDER, Museum of Science, Science Park, Boston, MA 02114-1099 ⟨csneider@mos.org⟩

ANITA SOHUS, Jet Propulsion Laboratory, 4800 Oak Grove Drive, MS 311-100, Pasadena, CA 91109 ⟨Anita.M.Sohus@jpl.nasa.gov⟩

ROBERT SPARKS, The Prairie School, 4050 Lighthouse Dr., Racine, WI 53402 ⟨rspark@prairieschool.com⟩

PEG STANLEY, STScI, 3700 San Martin Drive, Baltimore, MD 21218 ⟨pstanley@stsci.edu⟩

KEIVAN STASSUN, Vanderbilt University, Physics and Astronomy Department, VU Station B 1807, Nashville, TN 37235 ⟨keivan.stassun@vanderbilt.edu⟩

SIMON STEEL, Harvard-Smithsonian Center for Astrophysics, 60 Garden St, MS 71, Cambridge, MA 02138 ⟨ssteel@cfa.harvard.edu⟩

STEPHANIE STOCKMAN, SSAI NASA/GSFC, Code 921, Greenbelt, MD 20771 ⟨stockman@core2.gsfc.nasa.gov⟩

GEORGE STRACHAN, Solar Terrestrial Probes & Living With A Star, Code 460, Bldg. 6, Rm. S133A, GSFC/NASA, Greenbelt, MD. 20771 ⟨gstracha@pop400.gsfc.nasa.gov⟩

CAROLYN SUMNERS, Houston Museum of Natural Science, 1 Hermann Circle Drive, Houston, TX 77030 ⟨csumners@hmns.org⟩

SUZETTE SVOBODA-NEWMAN, Medical College of Wisonsin – Center for Science Education, 8701 Watertown Plank Road, Milwaukee, WI 53226 ⟨suzettes@mcw.edu⟩

JAMES SWEITZER, NASA Space Science Center at DePaul University, 990 W. Fullerton Ave, Suite 4400, Chicago, IL 60614 ⟨jsweitze@depaul.edu⟩

VALERIE TAYLOR, Northwestern University, 2145 Sheridan Road, ECE Dept, Evanston, IL 60208-3118 ⟨taylor@ece.northwestern.edu⟩

WILLIAM TAYLOR, SECEF/GSFC, Code 630, GSFC, Greenbelt, MD 20771 ⟨taylor@mail630.gsfc.nasa.gov⟩

Participants

TERRY TEAYS, Teays Consulting, Inc., 8811 Magnolia Drive, Lanham, MD 20706 ⟨terry.teays@comcast.net⟩

MICHELLE THALLER, IPAC, 770 S. Wilson Ave., Pasadena, CA 91125 ⟨thaller@ipac.caltech.edu⟩

JAMES THIEMAN, NASA/GSFC, Code 633, NASA/GSFC, Greenbelt, MD 20716 ⟨thieman@nssdc.gsfc.nasa.gov⟩

FREDERICK THOMAS, Sinclair Community College, 444 West Third Street, Dayton, OH 45402 ⟨fred.thomas@mathmachines.net⟩

VALERIE L. THOMAS, The LaVal Corporation, 2004 Clearwood Dr., Mitchellville, MD20721 ⟨vthomas@erols.com⟩

PAMELA THOMPSON, Lunar and Planetary Institute, 3600 Bay Area Blvd, Houston, TX 77058 ⟨thompson@lpi.usra.edu⟩

JENNY TIEU, Jet Propulsion Laboratory, 4800 Oak Grove Drive, M/S 301-451, Pasadena, CA 91109 ⟨Jenny.T.Tieu@jpl.nasa.gov⟩

KAY TOBOLA, Johnson Space Center, SA13/C23, Houston, TX 77058 ⟨kay.w.tobola1@jsc.nasa.gov⟩

NICK TOURNIS, Challenger Learning Center of Northwest Indiana, c/o Purdue University Calumet, 2300 173rd Street, Hammond, IN 46323

JOHN TRASCO, University of Maryland, Astronomy Dept., University of Maryland, College Park, MD 20742 ⟨jtrasco@astro.umd.edu⟩

ANN MARIE TROTTA, NASA HQ Office of Space Science, 300 E Street SW, Washington, DC 20546 ⟨atrotta@hq.nasa.gov⟩

CATHERINE TSAIRIDES, NASA Ames Research Center/Astrobiology, Moffett Field, CA 94035 ⟨ctsairides@mail.arc.nasa.gov⟩

PAIGE VALDERRAMA, ASU Mars Space Flight Facility, Moeur Bldg, Rm 131, PO Box 876305, Tempe, AZ, 85287-6305 480.965.3038 ⟨paigev@asu.edu⟩

DEBORAH VANNATTER, Reach for the Stars NASA/DePaul mini-grant, 310 Holly Hill Dr., Evansville, IN 47710 ⟨debvannatter@insightbb.com⟩

MICHELLE VIOTTI, Mars Public Engagement, JPL, 4800 Oak Grove Dr., MS 301-345, Pasadena, CA 91109 ⟨mviotti@pop.jpl.nasa.gov⟩

PROMOD VOHRA, College of Eng & Eng Tech, NIU, Engineering Bldg Room 331, NIU, DeKalb, IL 60115 ⟨vohra@ceet.niu.edu⟩

RICHARD VONDRAK, NASA GSFC, Goddard Space Flight Center-690, Greenbelt, MD 20771 ⟨vondrak@gsfc.nasa.gov⟩

MICHAEL WAGNER, Research Programmer, EventScope, Carnegie Mellon University, 5000 Forbes Avenue, Pittsburgh, PA 15213 ⟨mwagner@cmu.edu⟩

J. HUNTER WAITE, JR., The University of Michigan, Department of Atmospheric, Oceanic, and Space Sciences, Space Research Building, 2455 Hayward, Ann Arbor, MI 48109-2143 ⟨hunterw@umich.edu⟩

CONNIE WALKER, NOAO, 950 N. Cherry Ave, Tucson, AZ 85719 ⟨cwalker@noao.edu⟩

WILLIAM WALLER, Tufts University – NESSIE Broker/Facilitator, Museum of Science, Science Park, Boston, MA 02114-1099 ⟨wwaller@mos.org⟩

LINDA WEAKLY, SciTech Hands On Museum, 18 W. Benton St., Aurora, IL 60506 ⟨ronen@scitech.mus.il.us⟩

EDWARD WEILER, NASA HQ, Office of Space Science, Washington, DC 20546 ⟨eweiler@mail.hq.nasa.gov⟩

LISA WELLS, CFHT, 65-1238 Mamalahoa Highway, Kamuela, HI 96743 ⟨lwells@cfht.hawaii.edu⟩

ALICE WESSEN, Jet Propulsion Laboratory, 4800 Oak Grove Drive, Pasadena, CA 91109 ⟨Alice.S.Wessen@jpl.nasa.gov⟩

KRISSTINA WILMOTH, NASA Astrobiology Institute, MS 240-1, NASA Ames Research Center, Moffett Field, CA 94035 ⟨kwilmoth@mail.arc.nasa.gov⟩

JOHN WILSON, InDyne, Inc./NASA Stennis Space Center, Bldg. 1100, Room 209A, Stennis Space Center, MS 39529-6000 ⟨john.wilson@ssc.nasa.gov⟩

RONDA WILSON, Lunar and Planetary Institute, 3600 Bay Area Blvd, Houston, TX 77058 ⟨rwilson@lpi.usra.edu⟩

SHIRLEY WOLFF, Jet Propulsion Laboratory, 4800 Oak Grove Drive, 303-401, Pasadena, CA 91109 ⟨shirley.e.wolff@jpl.nasa.gov⟩

SIDNEY WOLFF, NOAO, P.O. Box 26732, Tucson, AZ 85726 ⟨swolff@noao.edu⟩

DAN WOODS, NASA Space Science, 300 E Street SW, Washington, DC 20546 ⟨dwoods@hq.nasa.gov⟩

BRYAN WUNAR, Adler Planetarium & Astronomy Museum, 1300 South Lake Shore Drive, Chicago, IL 60605 ⟨bwunar@adlernet.org⟩

RYAN WYATT, Rose Center for Earth & Space, 79th Street at Central Park West, New York, NY 10024 ⟨wyatt@amnh.org⟩

GILBERT YANOW, NASA/Jet Propulsion Lab., 4800 Oak Grove Dr., Mail Stop 264-370, Pasadena, CA 91109 ⟨gilbert.yanow@jpl.nasa.gov⟩

DON YORK, University of Chicago, 5640 So. Ellis Ave., Chicago, IL 60615 ⟨don@oddjob.uchicago.edu⟩

ESTHER ZIRBEL, CUNY & Yale, 42 Howard Ave, Branford, CT 06405 ⟨zirbel@astro.yale.edu⟩

Opening Remarks

Welcome Message
Edited transcription

Jeff D. Rosendhal

NASA Headquarters, 300 E Street, SW, Washington, DC 20546, Jeffrey.D.Rosendhal@nasa.gov

Just looking around the room, it's incredibly inspiring to see the set of people we've succeeded in gathering together for the next couple of days.

Words sometimes get overused, but this really is a singular event. It's a singular event because it's really a critical milestone in the journey that so far has taken us significant time, more than eight years, to put together a major program in education and public outreach. We spent the first three years or so developing very solid plans. We did some things in those plans that were very different than anything anybody has ever attempted to date: creating a support network, creating a set of theme-oriented centers for space science education, creating an institutional infrastructure that would enable us to connect with the world of education and understand what that world needed. You'll get a chance throughout this conference to meet and interact with many of those people.

We spent another year creating the infrastructure and developing important policies, and I think one thing that we need to recognize is that much of our effort is no more than five and a half years old. So that all of the activities that you'll hear about that OSS has sponsored, virtually all of them, have been created from a standing start a little over 5 years ago. I think that the number of people who are gathered here today, and the diversity of activities you'll hear about, are a testimony to the energy, creativity, enthusiasm and drive of the literally hundreds of people who are now participating in this program. I think we actually have what may well be, when you add all the pieces together, the single largest program in astronomy and space science education ever undertaken. That is just the tip of the iceberg of what's in the pipeline now and is yet to come.

The real aim of this conference is twofold. This is really the first gathering of the community that we have helped create and it really is an inspiration to look around the room and see who is here. We hope that in this first gathering of our community there will be a lot of effort made to understand what other people have done, to exchange ideas, insights, and experiences. One of our goals is to mutually enrich what we're all doing collectively by such an exchange.

But there's a second key reason for having this conference, which is associated with the whole issue of quality, effectiveness, and impact. The Office of Space Science is absolutely committed to doing the best that we possibly can in contributing to pre-college education and to broadening public understanding of science. We strive to give people an understanding of the meaning and role of science, engineering and technology in their own lives by using space science as a vehicle to address some of these larger themes. We want to make sure that we deliver programs that are of the highest possible quality and effectiveness. We want to do the right thing.

We have a number of programs underway and you'll see many of those outlined in the annual report and newsletter contained in your registration packets. Some of them are breathtakingly good. Some of them are OK. Some of them are well-intentioned but need work. But we really want to make sure that we are not going to rest on our laurels and say, "well, we've got this nice fat book of things that are going on." That's interesting but not necessarily good enough. We want to make sure that whatever efforts we undertake are informed by the best research in education and are done in collaboration with the education community. We want our programs to fill needs that have been defined by the education community. We want our programs to have regional or state impact, and we want to form as many collaborations with outside organizations as we possibly can. If you take a look at the annual report you'll find that approximately 600 outside organizations are now associated with the OSS education program. Every day we turn around and somebody new is approaching us, saying, "We're interested in what you're doing. We'd like to work with you. You have things you can bring to the table that will help us do our jobs better." And that is exactly what we want to hear. We're not doing this TO the education community, we're doing this WITH the education community.

One thing we've learned is that some of the personnel carrying out the OSS E/PO program have to deepen their understanding of education. So a very large part of this conference is meant to bring the participants together and, in sharing their experiences and ideas, to make sure that we are really doing the right things. That we are deepening our understanding of the needs of the education community and using this meeting as a way to really strengthen and enrich the dialogue that we've begun.

I look forward to sharing time with you, to drawing on this experience, to listening carefully. I hope all of you will get a chance to talk with me because we really are committed to continual improvement. So, I welcome you again, and I look forward to the next couple of days.

Thank you.

NASA Space Science –
Our Commitment to Education and Public Outreach
Edited transcription

Edward J. Weiler

Associate Administrator NASA Space Science, NASA Headquarters, Washington DC 20546, eweiler@mail.hq.nasa.gov

It's really a pleasure to be here for a lot of reasons. I grew up in Chicago about six miles southwest of here and if you had told me back when I was growing up in inner-city Chicago that someday I'd be standing in front of you as the Associate Administrator for NASA's Space Science Enterprise in charge of a $3 billion program, I would have thought you were worthy of a mental institution.

How did I get here? Two simple words: inspiration and education. These are two key words in my life. I was inspired when I was age 13. I remember getting up before grade school on three particular days to watch John Glenn, Alan Shepherd, and Gus Grissom go up into space. By age 13, I had decided I wanted to be an astronomer, attend Northwestern University to get my Ph.D. in Astronomy, and work for NASA someday. As I told a bunch of very high I.Q. 13 year-olds at a special school in Virginia just last week, you don't achieve your goals in life by just having a very high I.Q., because I don't have one. It also takes hard work, inspiration and education.

This is going to be a mixed up presentation. I'm going to tell you a little bit about a lot of things so I'm not going to give you a whole lot of meat in the presentation. That's going to come in the next two days. I will be giving you the hors d'oeuvre today and later you will get the main course, the salad and the dessert.

Sean O'Keefe has only been our Administrator for six months, but right away he saw the need for a new vision, something we could focus on. He could see that NASA was perhaps overly focused on finishing and controlling the cost of the space station. So he got the five Associate Administrators, I'm one of them, and the ten Center Directors together and locked us up in a little retreat-house somewhere in central Virginia for three days. There was a lot of interesting discussion about the future course for NASA and in the end, we came up with the following, not just by consensus, but by acclamation. It's basically three thoughts: to understand and protect our home planet, to explore the universe and search for life, and a very important one, to inspire the next generation of explorers as only NASA can.

Suddenly education was elevated to one of the three main thrusts of NASA. That's a very, very significant piece of news for all of us who are interested in education. As I speak, NASA is in the process of forming a new organization to bring all elements of education together, not to centralize it and undo all the good independent work that's been done by the NASA science enterprises and centers, but to bring it together with a clearer focus. It's a big task. That's why we're moving slowly and that's why you haven't heard of the new organization

Figure 1. *Birmingham News*, October 7, 2000

yet. But Sean O'Keefe feels strongly enough about this that he wants it to be a central focus of the Agency.

Why is education so important to NASA Space Science? Well first of all, a lot of us have always thought it was a part of our jobs. The US Space Act clearly says that one of our jobs is to explore the universe, but the sentence doesn't end there. It doesn't say "give all the data to scientists, give them all money and make them happy." We do that and scientists are happy, but it also says, "and share the discoveries with the American people." That's in the Space Act, except we haven't noticed it enough over the past 40 years. I think we're noticing it a lot more today.

I want to tell a story about the little girl in Figure 1. One day Jeff Rosendhal brought a newspaper into my office. I think it was front page, Section A of the *Birmingham News*. There was a little girl, a little blond girl touching a simulated Mars rock in a local museum. I guess it touched me for a lot of reasons. She could be the twin of my daughter Allison, who is now 16 but looked just like that when she was about 8 years old. But what got me was the look on her face. That was an especially tough day that day. Jeff brought this newspaper up to me during a rather difficult OSS review and it was like a shining star in a starless sky. I held this up to the rest of the staff and said, "People, this is why we come to work in the morning. This is what makes it worth it. Forget about the high pay as a federal civil servant and all the great benefits. Forget about all those wonderful things. This is what makes it worth our time."

Let me digress a little bit to say that what we seek to do in the Office of Space Science is obviously to share the discoveries with the public, with the people who pay for it, but also to enhance the quality of science, math, and technology, particularly at the pre-college level. When you saw the words "Next Generation of Future Explorers" in our vision, what do we mean by "Explorers?" We really mean mathematicians, scientists, engineers. I have a chart on that coming up. Why do we pick on those types of people? To help create the 21st century workforce.

The Office of Science Technology Policy, OSTP, which is the science advisory arm of the White House, put out a report under the previous President about the future of electrical engineering, and this applies not just to electrical engineering. We can put mechanical engineering up there or we can put physicists or astrophysicists up there. Basically, fewer and fewer undergraduates are completing degrees in the hard sciences and engineering in this country. Furthermore, the number of advanced degrees in science, math and engineering is continuing to go down. That includes the number of foreign Ph.D.'s who are graduating and don't stay here but go back to their home country. Why is that important? Why is science and technology important? Well, I know you probably don't remember this with the current situation in the stock market, but back in the '90s there was a technological revolution, an economic revolution. Cell phones, beepers, I mean it was endless. I guess some people think that all that stuff came from Santa Claus. But unless Santa Claus is wearing a slide rule, I maintain that the technological revolution of the 1990s was created by the engineers and scientists who were inspired in the '50s and '60s and became working engineers and scientists in the '80s and helped develop some of the technologies we all wear on our belt now or use at home. If this country's going to survive, it needs a strong military certainly, but it also needs a strong economy. Where is the next technological revolution going to come from with a situation like this? We've got to inspire more kids to get into physics and engineering and math and into teaching all those subjects. I think that's something we forget to mention. It's not just getting more physicists and engineers into this country. It's getting people who can train them.

So what can NASA do? Remember our vision says, "As only NASA can." NASA is not going to go out there and build schools or museums. We can't afford to build a new infrastructure to make this happen. We've got to work with and nurture the existing infrastructure. What can we add to it? We can add the raw material of discovery and inspiration. When I started working at NASA in 1978, I strongly believed that my education helped me get to where I was. When I arrived at NASA, I saw that NASA produced some really exciting stuff, but seemed to put a bushel basket over the shining light of its achievements. We weren't doing a very good job bringing our discoveries to the classroom. So when I became responsible for the Hubble Space Telescope, I made sure that HST engaged in a healthier education program. I could do that in the position I was in then as Program Scientist for HST. Now, the really neat thing about being the Associate Administrator for Space Science is I can make education and public outreach happen across all of the Office of Space Science programs. That's really fun. So we now have a policy in OSS that a fixed percentage, albeit a small percentage of 1 or 2 percent, of every single project in OSS is taxed to fund education and public outreach. I have a feeling, although I can't report it today, that that kind of "taxation," so to speak, is going to be seen more prominently across NASA very soon.

Let me tell you a little about NASA Space Science itself. The NASA Space Science vision was created prior to the new Administrator's vision. This was the vision that we created when I became Associate Administrator in 1998. As you can see, there is a remarkable symmetry with our vision and the overall Agency vision. So we feel very comfortable with the new Agency direction. I wrote this down one day, actually I wrote it in my head as I was driving on U.S. 50 toward my home in Annapolis where I do my best thinking for NASA. I was frustrated that we had all this great science, but we continued to describe it in technical or scientific terms. Wouldn't it be nice if we could describe our program not totally in scientific terms, but in terms the average human being could understand? These four questions flowed out of my thought process and it dawned on me that they not only describe everything we do in Space Science, but they also go beyond science questions. They're human questions. They're questions that the average person sitting on the plane next to you, like when I flew from Washington to Chicago today, is interested in. When people find out you're from NASA, they usually know what the Hubble Space Telescope is, so they're usually interested in if there is life in the universe and your plane ride is either ruined or enjoyable depending on the kind of person it is and depending on what you wanted to do on that plane ride, like look at your charts. How the universe began and evolved? Age-old question. Basic human question. How did we get here? Not "we" this group but how did "we" the human race, life, the Earth, the solar system, how did we get here after the Big Bang? Where are we going? What's the future of the universe? And perhaps the most fundamental and the most human question of all, probably asked for the first time when Fred and Wilma Flintstone looked at the stars: Are we alone? Some scientists think we put too much emphasis on that. I don't think we do. I think it's a very basic question and I think everything we learn in astronomy makes that question even more interesting.

The way we structure the Office of Space Science is in four thematic areas. One I used to be in charge of before I became an Associate Administrator is Astronomical Search for Origins – understanding the origin of galaxies, stars, and life itself. The missions involved in this theme include the Hubble Space Telescope, and future missions such as the Next Generation Space Telescope, and the Space Infrared Telescope Facility, which is HST in infrared. SIRTF will be launched next year.

The second theme is the Structure and Evolution of the Universe. This is really studying the high-energy universe and x-rays, gamma rays to the millimeter, and worrying about what are the forces that shape the universe. Are there physical laws we don't yet understand, even beyond Einstein? We seek to understand what happens at the event horizons of black holes, because if Einstein's laws are going to break down, they're going to break down somewhere in that vicinity. Missions involved in this theme include the Chandra X-ray telescope, which is the X-ray version of Hubble, and several gamma ray observatories.

Solar System Exploration is the third science theme. Basically understanding the history and diversity of the solar system including Mars, for which we have a very impressive program. Next year we plan to launch two – we call them SUVs in a bag – golf-cart-sized rovers. They're really roving geology labs, that will land on different spots on Mars and not just go up to a few rocks a few meters away like the Pathfinder mission a few years ago, but be able to go up to a kilometer each and last for about 90 days instead of one month. We intend to get that data, there is a lot of inspiration there, to kids on the internet as soon as possible. So that should be a very interesting time. Other missions in the solar system program include the Galileo mission around Jupiter right now, which has really, really helped us understand the moon Io and the possibility of life on Europa. Not *on* Europa but *in* Europa. It's pretty clear now that the moon Europa has an ocean underneath its ice. What we don't know is whether the ice is 20 meters thick or 20 miles thick or 200 miles thick. That's up to future missions. But Europa could have the conditions to support various rudimentary forms of life. And certainly Mars does.

Our final theme is the Sun-Earth Connection, featuring the Sun. By the way, something I've never understood, when I came to NASA 23 years ago the solar system division did not include the Sun. I still haven't figured out why. We have separate divisions and we have a separate theme for the Sun and its effects on the Earth. It's kind of ironic because without the Sun, the solar system wouldn't be much to brag about. Anyway, I think there's been a revolution in public interest in solar physics and the effects the Sun has on the Earth. To the great credit to the people at the Goddard Space Flight Center, we suddenly started getting our scientists in front of the cameras to talk about the fact that the Sun is not that constant yellow light you see up in the sky, but rather it has a profound effect on the Earth, its climate, and sometimes our everyday life, if your beeper or your cell phone quits working because of a solar storm. Suddenly pictures of the Sun like this were appearing on the Dan Rather nightly news and going to every TV in the nation. I'm very proud of the Sun-Earth Connection theme because for decades people who worried about space physics and solar weather and solar physics were sort of the backwater of NASA and now they're probably second only to Mars and Hubble in terms of generating excitement and interest in the American public.

Now I'm going to show you a plethora of examples of OSS education and public outreach activities we do. If the particular activity you do is not here, do not be offended. Especially don't blame me because Jeff picked these out. He'll be here for the next two days. I have to go back and be a bureaucrat, so take it out on him.

Scientists can contribute to education and public outreach. Again, emphasizing that unique resource that only NASA can provide, space scientists have a responsibility to get their discoveries out to the American people; especially school kids. OSS is aggressively searching for those space scientists who have the skills to effectively communicate their research to the general public. Now, I'm not criticizing scientists – I'm one of them – but some of them do not have the ability to talk to kids and you really don't want them going to a classroom of eight year olds because they might not inspire them to become scientists. You know who they are. There is, however, a growing number of scientists who recognize (1) they've got to speak in an understandable way, and (2) that education and public outreach is important. If you get federal money from NASA to do your science, doing your research and publishing your paper is not all you're paid to do. There's another step that again, is part of the US Space Act. As a scientist, you need to get your discoveries out to the American people, especially school kids. That's something we continue to emphasize to our scientists.

Traveling exhibits. The Space Weather Center is currently in Detroit. MarsQuest is a very popular exhibit that's currently here in Chicago. The Hubble Space Telescope exhibit is currently in Kansas City. So again we're not funding the museums, the infrastructure, or the planetarium to do this kind of stuff. We're working with people who know how to do education and create exhibits and providing them with our unique type of resource – science and scientists – to reach the American people. That's a key point that I want to keep driving across, is that we cannot compete with the existing infrastructure that is there but we can certainly feed it and work with it.

This next example is becoming very popular especially now that it's spring and summer in Washington, except now it's probably too hot to walk along the Mall. But I participated in the grand opening of this scale model of the solar system in Washington. It's a really, really neat idea. It's spread over 650 yards, about seven football fields. We put together, working with the space science community, a scale model of the solar system to one ten-billionth scale. So you start out with the Sun right outside the doors of the Air and Space Museum and you basically start walking. Very soon you get to Mercury and not too much later Venus, then Earth, then Mars. A series of long walks beyond Mars take you to Jupiter, Saturn, Uranus, Neptune and finally Pluto. When you get to each planet, you see a scale model of the planet and its moons along with a description. The exhibit opened in October 2001, and our previous Administrator thought strongly enough about education that he attended the event. So this is just a great idea. I thought it was really unique and I was happy to be part of it.

Why broadcast from the internet and television? I see Geoff Haines-Stiles is here. I worked with Geoff on the *Live from Hubble* broadcast about seven years ago. Through *Passport to Knowledge*, a "Live From" series, we've had *Live From Mars*, *Live From Hubble*, and *Live From the Sun*, which are all very popular programs that were broadcast on PBS nationally. Another really neat event is that we have been able to get solar eclipses covered live on the internet with student participation. Again, the use of the internet and television has led to a whole bunch of really innovative ideas for education and public outreach.

The Solar System Ambassadors Program consists of volunteers who are recruited and trained by JPL to bring OSS programs into local communities across the country. In 2001 we had 206 Ambassadors in 48 states. By next year we hope to have Ambassadors in all 50 states and Puerto Rico.

Here is another one of my favorites because it highlights the underdog Sun-Earth Connection theme right here in Illinois with the Illinois MagNet Project. This is something very simple to do and not very expensive. But it gets kids interested in science. It's a network of teachers and students who build magnetometers to collect data about the Earth's magnetic field. It began as a small teacher-led activity at Goddard, and with the help of Walter Payton(someone who I consider to be the best football player to ever play Illinois football) High School in Illinois, 100 teachers have been trained and have developed lessons with magnetometers. A 2002 workshop will extend that beyond Chicago.

SUNBEAMS is a local program in Washington, DC. Local sixth grade teachers in math and science spend five weeks at NASA Goddard in Greenbelt, Maryland, working with the technical mentors and developing lesson plans for their students. We have student workshops where each teacher brings 30 students to Goddard to be immersed for a week in science and math and then students actually develop their own webpage about their experience. With partners including Goddard and the District of Columbia schools, it's ultimately involving every sixth grade student in the District of Columbia.

Another one of my favorites is the SOFIA E/PO program. SOFIA is going to use one of the old SP-747's, the planes that could fly 10,000 miles nonstop, to conduct its science. We are working with United Airlines and a university consortium to basically put a hole in the fuselage of the plane. They tell me we don't actually cut a hole in the plane. What we do is tear off the skin and build the skin back up and a hole kind of appears. I made the mistake yesterday of asking when will the hole be cut in SOFIA and people kind of sheepishly explained to me that that's kind of the wrong way to look at it. So sometimes I guess I scare people into telling me the real truth. Anyway, SOFIA is going to fly a 3m infrared and 7mm telescope up in the stratosphere about 42,000 feet. Why do that? Why not just launch a telescope? Well, to launch a telescope it costs hundreds and hundreds and hundreds of millions of dollars. To do this kind of astronomy, all you have to do is get above most of the water in the atmosphere. When you're up about 41-43,000 feet like this plane we've got here, you can do really, really unique science. Because it's an airplane, because it's a very safe airplane with an excellent safety record, we plan to fly teachers. Not just for the ride but to actively work with the research groups, to develop experiences, and to take those experiences and real research back to the classroom. I think it's a really, really neat idea. We did a little bit of this on a C-141 airborne program about 10 years ago, but it's really going to take off, no pun intended, when SOFIA starts doing its research in 2004.

This one really touched my heart, especially after 23 years on Hubble. Everybody in the world who gets magazines and who can read a newspaper, see a newspaper, who can watch television, has seen the universe through the eyes of Hubble. Everybody in the world with one exception: the visually impaired. Somebody, one of the Hubble scientists, had an idea: wouldn't it be neat to use Braille techniques, tactile techniques to take the digital images from Hubble and put it into a format that the visually impaired could use and understand? The young lady in Figure 2 is touching a planetary nebula, feeling a planetary nebula, perhaps seeing a planetary nebula for the first time. That is a Braille explanation of the science they're actually feeling there. This has been produced into a book. We got the National Academy Press to print the book and they've already gotten numerous requests for the book. The price for this grant was $10,000. In my job we don't normally deal with $10,000 in a $3 billion program. Considering Hubble is a $4 to $5 billion program, this is quite a dividend for a very tiny amount of money.

Enhance the quality of education. None of this means anything if you can't find the OSS educational resources that are available. So we created an online educational resource directory. It's a convenient way to find all of our education products and your education products on the web right now.

Here is a subject that Jeff keeps reminding me of and I really appreciate this because I have not thought about it much. When we talk about educat-

Figure 2. A student from the Colorado School for the Deaf and the Blind examines a Hubble Space Telescope Image of the Eskimo Nebula (see also Grice et al. page 301 of this volume)

ing the next generation of explorers, engineers, mathematicians, teachers, and scientists, where is the next generation coming from? It may not be the traditional generation of people we're talking about because in the year 2030, if you take a collection of all minorities, Hispanics, African-Americans, Native Americans, Asians, they almost become a majority. We have to begin paying special emphasis to minority education, especially at the elementary school level.

The OSS Minority University Initiative recently allowed us to foster a strong linkage with historically black colleges. Fifteen institutions were selected to receive funding from OSS. We now have seven Space Science missions or suborbital experiments that have minority participation. We have eight new and ten redirected space science faculty positions, one new major and nine new minors in space science under development, and 32 new space science courses being offered around the country.

Jeff doesn't realize this because he didn't think I would use these charts but they're probably the most useful charts I've ever used. The irony is that this is a true credit to Jeff because Jeff only has a staff of three or four. He was Mr. Education at OSS for many, many years until recently. This is the education program that Jeff Rosendhal inherited (see Figure 3, left). Really

NASA Space Science – Our Commitment to E/PO   13

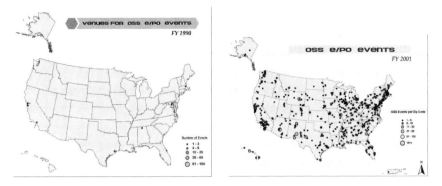

Figure 3.   Comparison of OSS-sponsored education events in 1990 and 2001

massive countrywide program, right? And if you're really sharp you'll notice a correlation to each one of the dots there. Anybody figure it out yet? NASA centers. This is what Jeff inherited. This is when our education program was in its infancy. This is 1990, about the time I started pumping money into the Space Telescope Science Institute, and then over the years through Jeff's leadership, next chart, this is where we are today (see Figure 3, right). We have venues for OSS sponsored education events, partnerships, etc., etc. I believe it's over 3,000 screened events in every state of the union and Puerto Rico. And that's just the beginning.

Actually, Figure 4 is one of my favorite pictures from Hubble. Most people think it would be the Eagle Nebula or the Hubble Deep Field. But you can have a lot of fun with 12 year-olds like I did last week with a chart like this. You can read the quote. I'm not going to read the quote. Anyway, what is that? That's a Hubble picture of the Milky Way galaxy. It covers one two hundred millionth of the sky. One over 200 million. So 200 million Hubble pictures would cover the whole sky. Quite a few stars. Now let's carry this out. I talked to a few physicists last week and I said, "let's do the math." 200 million times that in all the universe. That's just one galaxy. To make a long story short, if you count the number of galaxies that we think we can see and keep in mind each galaxy has 100 or 200 billion stars, the bottom line is that you come to a number of about 10 to the 22nd power stars. So I like to give each kid a homework assignment. I say, "if you really want to understand 10 to the 22nd don't go to your computer. Get a big long sheet of paper and write down 1 and write down 22 zeros after 1. That's how many suns are in the universe." We have found other planets around other stars – 90 to 100 of them. Biologists have found that no matter where they look on Earth, there is life. It isn't the Goldilocks theory like it was when I took biology, where life only existed on Earth because it was just right here – 70 degrees, sunny, lots of water. Mars was too cold, Venus was too hot. When I was growing up and going to school, nobody took it seriously that life might exist elsewhere in the universe. Life on Earth was considered special and unique. The bottom line is biology has been through a revolution. No matter where scientists look on Earth, if they just have water, energy, and organic material, material with carbon in it, they find life. 10,000 feet below the ocean, near volcanic vents. They don't find one microbe. They find tubeworms full of life when they're not quite sure why. No air. No sun. They've got organic material, they've got heat energy, they've got water. Antarctic ice. Yellowstone pools, boiling sulfur pools, if you're close to it they smell so bad, but life abounds. Why is that important? Water, organic material, energy. It's important because every place we look in the universe we find water molecules, we find lots of energy, and lots of organic material. And there's 1 followed by 22 zeros stars out there. Hubble has found that many of the stars we've looked at have a protoplanetary disk. The process of forming planets is going on as I talk to you. The kids get really excited about the possibilities of life in the universe, certainly on the science fiction side but now on the science side. Because yesterday's science fiction has become today's scientific fact.

We have a very exciting program going to Mars. Part of the reason we're going to Mars is not just to learn about Mars geology, but Mars' climate and surface mineralogy. The bottom line is we're also going to Mars to see if there's any life there or if there was any life there. If we go to Mars, the first place we will really study intensely for the business of life, and we find life on the planet, what does that imply for the rest of the universe, considering 1 followed by 22 zeros stars?

I like to think that the history of humankind has been one of a cosmic decentralization of humans. Humans have always tried to find a special place for themselves. If you don't believe it, remember we put ourselves at the center of the universe for a long time in human history. Galileo got quite a shocked response when he had the nerve to say that if everything revolves around us, why does this object Jupiter have things revolving around it? There were clerics who refused to look in his telescope because they couldn't handle the truth. So what did he do? He put the Sun in the center of the solar system. Only a few years ago, it was felt we were probably the only solar system because of the way the solar system must have been formed – a star went by our star, pulled off gas, which condensed into planets. That theory was around for a while, but it turned out to be wrong too. The process where planets form is very common. We're finding other solar systems constantly. What do we have left? What's the last crumb on the plate of human arrogance? We're the only life in the universe. That is the ultimate arrogance. So I'll leave you with that thought. But we've got a lot of raw material here to excite kids and I'm going to continue to support it as long as I'm Associate Administrator for OSS, because as far as I'm concerned it's even more important than the science we do. Thank you.

Biography

EDWARD J. WEILER is the NASA Associate Administrator for Space Science. Prior to this appointment, he served as the Director of the Origins Program at NASA Headquarters in Washington DC, from 1996-1998. He has served as the Chief Scientist for the Hubble Space Telescope since 1979. Dr. Weiler joined NASA HQ in 1978 as staff scientist and was promoted to the Chief of Ultraviolet/Visible and Gravitational Astrophysics in 1979. Prior to joining NASA, Dr. Weiler was a member of the Princeton University research staff. He

Figure 4. Hubble Space Telescope Image of the Sagittarius Star Cloud in the Milky Way Galaxy. Credit: Hubble Heritage Team (AURA/STScI/NASA).

joined Princeton in 1976 and was based at the Goddard Space Flight Center as the director of science operations of the Orbiting Astronomical Observatory-3 (Copernicus). Dr. Weiler received his Ph.D. in astrophysics from Northwestern University.

Knock, Knock ... Who's There? Looking into Our Schools

Maria Alicia Lopez Freeman

California Science Project, 3806 Geology Bldg., UCLA, Box 951567, Los Angeles, CA 90095, mafreema@ucla.edu

Education is a topic that people feel qualified (by virtue of their own education) to critique or praise. Frequently, comments reflect experiences that are long past and that have a museum-quality patina that tends to either idealize or demonize practices, teachers, classes and schools. Thus, many times cherished images call into place bright, sunny, well-ordered rooms; worn but usable books; extremely knowledgeable teachers; and classmates that were very similar in looks, language and experiences. Teachers were either very compassionate or mean; schools were great places with a well-ordered progression of classes, little disruption of schedules, and classmates and friends that were average to good students. Today's schools may in fact be very different. So let's knock on the classroom door and see what is going on.

A way to do this is by using the 2000 National Survey of Science and Mathematics Education to view science and mathematics classrooms in public schools across the country. This study is the fourth in a series funded by the National Science Foundation (NSF) since 1977. Topics include: a) science and mathematics course offerings and enrollments; b) availability of facilities and equipment; c) instructional techniques; d) teacher background; and e) needs for professional development. The survey elicited responses from almost 6,000 teachers, including all Presidential Awardees in over 1,200 schools. The questions and responses provide us with ample data to develop a much more realistic and robust picture of science classrooms today.

Who is teaching science? The survey indicates that the majority of high school science teachers are white, male teachers, while less than 8% of all elementary school teachers are white males. This, coupled with an overall percentage of white classroom teachers of more than 88%, indicates a very white teaching force with an increasingly diverse student population.

Of major concern to the science and mathematics education community is the lack of teachers qualified to teach mathematics and science. We know that a decreasing number of students are entering careers in mathematics, science and engineering. Of those students with degrees in science and mathematics, very few choose to pursue certification in teaching. While almost 94% of high school biology teachers have had six or more courses in biology, the corresponding number in physics is only 64%. When one looks at the science preparation of middle school teachers the numbers drop significantly, especially for physics, where only 39% of middle school teachers have had six or more courses. This lack of preparation, coupled with the fact that less than 30% of teachers consider themselves "very well qualified" to teach science, is extremely alarming!!

What is being taught if teachers are themselves not knowledgeable about the fundamental ideas and concepts of this rather magnificent body of knowledge called science? If teachers do not consider themselves "well qualified to teach science," then what is the quality of their science programs? How are teachers to judge the accuracy of instructional materials used to teach science? Do teachers then have high expectations for student learning or are they willing to settle for mediocre achievement given their own lack of knowledge? This lack of well prepared teachers with confidence in their ability to teach science, represents a major constraint in our quest for a scientifically literate citizenry. Teachers cannot teach what they do not know!!

While both the National Council of Teachers of Mathematics (NCTM) and the National Research Council (NRC) have developed national standards documents that are often quoted and are primary sources for mathematics and science policy at the national and state level, these documents have had a very uneven impact on schools and teachers. Although the 2000 National Survey has been able to capture much about who the teachers and students are, information about the implementation of standards is anecdotal and subject to a large range of interpretations. Many teachers do seem to indicate that they are implementing the standards in their schools and classrooms. And it only stands to reason that if a teacher agrees with the standards, this same teacher will indicate that they implement the standards ... not being very clear about what it means to "implement the standards." The national standards in mathematics and science identify the body of knowledge to be taught and the most current ideas and conceptualizations of that knowledge. In view of such, teachers were asked to select their professional development needs from a set of categories. The categories of need that are most interesting are a) the use of inquiry/investigations; and b) the deepening of content knowledge. It seems that both mathematics and science teachers indicate a need for professional development in the use of inquiry/investigation. This is not surprising, given the lack of coursework in science and the manner in which mathematics is taught in colleges and university.

Common practice in science is a laboratory program that has a set of cookbook experiences with outcomes as verification of theory, while most mathematics is taught in lecture courses that fail to have students follow a line of inquiry into a fundamental problem or assertion. However, when it comes to content knowledge, approximately 70% of the teachers of science at the elementary and middle school indicate a need for professional development compared to $\sim 40\%$ of teachers of mathematics at the same level. Such a difference may be due to the perception of what mathematical knowledge is needed based on mathematics programs. Most elementary and many middle school mathematics programs emphasize fundamental arithmetic operations. Only a small fraction of middle school students are enrolled in pre-algebra or algebra classes. These classes are highly coveted assignments for those teachers with strong mathematics backgrounds. Yet in spite of the glaring need for professional development either by self-identification or preparation, it is not clear that all the various professional development opportunities provide adequate support. This is in spite of tremendous efforts and a large infusion of resources by the NSF, National Institutes of Health (NIH), and state and private foundation programs.

Now let us look directly into the classroom. Time is the most precious element in the instructional day. The time we allocate to any particular subject is a measure of what is deemed most important or most critical in a student's education. The number of minutes allocated to the teaching of mathematics and science lags significantly behind the time dedicated to language arts. The impact of both decreasing NAEP (National Assessment of Educational Progress) scores and the ever growing achievement gap between still historically underserved populations and white middle class and Asian students have pushed districts and principals to mandate the time allocated to the various subjects. Thus we find that, on average, 60 minutes per day might be dedicated to mathematics and 30 minutes, at best, to science. There are many schools and districts across the country, that now ask teachers to dedicate 30 minutes once a week to the teaching of science, especially in those schools where low literacy and reading scores are the key preoccupation of parents, teachers and principals. This focus on literacy comes at the same time that more and more students with Limited English Proficiency (LEP) or students classified as English Language Learners (ELL) have become part of mainstream classrooms across the country. There has been a significant increase in LEP students in mathematics and science classes from 1993 to 2000. While the increase in LEP students has been most dramatic in the Southwest and Midwest, recent demographic studies indicate that both the Northeast and South are experiencing similar increases. Simultaneously, as reading tests rely on a knowledge of standard academic English, there are many non-LEP students that, by virtue of poverty, speak non-standard forms of English. These students consistently test very low on reading tests and provide another reason to support the increased allocation of time to the teaching of reading. If students going through the K-12 system arrive at our colleges and universities ill-prepared for the rigors of the study of mathematics, science and engineering, it should not be a surprise. We do not have a K-12 enterprise that supports a vibrant science and mathematics education.

We also see that even when high school mathematics and science programs provide the traditional college preparatory courses, non-Asian minority enrollments are very low. It is not acceptable that only 19% of students in physics are non-Asian minority students. Or that only 25% of the students in high school science classes are non-Asian minority students. Clearly the largest growing sector of our population is not enrolled in those classes that provide for a true choice at the post-secondary level.

This is our picture of mathematics and science classrooms in the year 2000. Yet, there is a new educational context that is applying another form of pressure on the K-12 system and that will bring a new set of actions for mathematics and science educational programs. The newly authorized education act "No Child Left Behind" is calling for every state to develop science and mathematics standards. These standards are to be used for the mathematics achievement test for every child, every year, and for science achievement tests at the elementary, middle and secondary levels. Schools will be asked to demonstrate 'average yearly progress' for all students. All achievement data is to be disaggregated by class and ethnicity and all students are to be tested. Those schools where the lowest achieving students do not meet the targeted 'average yearly progress' will be subject to various interventions and sanctions. All of this accountability will be done under very public scrutiny with scores and progress targets published in local newspapers for all to see and consider.

For the OSS community, the question is where are your efforts and resources best invested? How does the beloved CD created by many E/PO programs enhance the current efforts to assure that science and mathematics programs are effective? Standards-based? How does this CD develop and support the knowledge of science teachers in the field? Maybe the CD production line should stop and reconsider how it helps to make "highly qualified teachers." Does the CD enable the teaching of more science in elementary grades or just bring another glitzy show-and-tell moment to students? How does the content support and enhance standards-based programs and does it align with the textbooks and programs used in classrooms?

Identify a theory of action

Determine what you think is a key issue. Look to the various surveys and studies about the state of mathematics and science instruction, student achievement in mathematics and science, teacher certification and qualification, low performing districts and programs to guide your thinking about what to do. Find out what is known, look to the research to indicate efforts that have been successful or have very early results indicating effectiveness. Listen to schools, teachers and science and mathematics educators and consider their needs and concerns. And once you have a theory in place about what is needed and what to do. ACT ON IT.

Schools, classrooms and teachers can benefit from your knowledge.

Your community has a wealth of knowledge that needs to inform and benefit the educational community. Through joint collaborative work with those in science and mathematics education, your knowledge, insights and concerns will serve to enhance the teaching of mathematics and science. It may be the case that sometimes this is through a lecture for teachers or a classroom presentation. A more subtle but much needed participation is in study groups where teachers analyze student work and grapple with what science is understood and represented.

Partner with the schools

Look to a school where a partnership would greatly capture the strengths that you can provide and that address some critical needs. Schools do not always benefit from the support of science specialists. Be there to listen, to gently guide, and to support efforts in science and mathematics. Be there when science is not on the top priority list. It might just stay on that list because of your presence! Be there when standardized test scores arrive and average yearly progress did not take place. Be there for the good and bad times. Be there for the long term.

Efforts need to support learning!

Every one of your endeavors should be focused on learning. Your resources are squandered if you cannot say that either student or teacher learning are enhanced. If your efforts do not enhance but rather distract from what needs to be learned, then this huge investment represented by the OSS is for naught. Work with schools and those that have thought long and smart about this and continue to stay on target with learning as the ultimate goal of all of your efforts.

Biography

MARIA ALICIA LOPEZ FREEMAN is the Executive Director of the California Science Project and the Director for the Center for Teacher Leadership in Language and Status Issues. She holds both an MA and BA from Immaculate Heart College in Los Angeles, California. She has taught chemistry and physics in large urban inner-city high schools, teaching science, developing programs and serving as department chairperson. For the past nine years she has been working in science professional development, science education research, and educational change. She has published articles in both chemistry and science education equity. She is currently involved in the development of case studies that focus on the intersection of science and equity in urban schools. She has worked extensively at a science education policy level. She is on the Glen Commission for the Teaching of Mathematics and Science for the 21st Century, and the Expert Panel in Mathematics and Science Education for the US Department of Education. She also serves on the Committee on Science Education K-12 of the Center for Science, Mathematics and Engineering Education for the National Research Council and is member of the Board of Directors of BSCS. She was a member of the Working Group On Science Teaching Standards for the National Science Standards and also a member of the committee on Adolescence and Young Adulthood Science Standards for the National Board for Professional Teaching Standards. She was twice elected the Chairperson of the Curriculum Development and Supplemental Materials Commission for California.

Session on Science Education Research

Context for the Panel on Science Education Research

The organizers of the conference made the decision to ground the conference in an opening session on science education research. In plenary, a group of panelists gave presentations representing different perspectives on this subject. Following this, in facilitated groups of 10, conference participants discussed reactions and their own awareness and use of science education research.

Jon Miller, Director of the Northwestern University's International Center for the Advancement of Scientific Literacy, chaired the panel. Additional panelists were Shelley Lee, Science Consultant for Wisconsin's Department of Public Instruction, Philip Sadler, Director of the Science Education Department of the Harvard-Smithsonian Center for Astrophysics, and Michael Zeilik, Professor of Physics and Astronomy at the University of New Mexico.

This section includes papers by Miller, Sadler, and Lee. Those of Miller and Sadler emphasize the importance of developing rigorous research methods in science education comparable to those used in scientific research. Both authors provide instructive illustrations. Sadler traces the development of a line of education research related to materials and activities designed to teach the fundamentals of light and color. Miller describes the steps in the design and implementation of a careful evaluation plan by the staff of a hypothetical planetarium. Taking a slightly different approach, Lee speaks about the perspective that teachers of science bring to the area of science education research and the importance of understanding how teachers learn science content for those designing professional development programs.

According to responses on the conference evaluation form, participants found the session "thought-provoking," "informative," "interesting," and "important." One person, in responding to the question "What did you like best about the conference and why?" answered, "Persistent emphasis on science education research. It's important and those of us that aren't aware of the area should be." Another replied, "The importance placed on education research. It shows the importance of approaching all our endeavors from a scholarly perspective."

The Evaluation of Adult Science Learning

Jon D. Miller

Center for Biomedical Communication, Feinberg School of Medicine, Northwestern University, 303 E. Chicago Ave., Chicago, IL 60611, j-miller8@northwestern.edu

The extraordinary growth of science in the 20th century and its pervasive impact on the individuals, institutions, and society have combined to require a substantial effort to provide science information to adults, often after the end of an individual's formal education. This new obligation is driven by exponential rates of advancement in the basic sciences and by the development and wide use of new technologies in computing, communication, agriculture, medicine, and transportation. The last five decades of space exploration have been an important part of the growth of science and technology in our society.

Although the focus of this volume is primarily on the communication of space-related information, it is important to recognize that the evaluation of these efforts is a part of the more general task of trying to identify and understand the best techniques to foster adult learning about science and technology. The purpose of this paper is to discuss the need for rigorous evaluation of adult education efforts, outline some important aspects of designing and conducting evaluations, and encourage the building of multidisciplinary evaluation teams.

The Need for Rigorous Program Evaluation

A good deal of the current interest in the evaluation of outreach programs is driven by the demands of funders – government agencies, foundations, and individual donors – for a more rigorous demonstration of the impact of the programs that they support. Although this response is understandable, thoughtful program directors should recognize the importance of sound on-going evaluation procedures regardless of the funding source.

The harshest critics of adult education and outreach programs often assert that the primary driving force in many programs is the desire of the staff for regular employment and that there is little concern about the impact or effectiveness of program efforts. Having worked with a wide array of adult educators over the last 20 years, I reject that view. I believe that most program directors and staff believe in what they are trying to do, and many take these positions at lower salaries than they might earn in non-educational occupations.

There are substantial numbers of program directors and staff who are concerned about the results of their work, but often feel that they lack the skills or resources necessary to design and conduct an accurate evaluation of the impact of their program. This essay is addressed to these program directors and staff and will try to provide a context and a set of procedures that will be helpful in conceptualizing and implementing a regular and continuing evaluation.

The Design of Evaluation of Adult Science Learning

To be optimally useful, evaluation needs to be an integral part of normal program design and operation. Although the process of program development differs slightly among institutions and organizations, there is a core set of activities that generally occur in the development and implementation of adult science education programs. It may be useful to discuss each of these steps briefly and note how evaluation can become an integral part of the process.

Most adult science education programs begin with a purpose or a need. In some cases, individuals with a need may request educational services. In other cases, an educational program or institution may identify a need and decide to address it. Regardless of the origin, the first step of the process is the development or acceptance of some definition of need. In some cases and some institutions, this definition of need may involve a formal written statement with detail and explanation. In other cases, it may be a few sentences justifying the purpose of undertaking a program or course or activity. In broad terms, this original definition of need should provide a general definition of the outcome or outcomes that an evaluation should expect to measure.

> **Example:** The educational committee of a planetarium decides that it is important to expand the public understanding of the concept of the Big Bang and an expanding universe. Several members of the committee have read reports from national surveys that approximately a third of American adults reject the idea that the current universe originated with a cosmic explosion and think that it is important to try to improve public understanding of the concept of an expanding universe.
>
> For evaluation purposes, the major outcome variable is a measure of whether individuals understand the concept of the Big Bang and their interpretation of the meaning of that construct. In this example, the report that generated the original concern may have included questions that could be used in an evaluation of the planetarium's effort, but, in other cases, it may be necessary to look for other examples or studies that have tried to measure the outcome of interest.

In many cases, a specific need can be addressed by more than one approach. If an institution or organization considers more than one approach to meet a need, there may be an opportunity to compare the efficacy of two or more approaches.

> **Example:** Continuing the previous example of a planetarium that wishes to increase the public understanding of the concept of the Big Bang and an expanding universe, the education staff might begin by looking at the literature for examples of programs that have addressed this problem and may consult with other adult science educators about their experience in this area. Out of this review, the staff proposes two approaches. One approach would create a short film and provide a small seating and viewing area on the main floor of the planetarium. No staff would be present most of the time,

but some carry-away materials would be available. A second approach would create a freestanding exhibit and use a staff explainer to convey information to visitors. This conflict in approaches could be resolved by adopting one approach, or by trying both approaches and carefully evaluating the results.

For the purpose of this discussion, we will assume that the planetarium education committee had the curiosity and resources to develop both approaches and to design a systematic comparison of these two adult learning programs. During the development of the film and the exhibit, each development team might decide that they need to show some of their work to a small sample of visitors to get preliminary feedback. In the evaluation literature, these early assessments are sometimes called formative evaluations or pilot studies, and they often involve small numbers of individuals singly or in focus groups. Apart from the jargon, program development can be improved often by seeking some feedback from potential users at various points in the creative process. Some evaluators put more or less emphasis on this aspect of evaluation, but I think that the level of feedback needed can most often be determined by the design team and institutional leaders. I generally resist formula-driven solutions to this problem.

It is important to think creatively about testing a single intervention or to compare more than one intervention or program. In principle, every hypothesis about learning involves some assessment of change over time. The very notion of increasing public understanding of the concept of an expanding universe implies that some proportion of the public does not currently understand the concept and the objective of the program is to help some additional number of people acquire an understanding and appreciation of the concept. At both the individual and the group levels, learning implies change.

The measurement of change requires measurement at a minimum of two points in time. In most cases, the quality of an evaluation can be improved by a larger number of measurements over a longer period of time, but it is not possible to accurately measure change with a single measurement. Although virtually all science educators know and understand this principle when applied to the measurement of time, weather, or other variables, they disregard this central rule of measurement when it comes to evaluating their own programs.

The explanation of change involves making causal inferences. We want to be able to say that participation in a specific program led to – caused – a change in the level of understanding of a specific idea or construct. Statements of cause, however, require solid empirical evidence and a method that holds constant other factors that may have caused the observed change.

The best approach to causal inference is an experiment. Whenever the word experiment is used, a chorus of adult educators will immediately say that this is fine for a laboratory or for mice, but that it cannot be used in a public setting such as a planetarium. While there are many situations in which a true experiment cannot be conducted, there are far more opportunities for experimental evaluations than many adult educators recognize. Given the substantial inferential power of an experiment, it is an option that should be explored thoughtfully before moving to other options.

Example: Returning to the planetarium example, the evaluation needs to measure change in the proportion of adults who understand the concept of an expanding universe. Given the two approaches discussed above, it would be possible to think of three groups for evaluation: (1) adults who visit the planetarium and view the film, (2) adults who visit the planetarium and look at the exhibit and listen to an explainer, and (3) adults who visit the planetarium and do neither. Group (3) is obviously a control group.

Many educators and evaluators misunderstand the criteria for an experiment that require individuals to be randomly assigned to the three groups. They would be quick to point out that visitors would be resistant to being randomly (arbitrarily in their view) assigned to one of the three groups above. And, of course, that is a correct observation.

What they miss is that it is possible to randomly assign times or dates for the three groups above and to achieve the same experimental end. In this case, let us assume that the planetarium is willing to test the approaches with adult weekend visitors and that the staff and related resources needed for the evaluation could be available on any Saturday or Sunday. By making a list of available Saturdays and Sunday over a 12-month period and randomly assigning one of the three conditions to each day, the planetarium would have set up an experimental design to evaluate the two approaches.

On those Saturdays and Sundays randomly assigned to Group 1, a random sample of planetarium visitors would be asked to complete a short questionnaire at the time of entry into the planetarium and would be encouraged to visit the film area. Free parking could be offered to all of those who watched the film and completed an exit questionnaire. The level of understanding of the concept of an expanding universe at the time of entry to the planetarium would constitute a baseline measurement. The level of understanding at the time of exit from the planetarium would constitute a second measurement and the difference with the baseline would show short-term change in understanding. If each visitor was asked for a telephone number and e-mail address and queried about the same matters a month after the initial visit, a third data point could be obtained, which would show longer-term retention of the concept of an expanding universe. The film facility would be removed from the floor or closed on days assigned to Groups 2 or 3.

Similarly, on those Saturdays and Sundays randomly assigned to Group 2, a random sample of visitors would be asked to complete a short questionnaire at the time of entry into the planetarium and would be encouraged to visit the exhibit/explainer area. Free parking could be offered to all of those who visited the exhibit and completed an exit questionnaire. The level of understanding of the concept of an expanding universe at the time of entry to the planetarium would constitute a baseline measurement. The level of understanding at the time of exit from the planetarium would constitute a second measurement and the difference with the baseline would show short-term change in understanding. If each visitor was asked for a telephone number and e-mail address and queried about the same matters a month after the initial visit, a third data point could be obtained, which would show longer-term retention of the concept of an expanding universe. The exhibit/explainer facility would be removed from the floor or closed on days assigned to Groups 2 or 3.

On Saturdays and Sundays randomly assigned to Group 3, a random sample of visitors would be asked be asked to complete a short questionnaire at the time of entry into the planetarium and at the time of exit. Free parking could be offered to all of those who completed both questionnaires. The level of understanding of the concept of an expanding universe at the time of entry to the planetarium would constitute a baseline measurement. The level of understanding at the time of exit from the planetarium would constitute a second measurement and the difference with the baseline would show short-term change in understanding. If each visitor was asked for a telephone number and e-mail address and queried about the same matters a month after the initial visit, a third data point could be obtained, which would show longer-term retention of the concept of an expanding universe. Both the film facility and the exhibit would be removed from the floor or closed on days assigned to Groups 1 or 2.

By comparing the rate of short-term change and longer-term retention in the three groups of visitors, it would be possible to compare the impact of the two approaches and to experimentally control for all other extraneous background variables. Because there may be systematic variations in the backgrounds of visitors by season, it would be advisable to run the experiment over a period of a full year.

For a variety of reasons, an institution may not be able to randomly assign dates for alternative treatments or groups. In this case, the task of building a solid case for causal inference becomes more difficult, but still possible. In broad terms, there are several strategies to try to control the influence of extraneous variables statistically. Although it is 40 years old, Campbell and Stanley's Experimental and Quasi-experimental Designs for Research (1963) is still the best reference for thinking about research designs when random assignment is not feasible. It is clear, concise, readable, and available from www.amazon.com.

The Construction of Accurate Measures

In addition to a sound experimental or quasi-experimental design, an effective evaluation program requires accurate, valid, and reliable measures of the variables and constructs of interest. Although it may be tempting to measure the understanding of the concept of an expanding universe with a single question, it would be a bad measurement decision.

The existing literature is often a good place to begin. Many of the constructs of interest in adult science education have been measured in previous studies, and few of those measures are copyrighted.[1] Often, however, it is necessary to construct some original measures or scales, especially for programs that seek to influence a wide range of construct understanding, attitudes, or behaviors.

In general, the accuracy and reliability of a scale is higher than for a single question. During the last 50 years, there has been substantial progress in psychometrics – the science of test construction. There has been major improvement in both the theoretical basis for test construction and measurement and the software needed to construct and assess tests. Despite these advances, psychometrics is still not an amateur sport and program developers and leaders should seek professional advice and assistance in test construction. To a slightly lesser extent, there have been major improvements in question construction and attitude measurement. Although some program staff become competent question writers, few staff-constructed instruments reach professional standards. On balance, the cost of professional help in instrument and test construction is usually a good investment.

Equally important, a good evaluation design should include various kinds of measures. If the purpose of a program is to increase knowledge or understanding, one outcome measure would be a short cognitive knowledge test. But, there are other ways to measure knowledge and there are other attitudes and behaviors that may be associated with a change in knowledge that can be measured separately. The growing capabilities of small computers – PC tablets, palm computers, and related devices – and the falling prices of these instruments open interesting new measurement opportunities.

> **Example:** Continuing with our planetarium example, both the short questionnaire at the time of entry and at the time of exit could be collected on a tablet PC. This full power computer looks like a slightly thick tablet of paper and various models work by touch or by stylus. Most important, this device allows the construction of a questionnaire with logic and graphics, including dynamic graphics. It would be possible, for example, to display two or three models of expansion as a part of a question, allowing a respondent to use visual as well as verbal resources in thinking about the question.
>
> Beyond questionnaires, it is possible to use video cameras to monitor the behavior of a respondent during his or her visit to the film or exhibit area. From these data, it would be possible to code the amount of time spent in the area, the conversation patterns of the respondent (did he or she talk with other people about the film or exhibit?), and amount of attention to written information or other objects associated with the film or exhibit.
>
> It is also possible to ask each study participant to wear a special lapel pin during the time that they are in the planetarium and to

---

[1] As a general rule, it is useful to know that single questions cannot be copyrighted, but scales or sets of questions organized into a test can be copyrighted.

use the signal from that pin to create a time and location map of each individual's visit. This kind of data would allow measurement of the number of different exhibits visited during the day, the amount of time at each location, and the content of the exhibits selected for greatest attention.

More than 70 percent of American households have Internet access and it is likely that the percentage is higher for visitors to adult science learning facilities. It is reasonable to expect to do a follow-up questionnaire by e-mail or online for those respondents who have that service and by telephone for those who do not have Internet access. There are several relatively inexpensive software packages for the design and conduct of online surveys, and the resulting data usually cost less and are of higher quality than traditional printed questionnaires or telephone interviews.

A Reasonable Time Frame

The time frame for an evaluation should reflect the learning model and hypotheses that underlie the project. Often, the rationale for a new program will assert that it will produce an increase in knowledge that will be retained by participants and used for years into the future. And the same proposal will include an evaluation plan to test knowledge at the beginning of a program and then two weeks later. This disjuncture between the expected outcome and the measured outcome is obvious.

Earlier in my career, I used to teach an introductory university course in American government. Inevitably, sometime during the first class meeting, a student would ask whether the final exam would be comprehensive, that is, cover the whole course or just the material after the mid-term exam. My response was that I would give a mid-term exam, an end-of-course exam, and that I would re-contact each student every five years for the rest of their life and give them another test! I am not sure that most of the students got the point, but my objective was to prepare people to be citizens for the rest of their adult lives and not just to memorize a few facts for a 15-week period.

In practice, of course, I did not have the resources or the capability to follow and re-test students over a period of time, and few adult science learning institutions or programs will be able to do multi-year follow-up evaluations. Most institutions and programs, however, do have the ability to do follow-up measurements for periods up to a year or two, and the growth of the Internet makes this process easier and less expensive.

Knowing the answer to the length of impact should be important to adult science education organizations. If you knew, for example, that less than two percent of visitors would retain any information from a program one year after exposure, would you invest valuable resources in that program? Would you think that it was important to modify that program to develop a more effective content or delivery mechanism?

The value of a continuing evaluation program that uses sound measurement methods is that it can provide an institutional leader or a program manager with information about the relative effectiveness of various kinds of programming. If a planetarium director knew the relative effectiveness of six different programs operated by his or her institution, it would be possible to allocate resources to expand the most successful programs and either reformulate or drop unsuccessful programs.

The Challenge for the Field

Having spent a number of years surveying national samples of adults and evaluating various adult science education programs, I think that most of our current programs have a low level of long-term impact. Some are outstanding; most are not. The problem is that neither the program managers nor their funding agencies have accurate evaluation information about the effectiveness of current programs.

One of the most important challenges facing the adult science education community today is the development of evaluation programs that correctly identify outcomes of interest, use sound empirical measures over reasonable periods of time, and become a continuous part of program development and assessment. This kind of evaluation is not inexpensive and will require resources, but it is less expensive than the continued operation of programs that do not produce the outcomes we want.

The construction of good evaluation programs will require the development of evaluation teams that combine in-house technical expertise, some external content advisors, and some external measurement and evaluation expertise. It is unreasonable to expect that any individual will have the substantive knowledge to create a wide range of adult science education programs, the measurement knowledge and experience to build good tests and questionnaires, the technical expertise to fully utilize currently available measurement technologies, and the statistical and analytic experience to design the evaluation and interpret the results. Superman and Superwoman are, unfortunately, only comic book characters.

The first step is willpower. The leadership of adult science education programs and institutions – presidents, directors, board members, and funding agencies – has to demand sound and continuing evaluation programs. Far too many of today's evaluation programs measure process instead of outcomes, use inadequate methods, and are largely ignored by decision-makers. When leaders recognize that the price of continuing to support and operate ineffective programs is higher than the cost of good evaluation, we will see better evaluation and more effective adult science education programs.

References

Campbell, D. T., & Stanley J. 1963, Experimental and Quasi-Experimental Designs for Research (New York: Harcourt Brace)

Biography

Jon D. Miller has measured the public understanding of science and technology in the United States for the last two decades, and has examined the factors associated with the development of attitudes toward science. Jon is one of the few scholars in the United States that has studied both the development of knowledge and attitudes in adolescents and young adults and the attitudes of national samples of adults. His basic approach to the study of public understanding and attitudes has been replicated in approximately 30 countries. Presently, Jon is Director and Professor of the Center for Biomedical Communications in the Feinberg School of Medicine at Northwestern University. He is also the Director of the International Center for the Advancement of Scientific Literacy, now located at Northwestern University. He has published four books – *Citizenship in an Age of Science* (Pergamon Press, 1980), *The American People and Science Policy* (Pergamon Press, 1983); *Public Perceptions of Science and Technology: A Comparative Study of the European Union, the United States, Japan, and Canada* (Fundacion BBV, Madrid, 1997); and *Biomedical Communications* (Academic Press, 2001) – and more than 40 journal articles and book chapters. He received his M.A. from the University of Chicago in Political Science, and his Ph.D. from Northwestern University in Political Science.

Educational Research: the Example of Light and Color

Philip M. Sadler

Harvard Smithsonian Center for Astrophysics, Cambridge, MA 02138, psadler@cfa.harvard.edu

Introduction

Science education research can and should play an important role in NASA's education and public outreach efforts. This is not to say that it always does. This is not only a problem exhibited by scientists new to education, but to experienced teachers, museum educators and professors in the sciences. While research scientists make a habit of examining past efforts to help inform a new project and to avoid repeating mistakes, educational efforts often start without examining precedents. Museum exhibits, curricula, demonstrations, and laboratory activities are often developed "from scratch," ignoring earlier work. All too often, the content knowledge, schooling history, teaching experience, and enthusiasm of the development team will propel it prematurely into production mode, eager for a prototype or field test. Only when trying to evaluate the completed project's quality and impact do any previously relevant initiatives surface, but by then their influence on the project can only be minimal.

I advocate the rather mundane advice that educational projects follow many of the same steps as those that are scientific in nature. No corners should be cut. Few scientists will risk not undertaking an extensive initial literature review in a scientific domain to which they expect to contribute. They will seek out and value advice from earlier investigators, utilizing them in an advisory capacity. Although scientists mythologize scientific breakthroughs that occur without familiarity with past advances, such a breakthrough is quite a rare occurrence, and usually does not survive close scrutiny of the facts. Most scientists will carefully formulate goals that are measurable and that specify accomplishments that will constitute success or render the project a failure, with a drive to make a timely contribution to the research literature paramount. By way of illustration, this paper traces the development of a single line of research, that having to do with the development of materials and activities to teach the fundamentals of light and color, topics that are fundamental to understanding much of astronomy and space science.

It is nearly impossible for a scientist or a science teacher to imagine what is going inside a student's mind without concentrated effort and study. While we were all students ourselves, the conceptual frameworks that we now possess bear no resemblance to those that we once had. Reconstructed beyond recognition compared to initial ideas, these frameworks help experts make sense of the world around them in ways that only baffle students. Take, for example, the diagram in Figure 1.

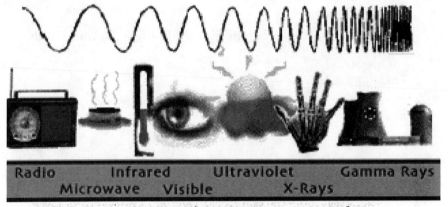

Figure 1. From the NASA website on the electromagnetic spectrum. URL: http://observe.arc.nasa.gov/nasa/education/reference/emspec/emspectrum.html

Figure 1 is an iconic representation of electromagnetic radiation. To a scientist, it shows the various accepted designations of bands of radiation from radio to gamma rays, with examples of objects that are associated with each. Some objects are detectors and some are emitters. All bands are shown covering the same width. The sine wave above represents increasing frequency, but only by a factor of 15 while in reality the range is closer to 1019. Students may make little sense of such an image or draw the wrong conclusions from its format. Without extensive explanation, using such an image can be counterproductive to many, while for those at the proper level it can be helpful. Key to understanding the impact of such objects is understanding the conceptual frameworks held by a particular audience.

Mining The Past

Most of the projects undertaken by our Science Education Department begin with a comprehensive review of the relevant academic literature and prior, published educational efforts. Often the very earliest educational works in the content area are informative. By way of example, light and color taught as spectroscopy and photometry surfaces as laboratory experiments for students in the works of Edward Pickering (director of the Harvard College Observatory for forty-two years from 1876). While a professor at MIT, Pickering published a manual to accompany the first physics laboratory created specifically for student use (Pickering 1874). Fourteen of the fifty exercises dealt with the properties of light, including use and calibration of a prism spectroscope to examine terrestrial and solar spectra and the creation, by hand, of intensity versus wavelength graphs, an example of Pickering's so-called "graphical method" (Hentschel 1999). These experiments spread rapidly throughout colleges in the U.S. and soon became commonplace.

Identification of student "misconceptions" about light and color can also be aided by examining ancient scientific works. Many students would be quite at home with Plato's (ca. 427-347 BC) view in Timaeus (Cornford 1937) that "Visual current issues forth [from the eyes]... (pp. 152-53)." Many students think of vision as an active process, that our eyes send out some sort of invisible rays in order to behold an object (Guesne 1985). In a study of twenty high school students, pupils were asked to explain "What is it that makes you see this object?" Most students never mentioned any linking mechanism between the object and the eye. Others explained that the act of seeing takes place by a "look" or "vision" going from the eye to the object (Anderson & Karrqvist 1983). Students who hold the latter view would have little problem thinking they could see non-luminous objects in the dark, since all they must do is "look" at an object to see it. It appears that many students believe that light is present only if there is enough light for visual effects such as shadows or bright spots to be noticeable (Stead & Osborne 1980).

Color is described as an innate property of an object, e.g. "the book is red" (Anderson & Smith 1986). In his study of fifth-grade students, Charles Anderson found that only two students of 125 described a green book as reflecting green light. Not a single student could accurately predict the appearance of an object when viewed in colored light. This way of thinking extends to the effect of transparent materials on a beam of light. Many people believe that colored objects can transform monochromatic light into another color, so that the color of light can be changed successively by a series of colored glass; white light can be turned into red by a red glass and then turned into green by a green glass (Watts 1985). Such reasoning could predict that prisms might actually just make all colors out of any color of light, not simply spread it out by differing photon energies (or wavelengths, as some prefer). Students are unaware that passing light of a single color through a prism a second time does not produce a rainbow. Marcus Marci (1595-1667) in Prague performed this crucial experiment in 1648, eighteen years prior to Newton's experiment, as shown in Figure 2 (Weise 1960).

Exploring Student's Conceptions and Test Item Construction

A literature review does not fill in all the detail of how students think about a scientific topic. With the view that "The unlearning of preconceptions might very well prove to be the most determinative single factor in the acquisition and retention of subject-matter knowledge (Ausubel 1968, p. 336)," knowing the initial knowledge state of learners is key in creating materials and experiences that foster learning. It is usually necessary to focus upon a specific age group and upon specific concepts. In our work, we usually generate protocols based upon "interviews about instances," setting up hypothetical situations for which students must predict what will happen based upon their own conceptual frameworks (Osbourne & Bell 1983). Perhaps the most well-known example of this method is *The Private Universe* (Schneps & Sadler 1988), a video in which Harvard students are asked to give their reasons for seasonal variations in tem-

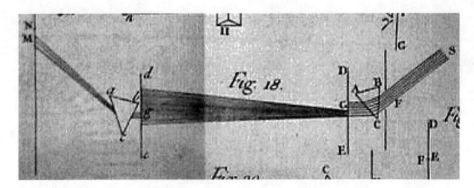

Figure 2. Newton's drawing of his Experimentum Crucis, proving prisms separate light into its component colors (Newton 1721).

perature. In the *Minds of Our Own* video, students likewise explain that vision is an active process, supporting the view that they can see in the dark, given enough time to adapt to the darkness (Schneps, Sadler, Whitney & Shapiro 1997). An additional illustration of this idea is that students claim that light from a lamp does not leave a lamp in the daytime, in spite of the fact that one can see that the lamp is lit up (Sadler 1992). Seeing a faint light in the distance does not mean that there is light actually reaching one's eye. For example, when shown the image below, many students say that the light illumination the road goes only as far as the end of the spot, adamantly denying that if they can see the headlights are lit means that the light is reaching the observer's eyes (see Figure 3).

Figure 3. Example scenario for interviewing students about the propagation of light.

After thoroughly exploring such ideas for the range of views that students express, it is possible to create multiple choice items that gauge the attractiveness of these ideas. An example appears below with the results from 1,414 high school students taking either astronomy or earth science (Sadler 1992).

You are in a completely dark room. There are no lights and no windows. Which group of objects do you believe you might be able to see?

A. bicycle reflectors, a cat's eyes 17%   D. more than one of these groups 16%
B. silver coins, aluminum foil 7%   E. none of these 38%
C. white paper, white socks 29%

Here one can judge that only a minority of students think that light is necessary to see, with different views expressed on the kinds of objects that can be seen in the dark. With a belief that vision is not a passive process, many astronomical concepts and topics become difficult to comprehend, if not impossible. Given that most students do not have a firm grasp on the functioning of vision, learning the workings of a spectroscope is probably built on a shaky foundation. Since many students do not even believe in the propagation of light, jumping directly to viewing Geissler tubes through a grating or examining black and white photographs of spectra may not be helpful (Antonucci & Sadler 2001).

Reaching Goals Through Development

With the goal of shifting students' understanding, abandoning misconceptions for more scientific conceptions, one can construct collections of test items that measure conceptual change. This should take place prior to the development of new materials. In this way one can avoid pursuing activities that may be interesting, but will not move one closer to the stated goal. For Project STAR, this meant creating activities and materials which would allow students to test their preconceived ideas against nature: that they can see in the dark, that objects "change" light, and that light just "exists" without propagating. Out of these efforts came inexpensive spectrometers, methods to project spectra in the classroom for demonstrations and experiments (Sadler 1991), and simple telescopes. Coupled with text readings and exercises, teachers had a whole set of new activities assembled into a cohesive whole. These "hands-on" activities displaced many of the paper and pencil exercises, text readings, and slide shows emulating the lecture-style teaching of astronomy seen in many introductory college courses at the time. Over the course of the project, yearly testing and teacher focus-groups helped to identify the most effective ways to change students' ideas. Through this formative evaluation process the curriculum eventually coalesced into its final form.

Our goals have been increasingly influenced by the NRC Standards (National Research Council 1994) and by the AAAS Benchmarks (American Association for the Advancement of Science 1993). Nearly every state has drawn upon these documents to construct teaching frameworks by grade range. Our aim is to develop materials, programs and assessment instruments that reflect these benchmarks and standards to aid the nation's teachers in fulfilling their states' frameworks. In this way, our efforts support the primary focus of classroom teachers and are not seen as superfluous or auxiliary. For example, this is one of the NRC standards for grade 5-8 Physical Science:

> Light interacts with matter by transmission (including refraction), absorption, or scattering (including reflection). To see an object, light from that object - emitted or scattered from it - must enter the eye (National Research Council 1994, p. 155).

The AAAS Benchmarks are an alternative collection of learning goals that have much in common (having been published two years prior) with the NRC Standards and are often more detailed. Here is an example of a relevant grade 6-8 Benchmark for the Physical Setting,

> Light from the sun is made up of a mixture of many different colors of light, even though to the eye the light looks almost white. Other

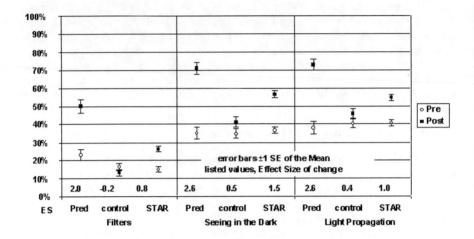

Figure 4. Pre- and post-test prediction from teachers and scores from Project STAR and control classrooms.

things that give off or reflect light have a different mix of colors (American Association for the Advancement of Science 1993, p. 90).

Pre- and post-testing of students engaged with a specific activity, program, or set of materials helps to establish the impact of the intervention. Yet just as in science, controls are very important to compare the gains between groups. For the example of light and color, I present in Figure 4 three questions responded to by 203 teachers predicting student outcomes, 336 students in traditional high school astronomy and earth science classrooms, and 737 students using the Project STAR text (Coyle et al. 1993).

Note that teachers predicted huge gains in students' ability to answer questions correctly. Students in control classrooms actually fared worse on one of the questions after a year of instruction and only marginally better on the other two. Students in Project Star classrooms improved dramatically, with an Effect Size = 1.10 SD (a gain of 1.1 standard deviations based on the sample distribution). The control group improved 0.21 SD overall. Teachers were fairly accurate in predicting the starting level of students, but grossly overestimated the gains they would achieve as a result of instruction, estimating that they would end the year at the level of what we found to be that of Harvard undergraduate astronomy majors (ES = 2.40 SD). For every test item we generated over the years, on average, teachers predict higher gains than are actually achieved in their classrooms.

Sometimes new findings arise from our research that were not initially imagined. Analysis of test results can easily separate easy from hard items, but one can also characterize the relationships between items in a hierarchical fashion. Using a technique developed by Peter Airasian (1993), the order in which concepts depend upon each other can be extracted from test data. One can interpret the resulting diagram (Figure 5) to conclude that very few students who do un-

Educational Research: the Example of Light and Color

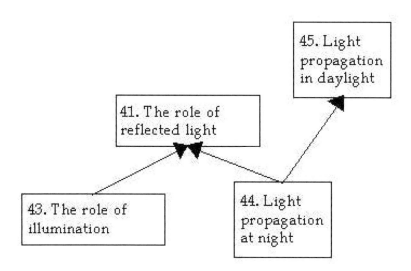

Figure 5. The ordered relationship between understanding light concepts.

derstand how light propagates in the daytime do not also understand how light propagates at night; the reverse is not true, leading to a causal inference. In this way, we found that the role of illumination and the propagation of light at night are the two most basic concepts on which other understandings of light depend (Sadler 1995). This information is important in curriculum design since it affects the optimal order in which topics should be presented.

Conclusions

Educational research is an intellectual domain that is very similar to that of the sciences. People get degrees in it. It has an extensive literature of journals and other publications. Conferences abound. It is highly valued by some and ignored by others. Educational research has methods and tools that work well for solving some kinds of problems, but not others. While it can be described as a field jealous of the success and universality of scientific research, it is less evolved, perhaps as science was three hundred years ago. Promising, but with far to go.

Yet educational research has much to offer those involved in space science and who wish to educate students and reach out to the public. Chief among these offerings is that scientific concepts form a hierarchy and that ideas are easier to understand if one has mastered the basics. Yet basic scientific ideas are always preceded by commonsense notions that are often in direct conflict with scientific beliefs. It is wise to be aware of how scientific understanding grows from that of novices to experts.

Building an educational research agenda into an E/PO project from the start helps to establish goals and provides rigor to the effort. By testing to see whether goals are achieved, formative assessment can guide a project in a productive direction. Moreover, to leave a formal record of the approach attempted for future researchers, as well as connecting to the efforts of past researchers, a useful edifice can be constructed that will serve scientists, educators, and the public. As Lord Kelvin (1824-1907) wisely observed,

> "If you can measure that of which you speak, and can express it by a number, you know something of your subject; but if you cannot measure it, your knowledge is meager and unsatisfactory."

Acknowledgments. I wish to thank several colleagues who participated in this research. Interviews on color and light were conducted by Jennifer and Paul Hickman, now of Northeastern University. Development of experiments on spectra was done by Ann Young of the Rochester Institute of Technology. Spectroscopes were designed by Sam Palmer and Andrew McFadyen of the CfA. Paul Antonucci headed our spectroscopy teaching project. Harold Coyle, Bruce Gregory and Irwin Shapiro of the CfA and William Luzader, now of the Plymouth-Carver School System, worked on many of the issues involving the teaching of light and color. Matt Schneps produced the superb video documentaries on children's ideas in science. The National Science Foundation supported our efforts with several grants (MDR-8550297 and 88504424, ESI-9553846, and ESI-9155229). Additional support was provided by the Annenberg/CPB Foundation and by the Smithsonian Institution.

References

Airasian, P. W. & Bart, W. M. 1973, Educational Technology, 13, 5

American Association for the Advancement of Science Project 2061 1993, Benchmarks for Science Literacy (New York: Oxford University Press)

Anderson, C. W., & Smith E. L. 1986, Educational Resource Information Center (ERIC), ED 270 318

Anderson, B. & Karrqvist, C. 1983, Light and Its Properties, Trans. by Gillian Thylander (Molndal, Sweden: University of Gothenburg)

Antonucci, P. & Sadler P. 2001, Light, Color and Spectroscopy for Every Classroom (Tonawanda, NY: Science Kit and Boreal Labs)

Ausubel, D. 1968, Educational Psychology: A Cognitive View (New York: Holt, Rinehart)

Cornford, F. M. 1937, Plato's Cosmology: The Timaeus of Plato Translated with a Running Commentary (London: Loeb)

Coyle, H., Gregory, B., Luzader, W., Sadler, P., & Shapiro, I. 1993, Project STAR. The Universe in Your Hands (Dubuque, IA: Kendall/Hunt Publishing)

Guesne, E. 1985, in Children's Ideas in Science, pp. 10-32, ed. R. Driver, E. Guesne, & A. Tiberghien (Philadelphia: Open University Press)

Hentschel, K. 1999, Physics in Perspective, 1, 282

National Research Council, National Committee on Science Education Standards and Assessment 1995, National Science Education Standards (Washington: National Academy Press)

Newton, I. 1721, Opticks, 3rd Ed. (London: William and John Innys)

Osborne, R. J. & Bell, B. F. 1983, European Journal of Science Education, 5(1), 1

Pickering, E. C. 1873, Elements of Physical Manipulation (Boston, MA: Houghton Mifflin)

Sadler, P. 1991, The Physics Teacher, 29(7), 423

Sadler, P. 1992, The Initial Knowledge State of High School Astronomy Students, Ed.D. Dissertation, Harvard University

Sadler, P. 1995, in Astronomy Education: Current Developments, Future Coordination, ed. J. Percy (San Francisco: Astronomical Society of the Pacific)

Schneps, M. H. & Sadler, P. M. 1988, A Private Universe (Santa Monica, CA: Pyramid Films)

Schneps, M. H., Sadler, P. M., Whitney, C. A., & Shapiro, I. I. 2001, Minds of Our Own (Cambridge, MA: Annenberg/CPB)

Stead, B. F., & Osborne, R. J. 1980, Australian Science Teachers Journal, 26(3), 84

Watts, D. M. 1985, Physics Education, 20(4), 183

Weise, E. K. 1960, in The Encyclopedia of Spectroscopy, ed. G. L. Clark (New York: Reinhold Publishing), pp. 188-199

Biography

Philip M. Sadler, Ed.D. is F.W. Wright Senior Lecturer in Harvard's Department of Astronomy and directs a team of 45 researchers and staff at the Science Education Department of the Harvard-Smithsonian Center for Astrophysics. His research interests include the development of students' scientific ideas (including astronomical conceptions documented in *The Private Universe*), developing curricula to encourage conceptual change learning (Project STAR, Project ARIES, DESIGNS), and creating computer tools that allow youngsters to engage in research (such as the WWW-based MicroObservatory telescope network). He has co-authored software that has won the American Institute of Physics prize four times, building simulations that allow students to study college physics topics at the high school level. Sadler won the Journal of Research in Science Teaching Award in 1999 for his work relating standardized testing and student misconceptions. His invention of the Starlab Portable Planetarium in 1977 has helped to reinvigorate the study of astronomy by an estimated twelve million youngsters worldwide each year. He admits to missing teaching middle school math and science.

Is the Teacher of Science a Scientist?

Shelley A. Lee

Department of Public Instruction, P.O. Box 7841, Madison, WI 53707-7841, Shelley.Lee@dpi.state.wi.us

Seemingly, the answer to this question is simple; but in fact, the answer resides within the heart of educational research on how students learn and specifically how students learn science. Research about how students are learning has increased dramatically in the past 25 years, beginning with the recognition and acceptance of Piaget's work that led to the constructivist philosophy widely used in education and education research today. Following Piaget's work, Gobbo and Chi (1986) examined the idea that students could be categorized as expert learners or novice learners based on their personal knowledge. Their research identified several important differences among expert learners and novice learners. Expert learners come into the classroom with more and deeper factual knowledge. Expert learners can think contextually, thus making it easier for expert learners to both group their knowledge and retrieve their knowledge when needed. Novice learners have great difficulty thinking deeply, and can only rely on knowledge that exists on the surface or in isolated facts (Lee 2002).

Lowrey (2000) agreed with Gobbo and Chi when he applied this research to the science classroom. He found that a typical student begins to learn science at the novice level and, through experiences in science and the learning process, the student can become an expert learner. He cited examples of hands-on learning to show how the student is able to use many of his or her own senses in the process of becoming an expert learner. This "hands-on" learning provides opportunities for the student to gain a deeper understanding of the science being learned, and leads to a greater retention of the science knowledge and content.

Learning and transfer studies (Bransford, Brown, & Cocking 1999) have examined the notion that students learn best when they are allowed to experience the learning, as cited by Lowrey, and then allowed to transfer that learning to a new and different situation. Within this is the idea that the context for learning is important for promoting and transferring knowledge. Equally important is the idea that the learner's own meta-cognition must be accounted for. This means that more learning takes place when the students become aware of themselves as learners (Donovan, Bransford, & Pellegrino 1999). When students are aware of how they learn science, they can more quickly become expert learners and can transfer that knowledge to new situations, thus retaining the knowledge they have gained indefinitely.

The National Research Council published summative research on how students learn and, specifically, how they learn science in *How People Learn* (Bransford, Brown, & Cocking 1999; Donovan, Bransford, & Pellegrino 1999) and *Knowing What Students Know* (Pellegrino, Chudowsky, & Glaser 2001). The National Science Teachers Association (NSTA) found that this summative research had strong implications for science education in the classroom and for professional development activities including the development of science education materials. To capture the essence of this research, the association published *Beyond 2000 – Teachers of Science Speak Out, an NSTA Lead Paper on How All Students Learn Science; and the Implications to the Science Education Community* (Lee et al. 2003). NSTA made the following recommendation in the publication:

> **The knowledge of how teachers learn science content, how to facilitate student learning, and how to access students learning should guide the policies, programs, and practices that establish professional development of teachers.**

That is, professional development for individuals that provide professional development for teachers must also include the research and understandings about how students learn, and recognize that that knowledge must grow and change as the research changes. The result is life-long professional learning. Equally important is an understanding of adult learning and an awareness that pre-service teachers of science, novice teachers of science, and experienced teachers of science need different professional experiences. Pre-service teaching must include extensive emphasis in field experiences that involve master teachers who model exemplary pedagogical practices and know the science content they are teaching. Novice teachers need professional development that includes increasing their mastery of science content, and expanding their understanding about how students learn the science content being taught, and how students are allowed to become expert learners (Loucks-Horsely et al. 1998; Berliner 1988). Experienced teachers should have opportunities to learn from their own practice, interact regularly with other teachers and researchers, and have opportunities to increase their mastery of science content (Bransford, Brown, & Cocking 1999). Clearly, the same needs are present for individuals providing professional development for teachers of science. These individuals must develop their understandings about professional development that ultimately allow them to become expert learners and have command of the professional development content they are using in professional development work.

Science curriculum programs must be both coherent and focused.

The basis for this statement by NSTA is from the NRC publication that presents resolutions to curricular issues that exist in many curricular products currently being used in science classrooms. The publication examines in depth the notion that science curricula should be both coherent and focused. A coherent and focused curriculum is one that allows all students to develop deep understandings of the science subject matter being covered. A coherent and focused curriculum has a clearly specified number of topics that allow the learners to make connections among the topics presented, has rich science content, allows all students to learn the subject matter science in depth, and follows a predictable path through the content (Bransford, Brown, & Cocking 1999; Donovan, Bransford, & Pellegrino 1999).

Curricular products must be based on current research and understandings about how students learn science. Each product should be strong in scientific content, contain appropriate pedagogical design, and be equitable to all learners. Assessment should be integral and embedded within the product and designed to reveal to everyone what the students are learning and how they are learning the content. Finally, curricular products should provide guidance about the type of professional development needed to implement the curricula (Lee 2002).

The second answer to the question "Is the teacher of science a scientist?" resides in what students should learn. Wiggins & McTighe (1998) coined the term "enduring understandings" as a means for establishing curricular priorities. The Wisconsin Department of Public Instruction (Lee 2002) specified that these enduring understandings in science were the state science education standards that were derived from the National Science Education Standards. Both sets of standards have the same goal in mind, achieving scientific literacy for all students, which is the fundamental "enduring understanding."

Teachers of science and scientists alike must recognize that each offers a unique perspective on science education and science content. Scientists have a depth of knowledge about their area of research and expertise that includes a vast repertoire of scientific facts and ideas resulting from asking and answering questions about the natural or designed world. The teacher of science not only must know science but also must understand how the students are learning the science.

Collectively it is simply not enough for a science teacher to understand or know science, because the knowledge needed to teach science is vastly different from the knowledge used by individuals engaged in scientific research. Teachers of science must ask and answer different questions from those of doctors, engineers, astronomers, or other scientists. Teachers must seek to understand how and what students are learning in the classroom, find errors in students' thinking and seek to correct them, and constantly search for new ways to reveal what and how students are learning in the classroom. Teachers of science and their students must work together to inquire about questions in science and seek to solve those questions. During this inquiry teachers of science must be able to use the knowledge gained from scientific research in ways that allow their students to make sense of that knowledge, to apply the knowledge gained to the natural and designed world; and to retain the knowledge indefinitely, hence becoming scientifically literate students.

References

Berliner, D. C. 1988, The Development of Expertise in Pedagogy (Washington: AACTE Publications)

Bransford, J. D., Brown, A. L., & Cocking, R. R. (editors) 1999, How People Learn: Brain, Mind, Experience, and School (Washington: National Academies Press)

Donovan, M. S., Bransford, J. D., & Pellegrino, J. W. (editors) 1999, How People Learn: Bridging Research and Practice (Washington: National Academies Press)

Gobbo, C., & Chi, M. 1986, Cognitive Development 1, 221

Lee, S. A. 2002, Planning Curriculum in Science (Madison, WI: Department of Public Instruction)

Lee, S. A., Badders, B., Barrow, L., Gadsden, T., Lang, M., Lopez-Ferrao, J., Pietrucha, B., & Pratt, H. 2003, Beyond 2000 – Teachers of Science Speak Out, an NSTA Lead Paper on How All Students Learn Science; and the Implications to the Science Education Community (Arlington, VA: NSTA Press) http://www.nsta.org/positionstatement&psid=17

Loucks-Horsley, S., Hewson, P., Love, N.,& Stiles, K. 1998, Designing Professional Development for Teachers of Science and Mathematics (Thousand Oaks, CA: Corwin Press)

Lowrey, L. F. 2000, Presentation at the Cutting Edge Conference, 13-15 July in Chippewa Falls, WI

Pellegrino, J. W., Chudowsky, N., & Glaser R. 2001, Knowing What Students Know: The Science and Design of Educational Assessment (Washington: National Academies Press)

Wiggens, G. & McTighe J. 1998, Understanding by Design (Alexandria, VA: Association for Supervision and Curriculum Development)

Biography

SHELLEY A. LEE is the science consultant for Wisconsin's Department of Public Instruction where she has recently completed the *Guide for Curriculum Planning in Science* and the *Wisconsin Model Academic Standards for Science*. In her role as science education consultant, she is responsible for assisting teachers and districts around the state in implementing science curriculum reform. She has been active in National Science Teachers Association by serving on the Board of Directors in many capacities, including President in 1995-96, and before that the Middle Level Division Director. She has served on numerous task forces and advisory boards for NASA and is currently serving on the National Research Council's Committee on Test Design for K-12 Science Achievement's Working Group of Science Supervisors. For 23 years she was a ninth grade science teacher for the Sand Springs, Oklahoma Public Schools. In that role, she was also the department chairperson for the district and served on many statewide science committees. She was the recipient of numerous awards, including the Teacher of the Year for her district. She received almost $700,000 in grant funds while teaching.

Sessions on Issues in Formal and Informal Science Education

Context for the Panels on Formal and Informal Science Education

The two sessions on issues in formal and informal science education – the first on the morning of June 13 and the second on the same afternoon – were designed to allow conference participants to engage in discussions on issues and challenges faced by those in the E/PO community. The discussions were stimulated by two means: plenary talks and panel presentations. For formal science education, George (Pinky) Nelson gave the plenary talk, and for informal science education, Paul Knappenberger. Papers by Nelson and Knappenberger are included in this section of the proceedings.

The panel sessions were an opportunity to feature exemplary E/PO programs and initiatives. In order to maximize the number of panel presentations while providing all participants opportunities for in-depth discussions based on common experiences, we organized four parallel panel sessions for both morning and afternoon. During the morning, each of the four panels included representatives, broadly speaking, from the areas of professional development, curriculum development, student involvement, and systemic reach. During the afternoon, the areas were large planetaria and museums, small planetaria, community outreach, and media. The focus during both morning and afternoon sessions was on issues that cut across all the areas.

The panelists were urged to focus on an issue or challenge that they encountered while developing, working in, or using a program. During the planning process, we also asked that all panelists address one or more of the following unifying themes: diversity and/or outreach to underserved/underutilized groups; the use of technology; and the involvement of scientists. We sought a balance from the four OSS theme areas and a balance of participants from different E/PO communities (including OSS scientists, OSS E/PO staff, representatives from other NASA divisions, and external community members.) In preparation for the conference, members of the organizing committee arranged and facilitated teleconference calls for the eight panels.

After the panel presentations, participants broke into groups of 10 for facilitated in-depth discussions. Recorders captured the main issues discussed and these issues have been summarized in papers by Isabel Hawkins and Terry Teays, which are included here. The summaries also reference papers by particular panel presenters included in this section of the proceedings.

Issues in Formal Science Education

Issues in Formal Science Education

Edited transcription of the keynote presentation given at the beginning of the session on issues in formal science education

George D. Nelson

Western Washington University, 516 High Street, Bellingham, WA 98225-9155, George.Nelson@wwu.edu

Good morning. This is an intimidating group to address, so I will ask for your forbearance. The topic this morning is about issues in formal education. That is a huge topic and you heard a lot about it yesterday. I hope to be a little bit provocative and stimulate some discussion. Your feedback and your comments will be appreciated.

What is the purpose of formal science education?

Let's start out by asking, what is the purpose of formal science education? I think if you ask some educators you might get this answer: that we are trying to foster science literacy for all students. Science literacy today has been pretty well defined by the scientific and education communities. I hope you have read *Science for All Americans* (AAAS 1989). If you haven't read it lately, please read it again. I would claim that *Science for All Americans* pretty well describes science literacy within 20% or so. There is certainly no claim that this is absolutely the set of ideas and skills that everybody ought to have to be science literate, but it was a careful effort to get a consensus from the communities and profoundly influenced the subsequent national goals documents. The important thing about the ideas that are in the Benchmarks of the American Association for the Advancement of Science (AAAS 1993) and the National Science Education Standards (NRC 1996) is that they include the nature of science. They talk about scientific inquiry. They talk about history, overarching themes, and habits of mind. These topics are generally not in the current curricula that primarily contain the traditional factoids and "scientific" vocabulary. Stepping back and looking at science as a whole is not in the current curriculum, but it is among the goals of the standards. Another key goal in the standards is knowledge of the interconnectedness of ideas. The *Atlas of Science Literacy* (AAAS 2001a) that AAAS published in 2001 is one attempt to show how these various ideas can be linked together to build a coherent picture of what science is like.

Schools, teachers, and administrators, and other groups have put forward other goals for science education. For this group, preparation of the next generation of scientists, engineers, and technical professionals is certainly high up there. Which means that there are some students who are going to go beyond basic literacy and should have the opportunity to do that. Aiming just at the standards for all students is a noble thing to do but we have to keep in mind that there is knowledge and are skills that go well beyond the standards and that those students who are capable ought to have the opportunity to achieve at that higher level, because they are the ones who are going to be us in the future. It should be obvious to us all that the population of future scientists and engineers can't look like us anymore, but should reflect the make-up of our population. We have to pay special attention to look out for those, the underrepresented groups, minorities and women, who have the potential to go beyond the basic literacy and join our ranks.

Another goal of science education is to help students get into the college of their choice. If you ask parents and administrators in the schools what is the most important thing you do in this school, they might answer, "Well, we give kids what they need to get into good colleges." For some bright students in K-12, the purpose of science education is to help fulfill requirements in high school because they don't want to take science in college. They take a lot of vocabulary courses that are called AP where they learn the words and formulas of science and how to show on tests that they know those words so they don't have to "learn" any more science in college.

Then a lot goes on in schools to promote the agendas of agencies and other groups under the name of science education. Grade school is probably the best (or worst) example. There are so many things added into the curriculum, shoehorned here and there. A lot of them fit under the general idea of science. Sex education, drugs, AIDS, health issues, and more, all looking for a place in the curriculum.

What have we learned from the research?

We heard yesterday from some of our best researchers. What does the research tell us? First, we seem to need to learn over and over again, that an outcome of teaching is not necessarily learning. Learning means gaining knowledge that's going to be retained for life. What is important is the residue that is left in students' heads ten years after the class. Also, we have learned that real learning takes significant time. One example of a curriculum unit that has been researched in terms of how it promotes student learning is a six-week, sixth-grade unit with the main goal to have students understand the idea of atoms and how atoms let you understand the conservation of matter. The research showed that in six weeks of instruction the material was able to help half the students in the class achieve the learning goals. This is twice the percentage of students that were successful with an earlier version of the material. So it's a great success. Half the kids got it. But it took six weeks to teach this one key idea.

Another research result that is very important is the notion that what is explicitly taught is more likely to be learned. For those of us who are engaged in the discussions about improving student understanding of scientific inquiry, this finding is key. A lot of times I hear teachers or I hear developers say that students are going to be doing great inquiry in a lesson and they claim the students are going to learn about inquiry by doing this lesson. Doing scientific inquiry is necessary but insufficient to help students learn about scientific inquiry, unless the curriculum actually comes around and helps the student address the ideas explicitly. If you want students to learn something, you have to focus on the specific concepts.

Elements of Effective Education (System Level)

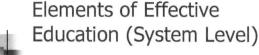

- Students ready, expected, and supported to learn
- Knowledgeable and supportive community (+Contractors)
- **Coherent, specific learning goals**
- **Well-trained and -supported teachers**
- **Aligned curriculum materials and assessments** (+ time to use them well)
- Coherent K-16 curriculum

Figure 1. PowerPoint Slide

Finally, maybe the most exciting research findings are those that are identifying the deep elements of successful instruction. Not the pedagogical tools, but the fundamental components that you use to construct the tools. We will talk about that a little more later.

What components contribute to student success?

I want to organize this discussion around this list (Figure 1). It is kind of an arbitrary list; you could organize the system in lots of ways, but this is one way of describing which pieces of the system have to come together in order to be able to have successful learning going on for students. At the top of the list is the requirement that students in classrooms are ready to learn, expected to learn, and supported to learn. If anyone knows how to do this I'm open to suggestions. This is a tough issue and one that we can work on in some ways, but this is a societal issue that takes a much broader community than we have here. You can't ignore it; you just can't do this without reaching out.

The second step is a knowledgeable and supportive community. This includes the parents. This includes the business world. I would throw the school administrators into this. The people have to know what it is you are trying to do and be on the same page with you. I put contractors on here because with the OSS projects it is one of the issues that I always found kind of strange. These missions aren't built in garages. They are built by huge aerospace contractors that have lots of money and are making lots of money on these satellites. Yet you see very little evidence of the contractor putting in his or her own resources to aid the NASA E/PO. I think it is important for you to work with your contractors to bring them on board. They are potentially powerful and helpful partners.

The third item on the list is coherent, specific learning goals. You have to know what it is you're trying to teach, what you want students to learn and when. So the notion of having clear standards is an important one. There will be more on this below.

Next are well-trained and well-supported teachers; the teachers are the interface between the student and the content. If they aren't knowledgeable and they aren't supported, then learning isn't going to take place. To support the teachers you need a curriculum that is aligned with the standards, which means that the activities specifically target the concepts and skills that are in the standards and that the assessments are aligned in the same way. Assessments not just to find out how well you've done, but to find out what you need to do at the beginning and how well you're doing along the way. So assessments should be embedded throughout the curriculum.

Finally, an important point that's often missed that I think the Atlas of Science Literacy points out well, is the idea of a coherent K-16 curriculum. What we want people to know when they emerge as adults, if they're going to be considered science literate, is the story of how the world works and how we have come to know how the world works. You get that by developing that story over time. Students don't become science literate by accumulating a set of disconnected facts and skills. Concepts and skills have to be assembled into a picture that makes sense to people. Go to a school and ask the 5th grade teacher what science they are doing in 3rd grade that she or he is going to build on in 5th grade. What is it that they do in 8th grade that's going to build on what is done in 5th grade? Where is it that the teaching that you're doing at this grade level is leading? Those ideas are ones that we need to think about as we do our work.

Standards

I'm going to spend a little more time on the three bolded bullets in Figure 1 and not so much time on the other three. First, clear goals. What are standards? There is a lot of nodding towards the standards but there hasn't been a lot of careful study of the standards. My view is that standards are what every child should know and the skills that every child should have. What every adult ought to know ten years after graduation. What the standards are not is everything the student should know. The science standards were intended to set a literacy floor. We want to bring everybody above this floor. Our measure of success could be a test, a valid test that measures knowledge of the standards. We would meet our goal if the average score of the bottom quartile of students on that test were above 90%. We want every student to be at that level. The Benchmarks and National Science Education Standards set a literacy core. For those students that reach that level, there's plenty of science that goes beyond the standards.

People talk about my standards or your standards or world-class or rigorous or high standards. Usually when someone puts adjectives in front of standards I know that they don't think about it the same way that I do. What makes them high standards is not that the material is hard, that there's more stuff there, but rather that we actually expect all kids to learn the standards. A lot of times

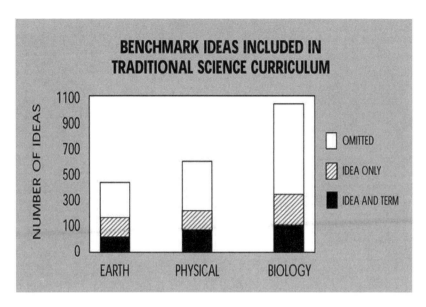

Figure 2. Comparison of the ideas found in science textbooks with the Benchmarks.

people will say we have this rigorous set of standards. When pressed about their expectations for student achievement, they say, "well we expect a bell curve, 10% of the kids will get all of them, 50% of the kids will get some of them, some percent will get only a few of them." That's their expectation. Our expectation with the Benchmarks and the National Science Education Standards is that we attempt to get every kid at that 90% level.

The standards demand that careful attention be paid to the content that is taught. At Project 2061, we went through a number of textbooks and compared the ideas in the textbooks with the Benchmarks. Figure 2 summarizes what we found. The white bars are the number of ideas that were in the books, but not in the Benchmarks. The striped band shows the ideas from the books that are in the Benchmarks but include the technical vocabulary that often substitutes for knowledge on common assessments and is not in the Benchmarks. The black band on the bottom is where both the idea and the vocabulary were in the books and the Benchmarks. So you can see that only about 40% or so of the content that is covered in schools today targets the ideas of scientific literacy. As an aside, one question that often comes up is, how much of the biology is omitted because we're not teaching evolution? We found that evolution is in all the books. You can't fault publishers for shying away from evolution, but you can fault them for the poor instruction that is built into their materials.

In a book AAAS put out in 2001 called *Designs for Science Literacy* (AAAS 2001b), there is a chapter called "Unburdening the Curriculum." The idea behind it is that if you are planning a literacy curriculum, you could go through your curriculum and eliminate the activities and assessments that do not target concepts or skills that are in the standards. To test this out, we went through books to find out whether the ideas that the books are teaching under common major topics are the ideas that are in the benchmarks and the standards. For example, in an Earth science book one finds solar features, stellar evolution, lunar features, and atmospheric layers, concepts not found in the standards. The idea of unburdening the curriculum is, if you're looking for ways to free up the time you need to have 6 weeks open to teach sixth graders about the conservation of matter, you're going to have to take something out. But you have to do that very carefully. There are some suggestions about how you might do that in the Designs book. We also made suggestions at the subtopic and vocabulary levels and discuss ways to eliminate needless redundancy that's in the materials so we don't keep doing the same thing over and over again. There are some amazing redundancies in the traditional math curriculum, for example. Kids get the chance to not learn fractions at 5th, 6th, 7th, 8th, 9th grade in exactly the same way. It makes so much more sense to take the 8 weeks or half a year at 6th grade and have kids really dig in and learn these things and save that time during the rest of the grades to build on that knowledge.

Teacher Preparation and Support

There are many issues surrounding teacher preparation and support. Yesterday, Maria talked about the tension between the need to produce more teachers quickly and efficiently, and the need to take the time and spend the resources necessary to produce highly qualified teachers. There is a shortage of not just competent teachers but a shortage of bodies to put in classrooms in many districts, especially in cities. So there is a big push from the Department of Education to get more teachers in schools. As an example, my own university is involved with the state in a federally funded alternative-path-to-certification program. Traditionally, a student going through the university can get a teaching certificate to be a secondary science teacher in 5 years. Even if someone comes to the university with a bachelor's degree in science, they will be there for at least 18 months earning a teaching certificate. The new pilot program takes people who have degrees in the disciplines, gives them a 3-week summer workshop, and places them into an internship program in the schools in the fall. They have the opportunity at the end of every quarter to demonstrate that they have filled enough squares to receive their teaching certificate. This is being done in spite of the research that shows that very few new teachers produced with this or similar shortcuts stay in the profession more than a couple of years. These kinds of tensions are going on now and we have yet to come to grips with them.

Another issue I'm facing now in my work and hoping to learn something about is: Why can't we graduate new teachers who don't need remedial professional development? Teachers, especially elementary and middle school teachers, coming out of the gate right now need professional development in both content knowledge and how to use the new curriculum materials. My goal in my new life as a social scientist is to try and figure out how to prepare novice teachers who receive their initial certification needing no remedial professional development. They will be able to go into the classroom, use good materials with high fidelity, and be able to use assessment in an appropriate way. That's a very tough challenge because the university curriculum, like the K-12 curriculum, is just packed. In my university elementary teachers take 12 methods courses.

And why are new teachers given the worst jobs? Retention of teachers is a critical issue and there is some really encouraging news from new mentor programs that we could learn from. UC Santa Cruz has a good example of a mentor program that is helping schools realize that if they're going to retain teachers more than 2 or 3 or 4 years, then they're going to have to take them in as novices, provide them with some real mentoring as they go along and help them grow into the profession to become participants in a professional community. It has been a long time coming and it's not there yet but districts are thinking about retention and the quantity of teachers.

Where is the time and support needed to become a professional? Time is the issue for everybody and I'm not talking about the time you can buy with your grants to provide professional development for teachers. I'm talking about how can we work with the system to institutionalize it so that teachers always have time to become better professionals. That is expensive. Any way you think about this in a country with the number of teachers we have, it's anywhere between a $20 and a $40 billion problem. Suppose you gave teachers a 12-month salary, let's say about a 30% raise to go from 9 to 12 months and said, "OK, in the summer months you're going to be paid to work as professionals on improving your teaching." Well if you add that up it is a $20-$40 billion commitment. K-12 is about a $500 billion enterprise, so we're talking about roughly 5-10%.

Curriculum materials

Another issue in teacher preparation and support is the availability and use of good curriculum materials. The tool that the teacher has to work with the students in the discipline is the curriculum material. How do you get good stuff in their hands that is going to support effective teaching and going to support students' learning? And how do you provide the necessary training to enable teachers to use the materials well? This is an issue that directly relates to NASA's E/PO activities.

Let me talk about materials for a bit. I don't know how many of you can say you are developing curriculum materials. The majority, I would guess. The content in your curriculum materials should be aligned with what you think is important for kids to learn. It should be targeted at that learning explicitly. Instruction that is in the materials should be such that it helps the teacher teach and it helps the student learn. The instructional support has to be there but it is missing in most materials today. As I mentioned earlier, there is generally too much content in materials. The content is there, if buried, in most materials, but the support for the teacher and students is seriously lacking. In curriculum materials, publishers haven't taken advantage of the research in how students learn and incorporate those ideas into their materials yet. The large publishers are struggling right now not to do that because it's expensive and time-consuming and the market that they have created and sustain has teachers who are comfortable with what they have. They want a book that looks like the book they have, the book they used last year. These are tough issues.

One approach to this problem is to do careful, credible, and public analyses of materials for their potential to help teachers teach and students learn.

Let me briefly describe AAAS Project 2061's curriculum analysis tool. The procedure looks at both the content and instructional support in materials. Of course we know content should be accurate. There are people who look at textbooks and find all these mistakes, and there are lots of mistakes in textbooks, and say, "OK, if we just fixed the mistakes, then kids would really learn." Well, if we just fixed the mistakes we'd have better books but kids won't learn from today's books. The content should be aligned with specific learning goals and, just as importantly, the content ought to be coherent in the books. There ought to be a story in there, not just a set of randomly assorted stuff. And if the books are part of a sequence, the story should run through the year and across the years.

When we analyze curriculum materials, we start with the ideas (Figure 3). Those boxes in the middle (marked with triangles) are the goals that we're looking for in the materials. They are statements from the benchmarks or the standards. Just as important as what is in the boxes are the arrows between them. The arrows between them are also content; the concept that this idea connects to that idea and that connection helps students understand the bigger picture. The boxes marked with circles are prerequisites, things that you should know before you take on these other ideas, and the boxes marked with stars are supporting notions. This map is around biology, the topic of cells. The middle two "goals" boxes are the notions that cells have lots of different parts and the parts do specific functions. The "supporting notion" boxes are mostly ideas about systems: the cell as a system and how do systems work, how does feedback work and things like that. Once you have them down like this you can take a textbook, and go through it and say, "OK, on page x I found half of one idea. On page y I found the other half. On page z I found this connection." You can map out where all of those pieces are in the textbook.

Figure 3 shows what reviewers looked for and Figure 4 shows what they found in today's high-school biology books in the topic of cells. You can see the ideas in the middle. What do biology books teach about cells when they teach this flat-looking thing? Cells contain lots of different things, and every one of them has a long name. The connections between the ideas are missing. Even though it looks like the content is there, it is not in a way that is useful to students. A lot of times the prerequisite idea will be on page 900, where the idea itself is on page 157. If you look closely at each book with a clear template in mind, you'll find some pretty disturbing things.

Given that the content is in a material, now how do you think about instructional support? We developed a set of criteria for exploring instructional support based on the learning research. It is important to note that we are not looking for hands-on. Hands-on doesn't automatically get high points. We are

Issues in Formal Science Education 63

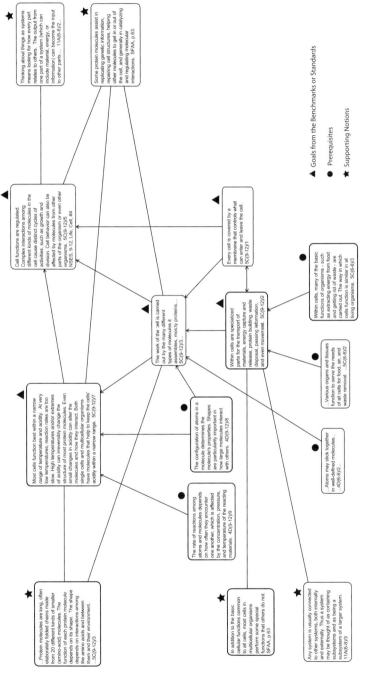

Figure 3. Key ideas and connections from the Benchmarks on the topic of cells. Project 2061 uses this map and other similar ones to analyze curriculum materials.

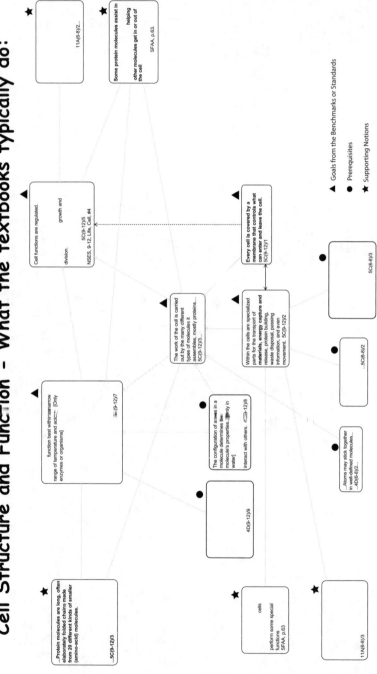

Figure 4. Project 2061's analysis of today's high-school biology books. The map clearly demonstrates that many important ideas adn connections described in the Benchmarks are not adequately represented.

not looking for any particular instructional model. We are looking for instruction that incorporates the fundamental elements of instruction shown to help students learn. The bottom line is a quote from *How People Learn* (Bransford, Brown, & Cocking 1999). "There is no universal best teaching practice." This game is about reaching the outcome that you want, that is to have students learn these ideas, not about the one way to get there. How you get there is like doing science. If you can dream the solution to the problem you have after a nice glass of wine at night and wake up and have the answer it counts just as much as if you think of it while wearing your suit and tie and sitting at a desk. The means to the end is wide open. There are a lot of ways to get there. The end is what's important. (Of course some ways are better that others. I'm not suggesting that we all start drinking wine before work!)

In a material we look for these instructional elements, and this is relevant to yesterday, because yesterday we talked about misconceptions a lot. One gets the impression that if we just knew the misconceptions we could confront misconceptions and be well on our way. That is one key point, but one of many.

The material has to provide a sense of purpose. One thing that you all talked about is space as a source of inspiration, stuff that's really exciting. But instruction should also have a sense of clear purpose. You ought to be able to walk into a classroom where students are engaged in some activity, ask the students what they're doing, what they're trying to learn, and they should be able to tell you. They should know the purpose of the activity. Remember Edison's quote, "Success is 10% inspiration and 90% perspiration." Education is the same way. Inspiration is important but it's not everything.

Taking into account student ideas is what we talked about yesterday and I will elaborate on that a little bit later. We also have to engage students with phenomena from the natural world. We're asking kids to learn concepts that come out of science. A lot of times they are just hard. They are not intuitive. It is not intuitive that the earth goes around the sun. In order to get them to make the scientific idea plausible, more plausible than some other ideas students have, they should engage with lots of examples from the world where this concept is relevant. In most materials today there may be one example from the real world to support a concept, a lot of times zero. Ideas or concepts are just laid out, here it is. Learn this, there's a formula you can apply to it. But no connection with real world is made. Once students have these ideas, then they need to use them to reinforce that they really do help describe the world.

Another key finding from the cognitive science research is that having students encouraged to think about their own thinking is one of the fundamental elements of instruction. As they learn that the scientific concepts are useful in explaining some things or in learning some things about the world, they should also be thinking "what do I think about this? How does this compare to the ideas I had before? Where else might I apply something like this? How does this new concept change any other ideas I had about the world?" Thinking about their own thinking is necessary and materials should guide the teacher and student to make it happen.

Very few lesson plans either included in or developed outside of textbooks have assessments in them to find out what students' ideas are prior to instruction, to look at progress along the way, to inform instruction so you can modify it as you go along, or to assess at the end of instruction whether students have learned the targeted goals. Quality assessment is important and missing in most materials.

Finally, materials should help teachers think about the learning environment. Schools in downtown Chicago are different than schools 100 miles out of town. The environments are different. The students have different backgrounds, different parents, and you have to make allowances for those sorts of things.

These are the categories we used to evaluate materials. The formal analysis process is grotesquely complicated. We didn't want to make it complicated, but in order to get reliability–having different people look at the same materials and come to the same conclusions for the same reasons–we had to get very detailed. Each one of these categories has a number of specific criteria. Each of the criteria has a number of indicators and there's a scoring scheme. It gets very complicated. But the basic idea is you look carefully for the important instructional elements in the materials and report the findings.

As an example of how the procedure was built, lets take a closer look at one of the categories, taking account of student ideas. *How People Learn* tells us that students come to the classroom with preconceptions. If you don't help students engage with these preconceptions you aren't going to change them or build on them. Teachers have to draw out and work with preexisting understandings. Formative assessment is essential. We have four criteria to look for in taking account of student ideas. 1) Does the material attend to prerequisite knowledge and skills? 2) Does it alert teachers to commonly held student ideas? You can't expect teachers to know all the research. 3) Does it assist teachers identifying their own students' ideas? If you know that some students may be thinking one way or another do more students think that way? 4) If they do, what do you do about it? We asked materials to do all of these things. Not just say these are the misconceptions. If you know them, how do you identify them in your students and what do you do about them?

Figure 5 shows the result of applying the criteria to a number of middle school earth science textbooks. On the left-hand column you can see there are six categories. We didn't include the last category of enhancing the learning environment but the other categories are there and below them are the criteria that we used. In the second row you can see taking account of student ideas and the four criteria we looked at as we scored these. You can see the books didn't do very well. Figure 6 shows how the same books scored in their instruction of selected physical science benchmarks. The right hand column has the scores for the unit I talked about that does conservation of matter. You can see how it does on the rating. It's not great, but it does a lot better than the textbooks because it was designed to take some of the research into account. It's a unit that came out of Michigan State University. This pretty much summarizes the state of textbooks.

What you want to hear

Here are some things you might like to hear. Astronomy and space science are great motivators. Kids love this stuff. There are wonderful new images

AAAS Project 2061 Middle Grades Science Textbooks

Summary of Instructional Analysis Ratings in Earth Science

Instructional Categories	Glencoe Life, Earth, and Physical Science (Glencoe/McGraw-Hill)	Macmillan/McGraw-Hill Science (Macmillan/McGraw-Hill)	Middle School Science and Technology (Kendall/Hunt)	Prentice Hall Science (Prentice-Hall)	PRIME Science (Kendall/Hunt)	Science 2000 (Decision Development Corp.)	Science Insights (Addison-Wesley)	Science Interactions (Glencoe/McGraw-Hill)	SciencePlus (Holt, Rinehart & Winston)
I. PROVIDING A SENSE OF PURPOSE									
Conveying unit purpose	■	⬓	⬓	⬓	■	⬒	■	■	■
Conveying lesson purpose	■	■	⬓	■	■	⬓	■	■	■
Justifying activity sequence	■	■	□	■	■	□	■	■	■
II. TAKING ACCOUNT OF STUDENT IDEAS									
Attending to prerequisite knowledge and skills	■	■	■	■	■	■	■	■	■
Alerting teacher to commonly held student ideas	■	□	■	■	■	■	■	■	■
Assisting teacher in identifying own students' ideas	⬓	⬓	■	⬓	⬓	■	■	■	□
Addressing commonly held ideas	■	⬓	■	■	■	■	■	■	⬓
III. ENGAGING STUDENTS WITH RELEVANT PHENOMENA									
Providing variety of phenomena	■	■	■	■	■	■	■	■	■
Providing vivid experiences	■	■	■	■	■	■	■	■	■
IV. DEVELOPING AND USING SCIENTIFIC IDEAS									
Introducing terms meaningfully	■	⬓	□	■	⬓	■	■	■	⬓
Representing ideas effectively	■	⬓	■	■	■	⬓	■	⬓	■
Demonstrating use of knowledge	■	■	■	■	■	■	■	■	■
Providing practice	■	■	■	■	■	■	■	■	■
V. PROMOTING STUDENT THINKING ABOUT PHENOMENA, EXPERIENCES, AND KNOWLEDGE									
Encouraging students to explain their ideas	■	■	⬓	■	■	⬓	■	■	□
Guiding student interpretation and reasoning	■	■	■	■	■	□	■	■	⬓
Encouraging students to think about what they've learned	■	■	■	■	■	⬓	■	■	■
VI. ASSESSING PROGRESS									
Aligning assessment to goals	■	■	■	■	■	■	■	■	□
Testing for understanding	■	■	■	■	■	■	■	■	⬓
Using assessment to inform instruction	■	■	■	■	■	■	■	■	■

■ = Poor (0-1); ⬓ = Fair (1.5); □ = Satisfactory (2); ⬒ = Very Good (2.5); ▣ = Excellent (3)

AAAS Project 2061 Middle Grades Science Textbooks Evaluation

Summary of Instructional Analysis Ratings in Physical Science

Instructional Categories	Glencoe Life, Earth, and Physical Science (Glencoe/McGraw-Hill)	Macmillan/McGraw-Hill Science (Macmillan/McGraw-Hill)	Middle School Science and Technology (Kendall/Hunt)	Prentice Hall Science (Prentice-Hall)	PRIME Science (Kendall/Hunt)	Science 2000 (Decision Development Corp.)	Science Insights (Addison-Wesley)	Science Interactions (Glencoe/McGraw-Hill)	SciencePlus (Holt, Rinehart & Winston)	Matter and Molecules (Michigan State University)
I. PROVIDING A SENSE OF PURPOSE										
Conveying unit purpose	■	⬓	⬓	⬓	■	□	■	■	■	■
Conveying lesson purpose	⬓	⬓	□	■	■	⬓	■	■	⬓	⬒
Justifying activity sequence	■	■	□	■	⬓	□	■	■	⬓	□
II. TAKING ACCOUNT OF STUDENT IDEAS										
Attending to prerequisite knowledge and skills	■	■	■	■	⬓	■	■	■	■	□
Alerting teacher to commonly held student ideas	■	■	■	■	■	■	■	■	■	▣
Assisting teacher in identifying own students' ideas	■	■	■	■	■	⬓	■	■	□	□
Addressing commonly held ideas	■	■	■	■	■	■	■	■	■	□
III. ENGAGING STUDENTS WITH RELEVANT PHENOMENA										
Providing variety of phenomena	⬓	■	■	■	□	■	■	□	□	□
Providing vivid experiences	⬓	■	□	■	⬒	⬓	■	□	□	▣
IV. DEVELOPING AND USING SCIENTIFIC IDEAS										
Introducing terms meaningfully	⬓	⬓	⬒	■	⬒	⬓	⬓	□	□	▣
Representing ideas effectively	□	■	■	■	■	■	■	■	■	⬓
Demonstrating use of knowledge	⬓	■	■	■	■	■	■	■	■	▣
Providing practice	⬓	■	■	■	⬓	■	■	⬓	□	⬒
V. PROMOTING STUDENT THINKING ABOUT PHENOMENA, EXPERIENCES, AND KNOWLEDGE										
Encouraging students to explain their ideas	■	■	□	■	■	□	■	⬓	□	▣
Guiding student interpretation and reasoning	⬓	■	□	■	■	⬓	■	■	⬓	▣
Encouraging students to think about what they've learned	⬓	■	⬓	■	■	■	■	■	■	■
VI. ASSESSING PROGRESS										
Aligning assessment to goals	■	■	■	■	■	■	■	⬓	⬒	⬒
Testing for understanding	■	■	■	■	■	■	■	⬓	□	□
Using assessment to inform instruction	■	■	■	■	■	■	■	■	■	□

■ = Poor (0-1); ⬓ = Fair (1.5); □ = Satisfactory (2); ⬒ = Very Good (2.5); ▣ = Excellent (3)

Figure 5. Summary of instructional analysis ratings of middle grades science text books in earth science (top) and physical science.

and results generated every day, great new science coming out, and amazing technology. Astronomy and space science are well represented in the benchmarks and standards. Tim Slater published a study of the astronomy content in the NSES and the Benchmarks (Adams & Slater 2000; Slater 2000). The Atlas literacy maps have a nice progression of how these ideas can grow through the grades. Even more important, I think, and one of the reasons that I'm fairly intimidated in talking to this group is that this is a terrific community that has been created through the NASA OSS E/PO office. There is incredible expertise and power out there with unlimited potential.

I'll tell you a quick story. When I was an undergraduate, my future wife gave me a little box with just a few buttons or something in it and she decorated it. Inside the lid she put a little Peanuts cartoon. It was Linus holding his blanket and sucking his thumb. The caption read: "There is no heavier burden than a great potential." This group has a great potential.

What you don't want to hear

But there are some things you might not want to hear. Most students are learning very little astronomy. There is some data to back that up. I find it fascinating to talk to my colleagues at the university because they have a split mind. They honestly answer basically the same question, two completely different ways. If you ask them what knowledge do you expect your students to have coming into your Astronomy 101 course, they will say, "I don't expect them to know anything." If you ask what ideas should kids learn about astronomy when they're in K-12, they'll lay out their Astronomy 101 class rather than say, "Well if they just knew a few things, if they really understood the geometry of the solar system and how light travels and what stars are. If they came in with that kind of knowledge, that would be great." But you don't get that response. Learning astronomical concepts is hard. We only figured out 500 years ago how the solar system works. This is not trivial stuff and there is some good research on how difficult it is to help kids learn those ideas, and again this learning takes a lot of time. Most current science teaching, not just space science teaching or astronomy teaching, is ineffective. *A Private Universe* (Schneps & Sadler 1988) is probably the best way to demonstrate that to folks. Most curriculum materials don't have the potential for helping students learn. You saw our analysis. Just as importantly for this group, there just isn't room in the curriculum for most of your stuff. You probably don't want to hear that. But there just isn't room in the curriculum for most of your stuff.

Let's think about that. How much time is available for learning in school? If you ask teachers how much time they actually have to teach when they're not going to assemblies or listening to the intercom box or taking attendance, it comes out somewhere around 100 hours per year. In typical astronomer fashion it varies between 50 and 150, depending on the optimism of the teacher. If we multiply that by the number of years that science is taught and assume that one-third of the time is available for Earth science, we get 300 hours or so through K-12. Earth science is conveniently divided into half the time talking about this little rock we live on and the other half talking about the rest of the universe. Divide that by the number of years and you get about 15 hours a year

**Space Science Missions
With E/PO Supplementary Curriculum
Components**

Magellan Yohkoh XMM Wind Voyager Ulysses TRACE SWAS Stardust SOHO SNOESAMPEX RXTE Polar Nozomi NEAR Mars Global Surv IMP-8 IMAGE Hubble (HST) HETE-2 HALCA/VLBI Geotail Galileo FUSE FAST Deep Space 1 Cluster II Chandra Cassini ACE ACCESS AIM CINDI CNSR Constellation-X Dawn Deep Impact Europa Orbiter Europa Lander FAME FIRST GEC GLAST HNX INSIDE Jupiter Ionosph. Mappers JMEX JOULE Kepler LISA Mag Const Mag Multiscale Mars 2003+ MESSENGER NetLander NGST Planck Pluto/Kuiper PRIME Rad Belt Mappers SDO Sentinels SIM Solar-B Solar Probe Space Tech 3 Space Tech 5 SPEAR SPIDR STEP Swift STEREO Titan Explorer TPF AMS ASPERA-3 CATSAT CHIPS CONTOUR GALEX Genesis Gravity Probe-B HESSI INTEGRAL MAP '01 Mars Odyssey Rosetta SIRTF SOFIA TIMED TWINS Mars Pathfinder Lunar Prospector

Figure 6.   PowerPoint Slide

for astronomy and space science, about three weeks. It could be distributed in lots of different ways but that's about the time that's available in schools for teaching astronomy and space science. You need to think very carefully about how you want to spend those three weeks a year. What are the important ideas we want kids to come away with in that short time? Our expectations for what most students can learn about astronomy and space science in school are probably way too high. The attempt we made to describe astronomy learning goals in the Benchmarks is probably over-ambitious in terms of what is actually possible given the time that we have.

Given that time, what is it we are attempting to do? Figure 7 is a collection of the NASA missions from a few years ago that had E/PO components, many with supplementary curriculum components. Where are they going to fit in the classroom?

What is needed?

We need good curriculum materials. You can see that this is a dilemma. I'm encouraging you not to develop lots of curriculum materials but then showing you that we desperately need good curriculum materials. We need good curriculum materials that target the learning goals we're after, that incorporate good instruction, and that are coherent across time and across the discipline. We need effective teacher preparation and professional education that incorporates the same models of good instruction as the materials. Yesterday, you heard that the most effective professional development is done around the content in the context of the materials the teachers are going to be using in their classroom. That is really important. We need a home for astronomy and space science in the K-12 curriculum.

We also need more research on how students learn the specific ideas that make up astronomical literacy. We talked a lot about misconceptions but they are always the same misconceptions. We know the misconceptions when we talk about gravity and light. There are a lot of other ideas where we really don't know what the student thinking is and how we might go about approaching that. As scientists here is an opportunity to contribute to the field by using your NASA resources to do some of this critical research.

What is not needed?

We do not need materials and activities that take up valuable classroom time that do not result in students learning important ideas in astronomy and space science. We do not need more stuff. We do not need materials and activities that take up teachers' professional education time and do not result in improving the ability of students to learn. My daughter is an elementary school teacher in Houston. She told me a joke one day about teachers. A teacher died and showed up at the pearly gates and St. Peter looked her up on the list and said, "Oh you're a teacher. Come with me and I'll take you over to a special place in heaven." They walked over and there is nobody there. St. Peter said, "Oh that's right, today's an inservice day, and they are all down in hell."

We do not need information technology for its own sake. There is no doubt that the technology is impressive and could be made useful. But it's not a straightforward issue. Using something doesn't mean it's going to enhance learning. Again you should try it, but it must be done in a research way as an experiment. The other thing we do not need is just more noise in the system. I read the article this morning about the Chicago schools, 81 out of 600 Chicago schools are on the failing list. They talked to some people in the three elementary schools that they are going to close this year, and they have been absolutely inundated with programs in the last ten years, one after another, on top of another, on top of another. We don't need that kind of noise in the system. Any reform needs to be careful and slow, targeted, and specific.

What can you do?

This group can approach education just like you do your science. You wouldn't take up a new science project without spending a lot of time in the library finding out what everybody else knows and talking to your colleagues and your peers. You can't read everything but you can interact with colleagues to identify what is most relevant to your work. You can pull in knowledgeable people from the education community who can help you get up to speed.

We often refer to the standards and once we have something we go to the standards see where we can fit it in. But very rarely do we take the time to actually study the standards to get a sense of what they're really after. Not just the vocabulary terms but what concept is this 5th grade standard really after? What isn't it after because that comes in 3rd grade or in high school? A lot of people will dump a lot of things into a standard because it has a topic match. So you should take the time to really read these documents and think carefully. Each time you think you're addressing a standard take the time to go away from what you're developing and look at the standards explicitly. Look around; what comes before it, what comes after, what are the words that are actually written in the essays and the probes, what's the research that's referenced that supports or talks about student learning? You can really help to sharpen your focus on the learning goals.

One of the things I think this group does pretty well is learn from teachers. We all need to spend more time in the classroom. There's a lot of wisdom and knowledge in the schools. Teachers should be involved as full colleagues in everything we do. Teachers know a lot about some things. They know a lot about kids. They know a lot about the realities of the classroom. Teachers don't know how to develop curriculum materials. We don't teach classes in curriculum development at the university. Teachers can assemble curriculum but they can't start with a concept and develop an activity that's going to help students learn that particular idea. There are some magic people out there who can do it, but generally teachers are not curriculum developers. Using them in that role is a misuse of teachers. Teachers should be on the team as materials are being developed to provide the kind of insights they can, but not be expected to develop activities and materials themselves.

What else can you do? You can do experiments. Science education reform is a social science experiment, and this social science stuff is really hard. Rocket science is relatively easy. These experiments involve people. We need to set the goals, gather data, and most importantly publish the results. I'm really excited about this journal that is starting, the Astronomy Education Review (see page 410). This is a terrific place to get this kind of information out so the rest of us can share it.

A final challenge

Finally I want to throw out one idea just to see if I can get you thinking about it. You saw in the reviews that the materials just are not there. This group is powerful, big, and talented. You might think about working together to develop one coherent K-12 astronomy/space science curriculum strand. You could even team up with one of the big publishers. Think about a way to use the collective resources of this group and pull one jewel together that can then stake claim to a permanent place in the curriculum. Starting with a coherent framework, you could find a natural home for the work done for individual missions. As missions come and go, details could be inserted and removed from specific curriculum materials but the specific learning goals, the targeted standards, would stay the same. It would be the best of both worlds, the excitement of NASA's mission and solid education.

My final comment is that this is an impressive organization and team. Every astronaut has to quote Gus Grissum. Let me close then by saying: "Do good work!" Thank you.

References

Adams, J. P., & Slater, T. F. 2000, Journal of Geoscience Education, 48, 39.

American Association for the Advancement of Science Project 2061 1989, Science for All Americans (New York: Oxford University Press)

American Association for the Advancement of Science Project 2061 1993, Benchmarks for Science Literacy (New York: Oxford University Press)

American Association for the Advancement of Science Project 2061 2001a, Atlas of Science Literacy (Washington, DC: AAAS and NSTA Press)

American Association for the Advancement of Science Project 2061 2001b, Designs for Science Literacy (New York: Oxford University Press)

Bransford, J. D., Brown, A. L., & Cocking, R. R. (editors) 1999, How People Learn: Brain, Mind, Experience, and School (Washington: National Academies Press)

Lee, O., Eichinger, D. C., Anderson, C. W., Berkheimer, G. D., & Blakeslee, T. S. 1993, Journal of Research in Science Teaching, 30, 249

National Research Council 1996, National Science Education Standards (Washington: National Academies Press)

Schneps, M. H. & Sadler, P. M. 1988, A Private Universe (Santa Monica, CA: Pyramid Films)

Slater, T. F. 2000, The Physics Teacher, 38, 538

Biography

George D. Nelson is the Director of Science, Mathematics and Technology Education, Western Washington University. Until recently, he served as the Director of Project 2061, the long-term science education reform initiative of the American Association for the Advancement of Science; he served as Deputy Director before he became Director. Prior to joining to AAAS, Dr. Nelson was Associate Vice Provost for Research and Associate Professor of Astronomy and Education at the University of Washington, Seattle. From 1978-1989, he served as an astronaut for the National Aeronautics and Space Administration where he flew three missions aboard the space shuttle. He received his B.S. from Harvey Mudd College and his M.S. and Ph.D. from the University of Washington.

Summary of Formal Education Discussions

Terry J. Teays

Teays Consulting, Inc., 8811 Magnolia Drive, Lanham, MD 20706, terry.teays@comcast.net

This is a summary of the discussions that were held on 2002 June 13, following the presentations and discussion on formal education subjects by panelists, and includes references to some of the papers submitted by the panelists and printed in this section of the proceedings. The summary represents a distillation of the common themes that appeared at many of the discussion tables. The information was obtained from the facilitators at each table and the flip charts that each table generated.

Partnerships between Scientists and Educators

Virtually every break-out group discussed how the success of a program/project/product was critically dependent on having meaningful partnerships between scientists and educators (The term educators is meant to include not just classroom teachers, but administrators, curriculum designers, informal science educators when engaged in K-12 activities, etc.). Scientists and educators come from different "cultures" and need to understand the issues and constraints that each must face in their respective milieus. Issues were cited that scientists needed to understand about the world of education, such as: The target audience and what is appropriate for it; the limited time that teachers have to devote to understanding a new product; the small allocation of time in the curriculum for astronomy/space science; the uses of interdisciplinary approaches that can expand the students' exposure to these concepts; the realities of the classroom, such as limited technology; and the nature of scientific inquiry. In his paper, Barstow (p. 107) recommends that NASA create a "Guide to Inquiry-based Learning" for scientists, mission specialists and others who create E/PO educational resources and learning activities.

Similarly, educators need to understand how to work effectively with scientists to use their expertise efficiently, and to identify the varying roles that a scientist may play. DeVore & Bennett (p. 77) report that teachers who flew on the Kuiper Airborne Observatory "observed science in action, learning that high-tech science is conducted by large, diverse teams that have many different skill sets" and that some teachers restructured their classroom science labs to reflect the teamwork they observed. Elitist attitudes by some scientists are still present, and came up for discussion, but times are changing and, as Shipman writes, "Those scientists who will really make a difference in preK-12 education are those who will work with groups of teachers in developing and implementing (these) materials." Madsen et.al. (p. 111) describe just such a program in *Astronomy in the Ice: Bringing Neutrino Astronomy to the Secondary Schools*.

Many people mentioned the need for ongoing professional development for both scientists and teachers in terms of E/PO. Teachers need to have their content knowledge updated on a regular basis in order to use the new discoveries in the classroom. In his paper, Brunsell (p. 115) describes the use of periodic telecon-based training sessions developed for the Solar Systems Educators Program during which a focussed content discussion by a scientist is followed by a question and answer session and a discussion of the educational use of the content. Scientists need to understand how to deliver information in a useful form. There were discussions of the trade-off between the need to reach a broad range of educators, versus the realities of working with the ones who were most interested in working with NASA. There definitely seemed to be interest in having NASA provide some professional development activities for educators (as well as scientists interested in E/PO).

There were several discussions about how to reach a broad base of teachers, rather than the self-selected minority that NASA has involved so far. A related concern was the efficacy of working at the local, even one-on-one, level versus expending more effort on working for systemic reform at the state level. (There did not seem to be a preference, more the sense that both were needed.) Pollock (p. 133) describes a state-wide program, Near and Far Sciences, that has served as a model for professional development programs for teachers in Illinois.

Another partnership that was mentioned as being frequently neglected is the large cadre of amateur astronomers in the country, many of whom would be interested in providing education support (see Bennett p. 246).

A recurring theme was the importance of establishing long-term, sustainable partnerships. The participants indicated that the long term financial support of NASA was crucial.

Multiple tables discussed the importance of long term commitment when addressing diverse communities of users. Diversity issues were understood to require time to build trust and mutual understanding. Participants did not seem to have a clear idea about how NASA materials fit into various cultural arenas, or how product developers addressed those challenges. In their paper *New Opportunities Through Minority Initiatives in Space Science (NOMISS)*, however, Kawakami & Crowe (p. 102) from the University of Hawai'i-Hilo describe a program that engages students in learning about the Hawaiian cultural context and its application and relevance to the study of space science. They have learned that "location matters," and that "Respect for the host community and culture is essential to reach and excite the populace about modern space science."

Coordination and Coherence of NASA Education Programs

The end-users' view of NASA as one, monolithic organization, was clearly not reflected in the experience of the meeting participants, most of whom interact more closely with the infrastructure. The myriad of products and services provided by the Enterprises, NASA Centers, and Education Division was frequently perceived as not well coordinated. The participants did not want each mission to be generating individual education products at random (the Forums and the Space Science Education Resource Directory or SSERD, notwithstanding), rather more collective efforts. Some people proposed that NASA establish specific "strands" of subject material and then produce products to fit those specific needs. Barber makes the case particularly well in her paper *Achieving Coherence, Effectiveness and Widespread usage in NASA-Related Standards-Based Curriculum Development* (see Barber, page 95).

A frequently voiced sentiment was, "don't produce ever more education products, but a smaller number of coordinated and effective ones." Emphasis on quality products versus the quantity of products was a common suggestion. There are many excellent materials already produced, which need to be better disseminated and organized into "packages."

Finally, it was mentioned that NASA should make it clear whether its main role is to inspire students or to support systemic change in education.

Identify Exemplary Products and Programs
Given the plethora of products from NASA, many tables discussed the need for identifying the exemplary products and activities. This was important for adoption decisions by educators, but also for identifying those products and activities that should be replicated or used as the basis for more extensive projects.

Basis for Developing Curriculum
There was considerable interest in emphasizing inquiry-based learning products from NASA. There were concerns expressed about focusing on covering content, to the exclusion of inquiry. Many participants indicated that the goals of inquiry-based learning were not always compatible with the standardized tests that they were faced with. Where does inspiration fit in? In general, people urged product developers to focus on student learning outcomes and science education standards for the target grade level, as the starting point for their product development. Challenges mentioned included the current emphasis on reading that is widespread, and often to the detriment of science education content. Student misconceptions were another hot topic. Product developers should be aware of them. Another challenge was what to do with all of the material that is in textbooks, but not covered in the content standards. It was also re-emphasized that teachers are swamped, and so it is a challenge to infuse new material into their existent curriculum. In their paper *A Research Based Approach to Developing Engaging Classroom Ready Curriculum* Prather and Slater (p. 125) say, "... it is clear that the development of effective education and public outreach efforts should begin with a thoughtful consideration of students' initial beliefs and furthermore efforts should strive to make use of innovative instructional strategies that provide a learner-centered environment if one wishes to reach the overarching goal of a better informed citizenry."

Maintain the Enthusiasm and Creativity of Grass Roots Efforts

Whereas the participants frequently wanted to see closer coordination of NASA E/PO programs, they were concerned about the imposition of a rigid, monolithic system. It was commonly recognized that the passion for E/PO is most often found in an individual or small group, and the discussion groups did not want to see that fervor blunted by overly restrictive management. They discussed the dilemma of aligning and coordinating these programs without squashing their creativity. Many of the grass roots efforts are multi-cultural and multi-disciplinary.

Lessons Learned
The sharing of "lessons learned" was considered especially important by many people. This was frequently coupled with the above-mentioned discussions on quality vs. quantity and identifying exemplary products and activities. The focus here was more on the needs of developers, who needed to know what had already been tried, and with what outcomes. In particular, the need to share information about what didn't work was mentioned often.

Professional Development For SOFIA's E/PO Program

Edna K. DeVore and Michael Bennett

USRA/SOFIA E/PO Program, NASA Ames Research Center, MS 144-2, Moffett Field, CA 94035-1000, edevore@seti.org

Introduction

NASA's Stratospheric Observatory for Infrared Astronomy (SOFIA) is a Boeing 747SP aircraft, modified as an airborne astronomical observatory; SOFIA is a joint project of NASA and the German Aerospace Center (DLR). Since inception, SOFIA has been envisioned as both a world-class infrared observatory and a "classroom in the sky" where educators participate in research missions onboard the flying observatory. SOFIA is designed to engage educators directly in the research environment, making SOFIA unique in the history of major ground and space-based astronomical observatories that are developed and operated for professional astronomers. Dedicated space in the forward cabin of the aircraft plus the E/PO consol in the general observing area on board provide ready access for educators and media representatives. Physical access to most observatories by the non-professional is extremely limited, generally to daytime tours on the ground, libraries of astronomical data (easy to difficult access), or necessarily via websites for space missions. As an airborne observatory designed to host non-research guests, SOFIA will allow intimate access to the research environment to a wide-variety of educators from both formal and informal institutions via the Airborne Astronomy Ambassadors program. The SOFIA professional development (training) program will draw upon the experience of prior airborne education programs of NASA's Kuiper Airborne Observatory. Further, professional development will be based upon best practices as described in the National Science Education Standards [1] and elsewhere. Complementary SOFIA E/PO programs reach additional educators as well as the general public. The SOFIA E/PO program will bring the science, technology and discoveries to the public via classrooms, science centers, community organizations, and homes of the nation and the world.

Airborne Astronomy and E/PO: Lessons Learned from the KAO

Scientists began experimenting with infrared astronomical observations from aircraft three decades ago. Why do research from an airborne platform? There are two major drivers: 1) to be in the right place at the right time to observe a short-lived phenomena such as an occultation or an eclipse, and 2) to rise above the water and carbon dioxide in Earth's atmosphere that absorbs almost all of the infrared region of the electromagnetic spectrum. On the ground, telescopes are often in the wrong place to observe an occultation or eclipse. Further, telescopes are blinded by the atmosphere when trying to observe beyond the near infrared. Hence, the need to be above the water in space or in an aircraft flying in the stratosphere where the amount of water vapor is vanishingly small.

Early airborne astronomy observations were carried out with equipment designed and constructed by the observers. Chasing eclipses across the Pacific, observers on NASA aircraft measured the temperature of the Sun's corona for the first time. These early experiments demonstrated the feasibility of aircraft as a platform for astronomical research, and the value of infrared astronomy as a tool to explore our universe.

In 1974, NASA christened the Gerard Kuiper Airborne Observatory (KAO). The KAO carried a 0.9m telescope, and was home-based at NASA Ames Research Center (ARC), Moffett Field, CA for 21 years. The KAO flew research missions around the world. Scientific discoveries included the rings of Uranus, water molecules in comets, complex molecules in the interstellar medium, early evidence of a black hole at the center of the Milky Way galaxy, and the study of objects in our solar system, starbirth regions and the ecology of galaxies. The KAO was engaged in education. More than 50 graduate students completed their Ph.D.s using research data gathered on board the KAO; many went on to work on space-based research missions. Over the two decades it flew, the KAO also invited pre-college educators and the media on board for research missions. *Sky & Telescope* reporter, Kelly Beatty, flew on the research mission where the rings of Uranus were discovered, and reported the experience as the astronomers argued and joked about whether the equipment was faulty, or whether they were making an unexpected discovery of the rings. They extended the flight, and obtained the data needed to confirm the discovery of rings; Beatty brought the story to the public.

Each summer, NASA ARC offered the NASA Education Workshop for Elementary School Teachers (NEWEST) and the NASA Education Workshop for Math and Science Teachers (NEWMAST) in conjunction with the National Science Teachers Association. (These subsequently became the NEW program.) About 30 teachers participated each year at ARC, and one was selected to fly on the KAO as a part of the summer workshops. Other educators, students and media representatives flew as guests of NASA and the teams conducting research with the KAO. From this informal outreach activity grew one of the first formal education projects of NASA's Office of Space Sciences (OSS). In 1992, the Flight Opportunities for Science Teacher EnRichment (FOSTER) project was initiated as one of the first OSS grants for an education project. It's purpose was to develop a program that trained teachers and brought them onboard the KAO to experience research missions; it operated from 1992 until 1995 when the KAO was retired from service. Teacher participants were pre-college educators: elementary, middle and high school teachers. FOSTER was funded by the OSS as an early experiment in education and outreach prior to the initiation of the OSS Forum-Broker system and the formal requirement for E/PO programs as a part of OSS Missions. Although the retirement of the KAO cut short the 5-year plans to expand FOSTER to a national program, during its three years of life the FOSTER project trained and flew 70 teachers from 14 states. Additionally, in 1993 the FOSTER project worked with NASA's Quest program on an early Internet education experiment where teachers and students could interact with the KAO team while it was on deployment in New Zealand. Scientists and mission staff emailed accounts of research missions, everyday experiences, and events; teachers and students sent back questions. This email-based communication brought KAO research from across the Pacific into classrooms at a time when email was new in schools. Operationally, FOSTER was directed by the author (DeVore) working in close cooperation with key individuals from Education, Operations, and Space Sciences at NASA ARC as well as the scientists from diverse institutions who flew on board the observatory. The lessons learned from this experience are discussed later in this article.

During the last year of KAO operations, education and outreach blossomed. NASA's ACTS (Advanced Communication Technology Satellite) was seeking demonstration projects for broadband communications, and the KAO was equipped with the experimental hardware (antenna and transmitter) to communicate during research missions via ACTS. In addition to the FOSTER project, in the final year of operations the KAO hosted the following E/PO projects:

- IDEAS Airborne Partnerships, four Texas teachers teamed with Dr. Dan Lester, University of Texas at Austin, for research flights.

- Exploratorium/Project LINK Broadcast: an experimental broadcast via the web and San Francisco's Exploratorium, a public science center.

- *Live from the Stratosphere*, a Passport to Knowledge production by Geoff Haines-Stiles, 13 hours over 3 different dates, including the last flight of the KAO.

Each of these projects utilized the broadband capability of the NASA experimental ACTS project. These projects point to the future potential for exciting public outreach from SOFIA when similar communications systems are installed on SOFIA for science, system upkeep (housekeeping data), and E/PO. In its initial configuration, SOFIA is not equipped for broadband communications.

At the close of the FOSTER project in 1995, the author conducted a survey of the participants seeking information about the impact of FOSTER on them, on their students, and on their schools. The results provide insight into planning future programs that embed educators in scientific research environments.

FOSTER Project: Lessons Learned

The KAO Flight–Positive Outcomes: Teachers valued the flight because it gave them experience in the research environment. An experienced FOSTER/KAO educator or scientist not participating in observing on that particular research mission always facilitated flights. FOSTER teachers observed science in action, learning that high-tech science is conducted by large, diverse teams that have many different skill sets: scientists, engineers, computer specialists, pilots, and graduate students. The notion that science is something done by isolated individuals evaporated, and some teachers reported that they restructured their classroom science labs into team projects to reflect the teamwork they observed. Other teachers reported surprise at learning that the data taken on the flight was just the beginning of the process; they had expected "results" at the end of the flight. The whole process of identifying the question, proposing, competing, observing, analyzing, and publishing–the life of astronomical researcher–became real through interactions between scientists and teachers on the ground and on-board during research missions. Several scientists grew to value their interactions with the FOSTER teachers as the scientists enjoyed a new set of eyes on their work: They enjoyed the teacher's interest in their work and the opportunity to share their research with teachers.

The KAO Flight–Challenges: The KAO was not constructed to comfortably accommodate teachers/visitors in the vicinity of the scientists and mission staff. (It was not particularly comfortable for the scientists or mission staff either.) On take-off and landing, seats for teachers were at the back of the aircraft, well away from the scientists, flight director, telescope operator, etc. To observe the team in action required that teachers stand for long periods of time unless invited to sit in as a team member. Some teams were inclusive; teachers were integrated into the team to assist with data collection. Other teams were happy to have teachers onboard as well-trained observers (of scientists), but not integrated into the team. And, at the extreme, a few scientists were concerned that the proprietary nature of their data would be compromised by teachers seeing it roll across the screen. SOFIA is designed to better solve the problem of teacher access by providing an E/PO console that allows teachers to sit and readily see all the activity onboard without compromising the mission. The overall design of SOFIA's observatory control room is both more accessible and flexible than the KAO's.

That said, the integration of educators with research teams is not architectural, but human, and will require careful pre-planning and cooperative science teams, a challenge for the SOFIA E/PO team. Finally, the more advanced science teachers (generally high school physics, earth science, astronomy, or chemistry teachers) were eager to have responsibilities during the flight and to obtain data from the KAO for their students to analyze. With the development of facility instruments and a data archive for SOFIA, there can be an opportunity for students, teachers and the public to access data that has been released by the observatory; such an archive did not exist for the KAO.

A final challenge related to flying on the KAO: The teachers also remarked, with humor, that their students thought they were astronauts flying in space because the KAO was a NASA aircraft. This misconception led to a "teachable moment" with good discussions of flight in the atmosphere vs. space as well as the depth and composition of the atmosphere.

Teacher Professional Development: FOSTER teachers applied as a team of two, generally from two schools in the same district or region. The objective was to build bridges between schools (e.g. a middle and high school), support mentor (experienced) teacher partnerships with young teachers, provide a partner for the FOSTER teacher in their home community for follow-on activities, and to generate community support for the teachers. One set of FOSTER teachers was selected and sponsored by their local Rotary Club. The FOSTER project provided a residential, summer workshop that grew to 10 days in length the last year of the program. All workshop activities were evaluated, and the results reported back to the group as well as the scientists, etc. who made presentations.

Interestingly, several scientists remarked that it was the first time they had received such feedback, and they appreciated the comments.

The teacher workshop encompassed general information about NASA and space sciences, fundamentals of the electromagnetic spectrum and astronomy, and airborne astronomy. Activities included scientific lectures, tours of NASA facilities, hands-on laboratory activities and lessons appropriate for the classroom, practical flight training, teacher developed and shared lessons, and an evening at University of California's Lick Observatory. Housing and per diem were provided; there was no stipend. At a later date, each team returned to fly on board the KAO. Again, expenses were covered by the FOSTER project; the school district provided substitute teachers as needed. Most of the respondents to the final survey remarked that although the flight experience was outstanding, they gained the most content information and practical (useful) materials from the workshop. This is not surprising as the workshop was designed and led by a scientist-educator team with the intent of enhancing teacher content knowledge as well as application of relevant science lessons in the classroom.

Media Coverage: For each flight, a news release was prepared on the upcoming flight plus background information on airborne astronomy. It was provided to the teacher's local news media, generally a newspaper. On several FOSTER flights, a local reporter accompanied the teachers. This resulted in a significant number of newspaper articles about the teachers and NASA's KAO in communities that otherwise had little connection to the space agency. The teachers came from urban, suburban, and rural schools, and articles appeared in newspapers throughout these communities. The FOSTER experience demonstrates the grass-roots appeal of a connection to NASA space sciences through community representatives, in this case, teachers.

Impact: The final teacher survey reported on several immediate outcomes for the program. Teachers integrated more information about space science and NASA into their existing courses/classes, initiated new classes in Earth Science (including astronomy), started new science and/or astronomy clubs, and worked with teams of students to enter local technology competitions. They reported a high level of interest from their students regarding their experiences with NASA; some reported that their students thought they were now astronauts!

The long-range impact of the FOSTER program could be obtained by interviewing the cohort of participants, most of whom continue to work in education. This study has not been funded or conducted. At the time of the final survey in 1995, only three years had elapsed for the first group, and as little as a few months for the last group. Thus, enduring long-range impact was not assessed. That said, anecdotal information offers insight into what might be called long-term outcomes. Eight years after she flew, a teacher shared with me that each year her new 7th grade students ask if she is "the NASA teacher;" she feels the FOSTER experience has had a long-term positive impact on her teaching and her school. Several teachers have become department chairpersons, principals, and one is the State Science Supervisor for Idaho (beginning in 2001).

Recently, a FOSTER teacher shared a letter from a high school student aiming toward a college program in robotics; the girl explained that she wanted to go into robotics work with NASA, and that her interest began with her 3rd grade FOSTER teacher. Did the FOSTER experience generate these successes, or did the program attract people who would naturally grow into outstanding educators and FOSTER contributed to their success? The latter is more likely true as many experiences contribute to teaching success.

That said, there is an enduring outcome for teachers affiliated with NASA programs that might be called the "halo effect." FOSTER teachers reported that the affiliation with NASA generated positive student and community attitudes toward both the school and themselves. The NASA affiliation enhanced the perceived value of the teacher and the school in their communities. This "halo effect" is enduring while hard to measure. The SOFIA E/PO program is ideally suited to develop and evaluate the long-term impact of E/PO programs such as the Airborne Astronomy Ambassadors because the observatory is planned to operate for 20 years. There is the opportunity for longitudinal studies of SOFIA E/PO programs not available for missions that are completed in a few years.

SOFIA's Education and Public Outreach Programs

The SOFIA program is dedicated to improving students and teachers' scientific, mathematical, and technological literacy. SOFIA's Education and Public Outreach Programs (E/PO) aim to increase public awareness and understanding of the importance and value of scientific research. To accomplish these goals, SOFIA scientists, engineers and E/PO staff will partner with educators in a number of complementary programs.

Although the Airborne Ambassadors program will gain the greatest public visibility, it is only a portion of the overall E/PO program for SOFIA. Other program elements include partnerships between SOFIA-affiliated scientists and teachers, symposia and courses for non-research college faculty (e.g., community college instructors), internships for educators and science journalists and other web-based resources for teachers and students. As planned, Airborne Ambassadors will be competitively selected to assure excellence as well as wide geographic and socioeconomic distribution of participation. The SOFIA E/PO team will sustain the Airborne Ambassadors as a national network of educator-leaders who continue to inform the public, educate other teachers, and support astronomy and space science education. Quality professional development for participants is required to meet the goal of trained educators who maximize the opportunity to participate in research missions and who sustain long-term affiliation with NASA.

The Airborne Astronomy Ambassadors Program: Once SOFIA is fully operational, approximately 200 educators will be selected each year to participate in one or more research flights. Training will prepare educators for the flights, during which they will work alongside scientists as they conduct astronomical research. Educator training will provide hands-on, minds-on classroom activities that educators can take back to their schools and communities. Together, SIRTF and SOFIA E/PO programs worked with Montana State University, Bozeman, to develop an online astronomy course (with extensive formative evaluation) for teachers as a potential training tool. And, SOFIA Airborne Ambassadors will be prepared to share their experiences with other teachers and the public. Following their flight experience, SOFIA's Airborne Astronomy Ambassadors will become part of a growing national network of master educators who present workshops, teacher in-service training, and other programs in their local school districts, science centers, and communities.

Community Involvement - The Education Partners Program: The SOFIA team is diverse and includes scientists, observers, instrument builders, engineers, technicians, flight crew, and educators located throughout the country. In their home communities, SOFIA team members will partner with teachers and youth group leaders to share astronomy with young people. SOFIA team members will be trained in appropriate hands-on, minds-on teaching techniques and will make multiple visits to their partners' classrooms or youth group events. Education Partners may also participate in research flights.

Enhancement for College & Informal Education - Science Literacy and Education Program: Periodically, educators and faculty involved in undergraduate instruction, science and technology centers, and planetariums will have the opportunity to participate in special summer workshops at SOFIA's base at NASA's Ames Research Center in Northern California. Participants will be encouraged to develop partnerships with SOFIA-affiliated scientists and may have opportunities to participate in research flights.

Staff Opportunities - SOFIA Visiting Educators: Each year, three or four experienced educators will have the opportunity to join the SOFIA Education and Public Outreach (E/PO) staff. Based at the SOFIA Science and Mission Operations Center at NASA's Ames Research Center, visiting educators will train and host airborne guests, develop classroom and web-based material, and support other SOFIA E/PO programs. Typically, visiting educator appointments will last for one to two years.

Public Outreach: SOFIA E/PO plans to develop partnerships with science centers and museums in order to best develop products and activities suited their needs, ranging from exhibits, workshops, webcasts, and planetarium programs to scientists as guest speakers. As a major observatory, SOFIA will be the focus of television, webcasts, and other media at the announcement of scientific discoveries. And, uniquely, SOFIA will become widely visible as educators from across the US train and fly on research missions.

Challenges for SOFIA E/PO

SOFIA, as an astronomical observatory, will be in high demand. Astronomers will propose and compete for use of the observatory, making observing time precious. The successful integration of E/PO teams into this environment will necessarily evolve as the observatory in commissioned and begins research flights. Lessons from FOSTER can inform this process, but, new experiences and tools onboard SOFIA will soon overtake the practical limitations of the FOSTER/KAO experience. SOFIA will be more accessible and comfortable, and offer better tools for teachers to see and experience the research process. Separate from the new facilities offered by SOFIA, the E/PO team will be challenged as they were on the KAO to create working relationships on the ground and onboard where the research team values the participation of the educator/guests on SOFIA research missions, and educator/guests have a role beyond that of a well-trained observer.

The SOFIA E/PO program can uniquely bring the high-technology, scientific research environment to schools, communities and the public through the flight and partnership programs, professional programs for college instructors, planetarium directors, and science center staff, and web and media presence. With careful planning, the SOFIA E/PO program will reach diverse communities throughout the nation in urban, suburban and rural areas. When broadband communications are installed in the observatory, broadcast events from SOFIA will reach a wide audience of schools and the public. The challenge for the E/PO team is to structure programs that complement observing, train participants and sustain their affiliation as ambassadors for NASA space science, and communicate research well to the public. In doing so, SOFIA E/PO will support NASA's goals and objectives for education.

The primary purpose and achievement of SOFIA will be astronomical discoveries, developing new technologies, and training future astronomers and space scientists. The SOFIA E/PO program will bring these discoveries and technologies to students, teachers, and the public, enthusing and inspiring the next generation of researchers.

The Committee for Action Program Services (CAPS)

Linda Davis
910 Bentle Branch Lane, Cedar Hill, TX 75104,
lindadavis@worldnet.att.net

The Committee for Action Program Services (CAPS) is a non-profit science oriented organization whose main offices are located in Cedar Hill, Texas, a suburb of Dallas, Texas. The main objective is to encourage students to pursue a career in the field of science and mathematics.

Although the National Science Foundation reports that the number of minorities enrolled and earning undergraduate science and engineering degrees continues to increase, there is still a disparity. The presence of minority scientists and engineers with diverse backgrounds, interests, and cultures assures better scientific and technological results. Statistics show, however, that the achievement gap between K-12 disadvantaged and minority students in the field of mathematics and science is widening from mainstream students. In addition, professional development for K-12 teachers is lacking, indirectly impacting the education of the disadvantaged and minority population. The shortfall of funding educational programs to address these deficiencies also contributes to the problem.

CAPS recognizes that to increase the number of minority students entering into the science field, a program was needed that would include after school tutoring and mentoring by peers. The second program would offer professional training and curriculum development for K-12 teachers. CAPS has plans to collaborate on the following programs with Rylie Academy of Dallas, Texas and the Dallas Independent School District (DISD):

1. Students who desire to be mentors and tutors will be trained by teachers

2. Peer Tutoring (older students to younger students)

3. Field trips to museums and other scientific workplaces (Space and Special Programs, NASA, NOBCChE, DEA Forensic Laboratory, etc.)

4. Students will participate in local, regional and national Science Bowl/science fair and robotics competitions

5. Teachers will attend educational workshops that will focus on teachers and curriculum development (NOBCChE conference, NASA workshops, etc.)

CAPS has already accomplished some of its goals of partnering with other agencies to meet the need of introducing science and mathematics to minority students. Some of those projects include:

- Partnering with the Lunar and Planetary Institute (LPI) "EXPLORE!" Program.

- Workshop sponsored by CAPS that was facilitated by members from LPI Explorers E/PO and hosted at the Drug Enforcement Administration, Dallas Field Division.
- Several field trips to the Space Science Museum, LPI, and Johnson Space Center in Houston, Texas.
- After School Mentoring Program at Rylie Academy
- Co-sponsored four students from DISD to participate in the Science Fair competition at the national meeting of the National Organization for the Professional Advancement of Black Chemists and Chemical Engineers (NOBCChE), held in New Orleans, LA.

CAPS will continue to work towards bringing more minority students into the fields of science and math, and to offer professional development to K-12 teachers.

The NASA Student Involvement Program (NSIP)

Farzad Mahootian
TERC, 2067 Mass Avenue, Cambridge, MA 02140,
farzad_mahootian@terc.edu

The NASA Student Involvement Program (NSIP) is NASA's national competition for students in grades K-12. NSIP rewards student research on NASA's missions of exploration and discovery, and supports national science, math, technology and geography standards. NSIP brings NASA into the classroom to support units on space, history, math, language arts, engineering, geography, and the sciences. Each competition features Resource Guides and rubrics designed to help teachers and students conduct research and prepare their projects for submission. Students may prepare entries as individuals, teams of 2-4, or as a whole class depending on the competition and grade level. Check the website for more details and to download entry packets and competition Resource Guides: http://www.nsip.net/.

NSIP 2002 features the six competition areas for grades K-12:

- My Planet, Earth ("Explore planet Earth in your backyard!")

- Science and Technology Journalism ("Share NASA's story of adventure and discovery!")

- Aerospace Technology Engineering Challenge ("Build and test a spacecraft structure!")

- Watching Earth Change ("Investigate the Earth using data from satellites!")

- Design a Mission to Mars ("Design a mission to explore the red planet!")

- Space Flight Opportunities ("Design, build, and send an experiment to space!")

Entry in NSIP is free of charge. All NSIP students receive a certificate of participation. High school winners receive a free trip to a NASA center to present their work to NASA scientists and engineers at the NSIP Symposium. Winners of the Space Flight Opportunities competition receive a free, week-long trip to NASA Wallops Flight Facility to prepare their experiments for flight on the Space Shuttle or on a NASA rocket. Primary and middle school winners receive a presentation at their school by a NASA representative. Middle school national winners receive a scholarship to attend Space Camp. NSIP is a rewarding experience for students and for their teachers. It is a great opportunity to watch your students shine!

Systemic Change in Delaware and Its Use of NASA-Generated Materials

Harry L. Shipman

Physics and Astronomy Department, University of Delaware, Co-Leader, Transformation of Energy Development Team, Delaware Science Coalition

Abstract. Since the mid-1990s, the State of Delaware has embarked on a program for systemic change which is not unlike the programs followed by other states (Adelman 1998). Professional development has been a key component of this program at all levels. Instead of asking how NASA-generated educational materials can incorporate professional development, this paper turns the question around and asks how a statewide, systemic professional development has used NASA-generated materials. I'll describe such a project. I will draw on this experience to suggest some ways in which curriculum developers like NASA can increase the chances that their materials will see widespread use in statewide systemic reform. In summary, materials which will be useful in K-12 education are those materials which take into account the curriculum frameworks or standards which have been developed in a variety of states. Furthermore, large-scale systemic change takes time. The days of the Lone Ranger approach, where a single individual enters the scene, spends a short time trying to fix things, and then leaves, are gone.

Stages of Systemic Change

The American educational system is currently in the midst of considerable systemic change. As this paper is written, one of the principal drivers of systemic change are the statewide, high-stakes tests which political leaders everywhere seem to place great faith in. While several authors (e.g., Black 1998) have questioned many of the uses of these high-stakes tests, they are a crucial part of the political landscape and cannot be ignored. For example, in states where science is not one of the subjects that is tested, it is not easy to get any time for any kind of science in elementary or even middle school.

In Delaware, the current wave of reform began in the early 1990s with the adoption of a set of standards in all four major curriculum areas: language arts, mathematics, social studies, and science (State of Delaware 1995). These standards were developed in concert with the National Science Education Standards (National Resource Council 1996). They reflect the same approach as the earlier effort by the American Association for the Advancement of Science (AAAS 1989, 1993). Development of the standards was quickly followed by the development of a set of performance indicators, which attempted to translate the standards into a set of objectives that curriculum developers could use. The same sorts of events are taking place in other states with some minor differences of detail and sequence.

In the late 1990s, two parallel efforts continued the reform beyond the standards. The state developed a statewide testing program which, most importantly, was aligned with the standards. Those of us, including the author of this paper, who wrote questions for the testing program, are a little disappointed that some of the new ways of assessment were not more thoroughly incorporated into the testing program. Still, students are often asked to provide verbal explanations of their answers to a particular question, and those verbal explanations are assessed by a fairly detailed and robust rubric.

The second of the parallel efforts was the development of a statewide curriculum. In the early grades, we could find a number of kits produced by the Full Option Science System (FOSS), the Smithsonian's National Science Resources Center and marketed by Carolina Biological Supply, and by Insights publishing, which fit reasonably well to the standards. As we pushed on to middle school, the available kits and texts fit our curricular needs considerably less well, and so a number of curricular units had to be custom-developed. The development of one of these units on Transformation of Energy is described later in this paper.

As Rodger Bybee pointed out to a group of us when he visited Delaware in the mid-1990s, this kind of curriculum reform takes time. After I had spent two years helping various groups develop parts of the state standards, I wondered if it was going to get any easier. Rodger pointed out that the next step, development of curriculum which incorporates the standards, was going to take the better part of a decade, not just a couple of years. And curriculum development is not the end of it. It's even harder, and takes several decades, to do what it takes to see the curriculum actually used in classrooms.

Astronomy and Space Science in K-12 Science in Delaware

A number of players in the science education reform movement have a particular agenda: the topic of this conference on space science education. We see astronomy and space science as curricular areas with a high degree of public interest. When K-12 students encounter astronomy, the persistent cries of "why do we have to learn this stuff?" become much more muted, because astronomy has an inherent appeal that some of the other sciences lack. Many people in NASA and a number of astronomers, including me, can appeal to some higher motives and not just our own parochial background in our efforts to undo what the Committee of 10 did in the late nineteenth century. This group selected biology, chemistry, and physics as the core sciences and banished astronomy, geology, and meteorology to the periphery of elective courses.

In Delaware our efforts paid off. I had met with several groups of teachers as the standards were being developed and "Earth in Space" became one of 8 major themes in the Delaware standards. When I was asked to write a set of performance indicators for 11th grade astronomy with 48 hours notice, I simply did it. I started teaching an astronomy course for middle school teachers in 1996, and one of the first students in that course is now teaching it. Others pushed for the use of the "Measurement of time" FOSS kit in the 6th grade. This unusually named unit includes a lot of astronomy including lunar phases and seasons. Astronomy is now back into K-12 classrooms in Delaware. The same general scenario could occur anywhere, and is occurring in many places. While different states write their standards in various levels of detail and place topics in different years, the same topics are found in almost all of them, and in the AAAS Benchmarks, and in the National Science Education Standards. But these standards should be seen as opportunities for the introduction of particular curricular areas rather than as mandates. Astronomy won't enter the classrooms unless someone – an astronomer – pushes for it and unless there are curricular units available that can make it possible. The Delaware energy story shows how it can work.

The Transformation of Energy Unit

Several years ago, I was approached by a number of teachers in the Capital School District who wanted to develop an 8th grade curricular unit on energy. Sherry Densler, Debbie Forest, Kristen Lupo, Mary Jo Moyer, Ellen Thompson, and I worked together for several days, including some days in the summer, to prepare this unit. The unit was then successfully used in one elementary school, Central Middle School in Dover, Delaware.

A year or so later, Rachel Wood, the state's science curriculum director, approached the University of Delaware's Mathematics and Science Education Resources Center (MSERC) and asked us to develop an 8th grade energy course. MSERC's mission, which grew out of some of the summer science courses that were taught for in-service teachers, is, broadly, to support mathematics and science education in the state in any way that is needed. I thought it would simply be a matter of taking the Capital School District unit and adapting it slightly. However, after meeting with a relatively large group that included many of the state's lead teachers, it became clear that much of the content of the Capital School District unit was being taught at other grade levels. We had to develop a lot of new stuff.

So we met, and discussed, and argued. The core group that developed the unit consisted of Gwyneth Sharp, a chemistry and biology teacher at Cape Henlopen High School; Anu Dujari, a science education professor at Delaware State University, and myself. Of the others at these meetings, not one was a space scientist and only one (Barbara Duch, Associate Director of MSERC) was a physicist. I won some arguments – succeeding in presenting heat as heat transfer, following much of the misconceptions literature (Solomon 1992, Stepans 1996). I lost some arguments; we are including much more detail about color than I thought was necessary in the 8th grade. And when Gwyneth Sharp showed me how photosynthesis is taught in the 8th grade, with students doing bottle biology experiments and illuminating plants with different colors of light, I then understood why our unit needed a reasonably detailed explanation of color. I learned a wonderful new teaching technique when Anu Dujari shared her knowledge of an activity on "colored shadows" which teaches the physics of color so much better than any other activity I've seen. (This activity was developed at the Exploratorium and can be found on their website.)

The implementation sequence of the unit builds on Delaware's experience with introducing kits into the elementary school classroom. Delaware borrowed this scheme from Montgomery County, Maryland. Teachers wishing to use our unit first take a 3-credit graduate course which is offered jointly by the University of Delaware and Delaware State University. The course has an intensive week of instruction in the summer and 3 follow-up sessions in the fall. Once they finish the summer course, they will receive a kit of materials, three boxes which contain everything they need in order to teach the unit. They will be eligible to receive this kit in subsequent years. Sharp, Dujari, and I decided what was to be in the kit, and the Delaware State Department of Education's Science Coalition purchased, built, and distributed the kit.

Use of NASA Materials

Where do NASA materials enter the scene? One of the topics we treat is the electromagnetic spectrum. At an AAS meeting, I saw Michelle Thaller demonstrate the existence and properties of infrared radiation using an infrared camera. She and the public outreach program of the Infrared Processing and Analysis Center at JPL now distributes a short, 5-minute videotape showing Thaller and her camera. The video covers the electromagnetic spectrum in just the right amount of depth for our purposes. It also conveys Thaller's enthusiasm, showing students that scientists are not confined to the shy, introverted, male stereotype.

Does It Work?

Feedback from the teachers who used the unit in its first, pilot year indicates that the unit is well on its way to becoming very successful. Ellen Mingione of Milford Middle school wrote "I can now see more clearly how the scope and sequence is going to fit. After exploring energy chains, students can now launch into a deeper understanding of transfer and transformation. Eureka!" Many other teachers echoed her sentiments.

We have more quantitative data on other parts of the reform. Lyn Newsom's gave her 9th grade science students at Concord High school the same test in two successive years, but the second group of students was impacted by the reform. Before the reform there were 23% A's, 36% D's, and 9% F's. After the reform there were 36% A's, 25% D's, and only 4% F's (Dillner and Melvin 2000). On the statewide testing program, the number of students who scored below the standards at William Penn High School dropped from 295 (out of 810) to 140, after several reformed courses were introduced.

First Lesson Learned: Curricular Materials Should be Designed with the Standards in Mind

A generation ago, a teacher doing 9th grade, or 3rd grade, or 1st grade science could really cover just about any topic that he or she wanted. Standardization did not start until 10th grade biology. But now in school district after school district, there is a curriculum structure, dictated by state standards and high-stakes tests. If a proposed curricular unit doesn't fit within this structure, it won't be used. There is enough commonality across states that units can be designed for the entire country and fit reasonably well. For example, I know a great deal about white dwarf stars since my astrophysics research has concentrated on them. It would be easy and fun for me to design a unit on them for the 4th grade. But I won't do it. Such a unit would never be used. White dwarf stars are not, and should not be, in the K-8 standards. They may have a very small place in high school astronomy as part of the treatment of stellar life cycles. While curriculum developers, including those funded by NASA, are beginning to pay more attention to the standards, I still see a fair number of activities which are developed in isolation and then tossed into the marketplace of the world wide web in the hope that someone will make use of them.

Furthermore, people who develop and promote a curricular unit have to be aware that even the language that is used in K-8 science is different from the language in the research community. What's taught at this level is called "Force and Motion" and "Transformation of Energy," not "physics." Delaware's curriculum strand is "Earth In Space," not "Astronomy." There has been an increasing effort to index some of the wonderful stuff that's out there on the web using terms that will permit teachers and curriculum developers to find it, but considerable work remains to be done.

Second Lesson Learned: The Lone Ranger Approach Won't Work

Some scientists see their contribution to elementary education as visiting a few (one?) science classrooms, leaving some materials behind, and going back to the lab. It's better than doing nothing and it does mean you get to meet some teachers. But deep down, this approach is very similar to what the "Lone Ranger" did in some TV (and radio!) shows that were broadcast several generations ago. A scientist developing a curricular unit with little or no input from teachers is also being a Lone Ranger. The energy unit which we developed is enormously stronger as a result of the contributions of a large number of people. Those scientists who will really make a difference in preK-12 education are those who will work with groups of teachers in developing and implementing these materials. As with any collaborative enterprise, teachers are not necessarily going to trust university professors at the outset of a project. In my case, it helped that Delaware is a small state and that I've been visiting classrooms forever. For example, the astronomy for middle school teachers was taught last year by Bill Phillips and Sherry Densler. I first met Bill when I visited his classroom in the 1970s. I visited Sherry's classroom in the 1980s. Now I'm working with Bill, and Sherry, and Gwyneth, and Anu, and many other people on these various teams. We trust each other.

It's probably pushing the Lone Ranger metaphor too far, but . . . Readers who may remember the content of Lone Ranger shows may recall that the mysterious, masked Lone Ranger came in, killed the bad guy with a silver bullet, and then rode off into the sunset. The conditions which led to the problems with the bad guy were still the same. Real change, long-range change, requires something more lasting, some changes in the system which will outlive any particular individual. The Lone Ranger didn't accomplish systemic change. We educational reformers must do better.

Acknowledgments. My work on this unit has been supported by the National Science Foundation (DUE-97-52285). My colleague Anu Dujari has been supported by the U.S. Department of Education through one of the last Title II ("Eisenhower") grants to be awarded. Considerable support to the energy unit has been provided by the Delaware Science Coalition, which is supported by a combination of state, federal, and private funds.

References

Adelman, N.W. 1998, in SSI Case Studies, Cohort I: Connecticut, Delaware, Louisiana, Montana, ed. A. A. Zucker & P. M. Shields (Menlo Park, CA: SRI International)

American Association for the Advancement of Science Project 2061 1989, Science for All Americans (New York: Oxford University Press)

American Association for the Advancement of Science Project 2061 1993, Benchmarks for Science Literacy (New York: Oxford University Press)

Black, P. 1998, Testing: Friend or Foe? Theory and Practice of Assessment and Testing (London: Falmer Press)

Delaware, State of 1995, Delaware Science Education Standards (Dover, DE: Delaware State Department of Education)

Dillner, H., & Melvin, K. 2000
http://www.k12.de.us/science/scivan/studentachievment.htm

National Research Council 1996, National Science Education Standards (Washington: National Academies Press)

Solomon, J. 1992, Getting to Know About Energy-In Schools and Society (London: Falmer/Routledge)

Stepans, J. 1996, Targeting Students' Science Misconceptions: Physical Science Concepts Using the Conceptual Change Model (Riverview, FL: The Idea Factory)

Building a Technology Culture in 29 Chicago Schools

Don G. York
The University of Chicago, 5640 South Ellis Avenue, Chicago, Illinois 60615, don@oddjob.uchicago.edu

The Chicago Public Schools / University of Chicago Internet Project (CUIP) is a program to create a sustainable technology culture in 29 public schools on the Southside of Chicago. After four years of working in partnership with principals and with the Chicago Public Schools (CPS) system, we have trained hundreds of teachers in the rudimentary use of computers and have trained 160 teachers in use of technology to create curriculum. Individuals in the latter group are now the technology leaders in their respective schools. We have created a program to provide low cost computers to each school. We have worked with CPS to provide Internet access to all classrooms and plugs to power the computers. We have established a support network to deal with maintenance issues, beyond that provided by CPS. Finally, we have initiated three technology integration programs to provide peer-reviewed, teacher-written material for all CPS schools. These are the digital library (e-CUIP), a set of instructional modules based on material in Chicago museums (Chicago WebDocent, CWD) and a program to search out, grade and organize websites appropriate for certain subjects and grades (Websift).

Achieving Coherence, Effectiveness and Widespread Usage in NASA-Related Standards-Based Curriculum Development

Jacqueline Barber

Great Explorations in Math and Science (GEMS) Program, Lawrence Hall of Science, University of California, Berkeley, CA 94720, jbarber@uclink4.berkeley.edu

Introduction

As we enter the 21st century, the criteria for exemplary science curriculum development have undergone significant transformation. These new criteria have been deeply influenced by the development of the National Science Education Standards and Benchmarks for Science Literacy, the findings of the Third International Mathematics and Science Study (TIMSS 2000), curriculum design strategies (Wiggins & McTighe 1998), and a general climate that demands both school/teacher accountability and research-based evidence of educational effectiveness. Within this context, as NASA seeks to maximize the quality, effectiveness, and use of its Education and Public Outreach efforts, some important lessons have been gained through collaborative work with the Great Explorations in Math and Science (GEMS) program at the UC Berkeley Lawrence Hall of Science (LHS). The following will highlight some of these lessons and propose an innovative model for a space-science/astronomy curriculum strand for grades 4-8. In the context of the desire for increased curricular coherence, we describe the issues that drive curriculum choices, offer a few clarifying definitions, and identify a viable role for NASA Space Science.

The GEMS of Astronomy

The GEMS Program creates curriculum modules of from 2-6 weeks duration for preK-8th grade, and was named a Promising Science Education Program by the U.S. Department of Education Science and Mathematics Education Expert Panel. Its growing national network of more than 50 professional development centers and thousands of teacher-associates establishes a powerful base for educator professional development and national dissemination. There are now more than 60 GEMS curriculum units, with a solid core of space science activities for upper elementary and middle school, including: *Earth, Moon, and Stars* (Sneider 1986); *Moons of Jupiter* (Sutter et al. 1993); *Messages from Space* (Beals, Erickson, & Sneider 2000), *The Real Reasons for Seasons* (Gould, Willard & Pompea 2001) and *Living with a Star* (Glaser et al. 2003). The latter two were developed in partnership with NASA's Sun-Earth Connection Education Forum (SECEF). As part of the E/PO effort for the SWIFT Mission, a GEMS guide, *The Invisible Universe* (Pompea & Gould 2002), has also been published. Several other GEMS units are outlined in pending E/PO proposals. One of the flagship GEMS modules, *Oobleck: What Do Scientists Do?* (Sneider 1985) also has a strong space science component.

Lessons Learned about Developing Effective Curriculum Units

Curriculum development requires a range of skills and knowledge that is best addressed by a collaborative team. Successful curriculum development projects will assemble a development team composed of teachers, scientists, and curriculum developers that can bring broad expertise to the task. The first step is to start with well-defined student outcomes (what we want students to know) that are aligned to appropriate science education standards. An assessment system to determine how well students are achieving learning goals is also necessary. Subsequently, the team will put together a sequence of inquiry-based experiences that guide students to these outcomes, taking into account research-based knowledge about conceptual development and a context of motivation and relevance to students. Several rounds of field testing of the curriculum units are essential, conducted at local and national levels. Such field testing should be conducted with an audience that reflects the cultural and socio-economic diversity of the ultimate audience expected to use the final materials. Effective units have depth of content, but are not too long to fit comfortably in the school curriculum. One of the challenges in curriculum development is how to modify and refine the field test materials while balancing the breadth of content, depth of content, and length of the unit. Robust curriculum materials pay special attention to the needs of the full continuum of learners, and are relevant to a multiplicity of audiences, as decisions are made to modify and refine the field test versions.

Lessons Learned about Supporting Widespread Usage of Curriculum Units

As a supplemental enrichment program, GEMS units are used very widely. Their high level of scientific and pedagogical integrity, inquiry-based approach, accessibility, flexibility, and teacher-friendliness are widely recognized. Creating a strong and effective curriculum unit is the first step to ensuring that it gets used, for units that do not meet this high standard are not even candidates for a high level of usage. But even great units will sit on the shelf if their use is not supported. Developers should plan on spending substantial effort supporting the use of curriculum units, working closely with teacher-leaders to enable them to become expert in the use of the curriculum units with students. It is important to build the capacity of teacher-leaders to support other teachers in the use of those curriculum units within their specific regional and local context. Establishment of regional sites provides training and implementation support for the curriculum units, especially when these sites are based at institutions committed to supporting teachers in the use of inquiry-based science for greatest sustainability. This type of infrastructure and support network is required to achieve widespread usage.

Towards Promoting Greater Curriculum Coherence

While depth can be achieved in a single unit, coherence must be built over multiple curricular units, namely, through a curriculum sequence. Many excellent teachers and schools build curriculum sequences with depth and coherence, however, it is not uncommon for supplementary units to be taught out-of-context or tacked on as a "break" from the core program. Curriculum developers can contribute to promoting coherence by developing carefully constructed sequences of curriculum units based on a conceptual framework.

In trying to define a position for NASA Space Science relative to curriculum development and increased coherence, it is necessary to think through the issues that drive curriculum choices, and to define key terms that are often used in this context. Words like standards, framework, scope and sequence, strand, and even curriculum get used differently by different groups. What follows is an attempt to define some of these words for clarity, and to identify potential roles that NASA Space Science could play related to each.

Standards: Content standards outline the details of what students should know, understand, and be able to do over the course of K-12 education. Standards (sometimes referred to as benchmarks) are typically not listed grade-by-grade, but rather represent what students should know, for example, by the end of 4th grade, 8th grade, and 12th grade. The National Research Council's National Science Education Standards (NRC 1996) are considered to be "the national standards." They are highly correlated with the AAAS Benchmarks for Science Literacy. All states (except Iowa) have their own state standards. Many states' science standards are derivative of the NRC Standards, but a notable number are not (like California, whose standards are quite different). Even many districts have their own standards (some derivative of national standards, some derivative of their state standards). This makes "aligning with standards" more challenging for groups with a national reach, like NASA and Lawrence Hall of Science, since, what's aligned in Waco, Texas, may not be in Gainesville, Florida. Whatever NASA does relative to standards needs to have some degree of flexibility if it is to be used broadly.

Standards are usually advisory, technically speaking. But with high stakes testing on the rise, the testing drives the teaching. Increasingly, schools are revising their curriculum content to better address whatever standards their district/state consider as "theirs." Alignment is happening more quickly in states where science is tested, but many states only test literacy and mathematics. This will change as the No Child Left Behind (NCLB) legislation phases in required science testing at the state level. Even when students are tested in science, few states' standardized tests are very closely correlated to their own science standards, since many states are using off-the-shelf tests. The bottom line is that curriculum and instruction that are not aligned to content in "the standards" fall seriously low in teachers' curriculum priorities. This state of affairs seems to be increasingly true in many districts.

NASA could certainly improve on the NRC standards in space science and astronomy (as many space scientists point out), and this could potentially influence the development of future national, state, and district standards. However, creating another set of content standards is not likely to be viewed favorably by the K-12 community. Instead, by building a framework based on existing standards, NASA could both position its work in education in the main stream, and influence instruction through a strong statement about space science teaching and learning.

Framework: The word "framework" is used in several different ways in education – often quite different from one another. For the purposes of this discussion, a framework is a roadmap that organizes standards-aligned sequences of space science topics by grade level. It provides recommendations of a sequence of content that would enable students to achieve progress towards standards. So for instance, the NRC content standards specify what students should have achieved by the end of 4th grade, or 8th grade or 12th grade. A framework would provide a more detailed roadmap of how students could get there.

Creating a NASA Space Science Framework could have some important impacts.

  a. It would enable NASA to make a statement about what they think is important for K-12 students to know in space science. To be credible, any framework would need to be based mostly (85-90%) on the NRC Science Education Standards, however, there would be room for NASA to vary the content somewhat – essentially to have a NASA brand on it. Being highly correlated with the NRC Science Standards would enable the use of the framework by schools, districts, and/or states that chose to align with the NASA Space Science Framework. While based on the NRC Science Standards, such a framework could effectively evolve those "standards" a bit.

  b. Users could find K-12 space-science-related content standards in one place, an improvement on the arrangement presented in the NRC Standards. Different content standards appear in different sections, not just Earth Science and Physical Science, but Science & Technology, and Science in Personal and Social Perspectives. Unless you know your standards well, it's possible to overlook whole sections that describe relevant content. The call for E/PO groups to align their products to space science standards is easier said than done.

  c. It would provide educators, NASA scientists, and other personnel working on E/PO efforts with more detail about how a specific science standard to be achieved by the end of 4th grade, for instance, might develop in earlier grades by outlining a conceptual path.

  d. It could provide more cohesion to the Space Science Education Resource Directory (SSERD), by adding standards-alignment information, in the context of the Framework, to the Directory's database records. This would provide a context for both coherence (among NASA-created products) and alignment to standards for sequences of products, or even solitary, stand-alone resources. The Framework would allow SSERD resources to relate to each other, to the topics and concepts they teach, and to the standards.

  e. In addition, such a framework could provide a good tool for product developers to map where there are gaps (a lack of materials that address a specific topic or concept or grade level) and/or gluts (lots and lots of existing materials). One could even imagine that NASA might want to require E/PO proposals to refer to existing materials and describe how the proposed materials are different and/or meet an unfilled niche.

Additional work is needed to further evolve the idea of a NASA Space Science Framework, keeping in mind the necessary attributes of flexibility and ease of use, while considering the level of detail to include. It's not a trivial task.

Scope and Sequence: A scope and sequence is similar to the framework described above, but is usually less comprehensive. Scope and sequences always suggest a path (sequence of content) to achieving progress towards one or more content standards. Scope and sequences usually focus on one topic at one grade level. Scope and sequences sometimes mention specific instructional materials or learning activities. Technically, what is described above as a framework could be called a scope and sequence, but given the range of meaning of the term "scope and sequence," it might be advisable to use the broader term "framework."

Curriculum Strand/Curriculum Sequence: Typically, curriculum strands or sequences refer to a series of specific lessons that together form more than one "unit" of specific study. Strands and sequences are by definition coherent in that they build deeper, connected knowledge over time. They provide a teacher with ready-made coherence. Teachers can do the work of building coherent sequences on their own (by sequencing a variety of resources)-and many schools and teachers do this very well. But constructing coherent curriculum sequences requires more work and coordination for teachers, and is dependent on an understanding of the bigger connected picture of space science. Curriculum sequences are tremendously useful tools for one large subset of teachers who would rather have that construction done for them. Providing curriculum strands/sequences increases the number of teachers who are able to teach a coherent program, even in subject areas where they have little prior knowledge.

The GEMS Program has begun the development of a 3rd-5th grade sequence and a 6th-8th grade sequence in Astronomy and Space Science. It is not our belief that there should be only one space science curriculum sequence or strand for K-12, or even for a narrower span such as middle school. No curriculum sequence can address all of what the standards suggest need to be taught at a specific grade span, and still fit within the time teachers have available. (There is too much content defined by the national standards for any teacher to be able to teach it all in one year.) Teachers and schools need to make choices about where they want to focus their efforts in deep ways. Ultimately, providing the K-12 education community with a choice of space science curriculum sequences, each focused on different collections of specific topics, but all responsive to standards and a Space Science Framework, will make powerful contributions to increasing the quality and coherence of space science education in our nation's classrooms.

Assessment Tools: With the increased emphasis on accountability in schools, teachers are being pressed to demonstrate not just that they are presenting specific content, but that students' knowledge and understanding of that content is increasing. A look at the "misconceptions" literature reveals that even students who score well on traditional tests of knowledge, often don't understand basic concepts in the ways we would want them to. Measuring actual student understanding (vs. regurgitated knowledge) requires different and more sophisticated kinds of assessment instruments than are typically developed and used by teachers. Most teachers have neither the time nor the expertise to develop assessments that will enable them to see what their students are truly understanding, and what they are not. Armed with this information, teachers are able to adjust and modify their instruction to better address students' misunderstanding.

NASA could make a powerful contribution to the K-12 education, and the effective teaching of space science, by sponsoring the development of a collection of assessment instruments. These "test items" could be aligned to the concepts in the NASA Space Science Framework, and used by teachers regardless of the specific instructional materials used. In this way, all teachers could have access to high quality assessment tools to use in measuring their students' understanding of key space science content.

Conclusions

The creation of a NASA Space Science Framework would have tremendous value in defining the conceptual base of K-12 space science and establishing increased context and coherence among existing and future NASA-created educational products and materials-both standalone resources and curriculum sequences. Evolution of the SSERD based on this Framework would serve as an educational tool in and of itself- increasing accessibility to, and knowledge of, the wide range of NASA educational products and demonstrating their connection to each other and to the standards. Additionally, providing the K-12 community with a bank of assessment instruments, enabling teachers to measure students' knowledge and understanding, would make a valuable contribution to K-12 space science education.

NASA's ability to encourage and support the development of a number of practical and rigorous space science sequences, focused at a variety of grade ranges, would make a significant contribution to the state of space science education and go a long way to increasing the depth and coherence of space science taught in K-12 classrooms. Future E/PO efforts should focus not just on creating additional single space science resources, but should build coherence between existing resources, thereby elevating stand-alone supplements to important core curriculum resources. It's worth noting, however, that curriculum sequences are just one important tool. There is still a tremendously important role for stand-alone supplementary materials – to enhance a text-based program, as a first foray into using unfamiliar materials, for teachers who want to create their own custom coherent sequence, or to pinpoint a very specific topic or content area – as well as their importance in informal education. It would be a mistake to swing too far towards creating only deep and coherent approaches at the expense of eliminating the importance of the awareness level and focused stand-alone resources that NASA has historically created.

References

Beals, K., Erickson, J., & Sneider, C. 2000, Messages From Space: The Solar System and Beyond (Berkeley: Lawrence Hall of Science)

Glaser, D., Beals, K., Pompea, S., Willard, C. 2003, Living with a Star: From Sunscreen to Space Weather (Berkeley: Lawrence Hall of Science)

Gould, A., Willard, C., & Pompea, S. 2001, The Real Reasons for Seasons (Berkeley: Lawrence Hall of Science)

National Research Council, National Committee on Science Education Standards and Assessment 1995, National Science Education Standards (Washington: National Academies Press)

Pompea, S., & Gould, A. 2002, Invisible Universe, (Berkeley: Lawrence Hall of Science)

Sneider, C. 1986, Earth, Moon, and Stars (Berkeley: Lawrence Hall of Science)

Sneider, C. 1985, Oobleck: What Do Scientists Do? (Berkeley: Lawrence Hall of Science)

Sutter, D., Sneider, C., Gould, A., Willard, C., & DeVore, E. 1993, The Moons of Jupiter (Berkeley: Lawrence Hall of Science)

Third International Mathematics and Science Study (TIMSS) 1999 http://isc.bc.edu/timss1999.html

Wiggins, G. & McTighe, J. 1998, Understanding by Design (Alexandria, VA: Association for Supervision and Curriculum Development)

New Opportunities Through Minority Initiatives in Space Science (NOMISS)

Alice J. Kawakami and Richard Crowe

University of Hawai'i at Hilo, 200 W. Kawili St., Hilo, HI 96720, alicek@hawaii.edu

Introduction

New Opportunities through Minority Initiatives in Space Science (NOMISS) is a program designed to engage minority students (particularly those of Native Hawaiian ancestry) in learning about the Hawaiian cultural context and its application and relevance to the study of space science. The University of Hawai'i at Hilo is located at the foot of Mauna Kea, acknowledged as the premier site for viewing the sky from Earth. The purpose of this grant is to engage a broad spectrum of participants, K-12 students and their teachers and undergraduate university students, in activities that bring together the concepts of modern space science, the history of Hawaiian celestial navigation and traditions of the land.

NOMISS, one of 15 grants awarded by NASA's Office of Space Science Minority University Partnership, is a partnership of two university departments, Hawaii's public and private K-12 schools, the observatories atop Mauna Kea and NASA's outreach scientists. Principal Investigator Dr. Richard Crowe, Professor of Astronomy, is responsible for the Astronomy major undergraduate curriculum and leads efforts to collaborate with the observatories and astronomers. Co-Investigator Dr. Alice Kawakami, Chair of the Education Department, is responsible for the K-12 outreach component that focuses on collaborative projects with teachers in public and private schools. The dual focus of this grant establishes the context for successful initiatives to address issues of diversity and systemic change. Unfortunately, previous efforts have not been able to establish and institutionalize curriculum that capitalizes on the complementary nature of modern space science and traditions of Polynesian voyaging and settlement. Our aim is to find the right combination that could hold the key to local students' success in the study of math, science, and their own culture.

Tensions and Challenges

In the island environment, identity and place are powerful determinants of any successful collaboration. On the island of Hawai'i, there is tension between the native Hawaiian community and the observatories atop Mauna Kea. An example of this tension is the lens through which the mountain is viewed.

A website of the Office of Mauna Kea Management's Astronomy Education Committee (Office of Mauna Kea Management 2003) describes Mauna Kea in the following terms:

> Many experts consider Mauna Kea to be the finest location for land-based astronomy in the world – or at least the best in the northern hemisphere. Its summit rises 13,796 feet above sea level – well above the clouds, which tend to settle at around the 9,000-foot level. As a result, astronomers are afforded clear viewing for more than 300 days out of each year. Situated on an island in the middle of the Pacific, Mauna Kea is well buffered by thousands of miles of ocean with little of the air and light pollution that plague other locales.

This description focuses on the value of Mauna Kea as a location for observing the sky. A perspective more typically held by native Hawaiians and citizens of the Island of Hawai'i is provided by a website created by a student named Kanoelani (Sakihama 1999). Her website on Mauna Kea opens with the following paragraph:

> Rising 31,796 feet from the ocean floor, Mauna Kea is the tallest mountain in the world. There are many legends that surround this white-capped mountain. One such story tells of a small lake near the summit named for one of the Hawaiian snow goddesses, Waiau. The lake's water comes from the annual snowfall which makes up most of the annual precipitation. According to Hawaiian legend, the lake was originally created for another snow goddess, Poliahu, by her godfather Kane as a swimming pool. Mo'oinanea, the guardian of the pool, numbed with cold any who happened to find the lake causing them to fall into a frozen sleep.

Hawaiians view Mauna Kea as a sacred mountain, valued and valuable historically, spiritually, geographically, and environmentally. This indigenous view has led to a Hawaiian community embroiled in debate about observatories built and operated on land that during ancient times was reserved for only a privileged few. Space scientists and researchers from around the world value Mauna Kea as the perfect site for exploring space beyond Earth. Hawaii's youth are caught between the two perspectives, looking outward into space and looking inward to the land and to the traditions of the people who inhabited this place long ago. In response, many students choose to disengage from learning through opportunities presented by Western models of schooling in this very unique Hawaiian place, at this very unique time.

One of the biggest challenges facing NOMISS is recognition of the cultural traditions of Hawai'i by educators, researchers, and undergraduate Astronomy majors at the University. Another equally formidable challenge is making learning about space science relevant and engaging to young students in Hawai'i's schools. In order to address these challenges, NOMISS is working very hard to establish positive, effective, working relationships among partners who have a stake in the future of Mauna Kea.

Addressing the Challenges

Through epistemology, the study of knowledge and ways of knowing, attention is given to both the indigenous perspective with a rich heritage of celestial navigation, voyaging and a reverence of the land and the astronomers' perspective of technological support to search for answers about the origins of the Solar System and the Universe.

We believe that teachers and professors must be provided with learning opportunities that reflect multiple views of knowledge. We have held two summer retreats and three one-day sessions (Protocol 101) to learn about Hawaiian traditions and protocol. Our summer retreats focused on space science, traditional knowledge and Polynesian voyaging. These retreats included visits to Mauna Kea, other significant cultural sites, and the observatories. Our Protocol 101 sessions led to discussion of spirituality and cultural practice by both native Hawaiian and non-native teachers. All of this yielded exciting and engaging curriculum innovations in the K-12 classes and additional experiences for undergraduate Astronomy students. Partnerships with the Hawaiian community, the Polynesian voyaging community, Mauna Kea observatories, Jet Propulsion Laboratory, local businesses, public and private K-12 schools, and the university provides critical support for our efforts. The impact of these experiences for teachers and students, professors and astronomers has been nothing short of life-changing.

The NOMISS teachers, from both public and private schools on Hawai'i and O'ahu, have enthusiastically met the challenge of making learning about space science relevant and engaging to young students in Hawaii's schools. In the lower elementary grades, a focus on multi-sensory and naked eye observation provides entry into the study of the Hawaiian moon calendar, traditional agriculture and aquaculture, and voyaging based on non-instrument navigation. It also provides the basics of Earth, sun, moon, and planetary system knowledge base. In the upper grades, the theme of Polynesian voyaging integrates language arts, social studies, math, and science in year long projects that culminate in a "mock-voyage" – complete with crew assignments, sail plans (including personal genealogies, maps, meal plans, and provisions), and positioning of sails and steering paddles as part of the final exam.

NOMISS curriculum development for grades Kindergarten through 12 is compiled by teachers who field-test activities as part of their year-long instructional program. They sometimes face barriers in trying to provide authentic learning experiences through field trips and day-long simulation activities. In many cases, their classes run contrary to institutional rules and regulations. The NOMISS project is instrumental in lobbying for special arrangements and providing support through partnerships. Observatory staff and cultural experts have been very generous in their support so that students may partake of both knowledge-based learning activities with astronomers and outreach scientists and experience-based learning at cultural sites and on voyaging canoes.

The undergraduate Astronomy major at UHH is also changing with a new text book and curriculum designed to address the issues of the cultural and historical context of Mauna Kea as well as the most current research and technology. Internships with the observatories and NOMISS participation on the observatories' outreach committee begins to bring an awareness of space science and the special context of the Hilo community.

Lessons Learned

Perhaps the most important lesson we have learned is that location matters. Generic programs designed in far-away places do not capitalize on the wealth of local knowledge and resources so necessary to engaging today's youth. Key determinants in effective education and outreach are "where we are" and "who we are." Respect for the host community and culture is essential to reach and excite the populace about modern space science. The need to acknowledge the people, place and history of Hawai'i is essential to developing individuals who are at peace with both ancient traditions and the futuristic technology of observatories on Mauna Kea. Outreach scientists, curriculum developers, and K-12 educators must experience traditional learning in order to be humbled by the power of knowledge that cannot be expressed through words or numbers. Likewise, professional development must address both Western and indigenous perspectives.

Teams of teachers, who know their students and their communities, must develop curriculum with support from content experts in space science, Hawaiian culture, and curriculum. Their efforts to design relevant learning experiences also need partners to provide field experiences and hands-on observational coursework. Educational systems are not often structured to honor multiple perspectives and grants such as NOMISS provide much needed encouragement for such innovations.

A successful team effort is based on productive partners. NOMISS addresses the issue of diversity starting with the partnership of the PI, an astronomer and former observatory employee, and the Co-I, a native Hawaiian educator focused on infusing cultural practices into existing academic settings. The K-12 teachers and undergraduates come from a variety of different heritage groups including indigenous groups and immigrants to Hawaii. Our partners include a broad spectrum of the community including educators creating curriculum in the Hawaiian language, navigators of voyaging canoes, business owners, technicians, telescope operators, outreach scientists, and astronomers. NOMISS continues to convene groups of individuals with different perspectives, dedicated to making our community a better place to live because of Mauna Kea.

Conclusion

The following list summarizes the important lessons NOMISS has learned:

- Location and context matters.
- Know your place, its people, its history.
- Partnerships must be based on mutual goals for long-term benefits to the community.
- Professional development is a must for everyone involved.

Finally, although all knowledge is not always visible or tangible, it is still worth learning. We believe that to effect systemic change, we acknowledge the following view expressed by Japanese researchers, Nonaka and Takeuchi, cited in Fullan's (2000) book on systemic change: [Japanese companies] recognize that knowledge expressed in words and numbers represents only the tip of the iceberg. They view knowledge as being primarily tacit – something not easily visible and expressible. Tacit knowledge is highly personal and hard to formalize, making it difficult to communicate and share with others. Subjective insights, intuitions, and hunches fall into this category of knowledge. Furthermore, tacit knowledge is deeply rooted in an individual's action and experience, as well as in the ideals, values, and emotions that he or she embraces. (p.8)

Ua mau ke ea o ka aina i ka pono. – The life of the land is perpetuated in righteousness. (Pukui 1983)

References

Fullan, M. 2000, Change Forces: The Sequel (Philadelphia, PA: Falmer Press)
Office of Mauna Kea Management, Astronomy Education Committee 2003
    http://www.malamamaunakea.org/astronomy.php?article_id=24
Pukui, M. K. 1983, 'Olelo no'eau: Hawaiian proverbs and poetical sayings (Honolulu: Bernice P. Bishop Museum)
Sakihama, K. 1999, Journey to Mauna Kea,
    http://tqjunior.thinkquest.org/5380/

Additional references not cited in this article

Kawakami, A. J. & Aton, K. K. 2001, Pacific Educational Research Journal , 11(1), 53 (Provides a qualitative study of Hawaiian educators' view of essential components for successful educational programming for Hawaiian students.)
Kawakami, A. J. 1999, Education and Urban Society, 32(1) 18 (Provides an overview of recent educational reform to integrate Hawaiian cultural elements into established educational institutions)

Guide to Inquiry-Based Learning Using NASA Resources

Daniel Barstow

Center for Earth and Space Science Education, TERC, 2067 Mass Avenue, Cambridge, MA 02140, Dan_Barstow@terc.edu

Inquiry into authentic questions generated from student experiences is the central strategy for teaching science

– National Science Education Standards

The National Science Education Standards (and virtually all state curriculum frameworks) place a major emphasis on science as inquiry. Inquiry-based learning engages students in DOING science. Students ask questions, design experiments, make observations, analyze data, draw conclusions - all in pursuit of answers to questions of real interest and relevance. Inquiry-based approaches help students develop crucial skills of scientific thinking and learn science content more deeply. Inquiry-based learning sparks student interests and is driven by the same spirit of exploration and discovery that is at the heart of NASA.

Ironically, inquiry-based learning is often left out of current classroom practice. Only 33% of the nation's science teachers use projects and extended investigations as part of their standard teaching practice (Report of the 2000 National Survey of Science and Mathematics Education, Horizon Research 2001). The Third International Mathematics and Science Study (TIMSS), National Assessment of Educational Progress (NAEP) and state tests scores confirm that students are not developing the skills of inquiry and research that are so essential to learning and doing science. This is a serious deficiency in our national implementation of the science education standards for all students. It also jeopardizes efforts to improve the "pipeline" of scientists who have the content knowledge and scientific thinking skills vital for our nation's future.

NASA is powerfully positioned to make a major contribution to solving this problem. More than any other federal agency, NASA is driven by inquiry, exploration and discovery. NASA scientists relentlessly ask questions about Earth, the solar system and the universe beyond. To find answers, they develop and use cutting-edge tools, such as robotic space probes, advanced telescopes, visualization software and take advantage of the human presence in space. This same spirit of inquiry should infuse NASA's education and public outreach efforts as well.

NASA's OSS program has created some stellar examples of inquiry-based resources. For example, the NASA Astrobiology Institute designed learning activities that stimulate and respond to students' natural curiosity about life on other worlds. Students design classroom experiments to explore the range of life on Earth, search for evidence of water on Mars and Europa, and peruse the latest images from Hubble space telescope. Students ask questions and pursue answers in ways that parallel the work of scientists. The Mars exploration program even provides the opportunity for students to do their own primary research, selecting targets for instruments on Mars missions, interpreting the data and presenting the findings to the scientific community.

There is a growing awareness of the need to more effectively and consistently integrate inquiry based learning into the formal educational products, materials and activities developed in conjunction with NASA's space flight missions and research programs. Because few scientists are familiar with inquiry-based approaches to support student learning, many of NASA's K-12 efforts have taken the approach of simply delivering mission specific content knowledge. Although such content knowledge is important, scientists could extend their contribution, by viewing their space science missions as powerful opportunities to stimulate student inquiry and model the process of space science research. This "perceptual shift" in how scientists view K-12 science education has the potential to set the stage for a quantum leap in NASA's contribution to improving science, math, and technology education.

Hence, we recommend that NASA create a "Guide to Inquiry-based Learning" that will facilitate the development and delivery of effective K-12 educational resources "as only NASA can." In fact, there really is a need for two guides - closely related, but targeted for two different audiences.

Guide for Scientists – The first guide is for scientists, mission specialists and others who create E/PO educational resources and learning activities. This guide should briefly and clearly present the rationale for inquiry-based approaches and a common framework for designing inquiry-based resources and activities. The guide should include a "taxonomy" of inquiry-based approaches, with concrete examples, ranging from pre-defined questions to open-ended exploration to engaging students in authentic research.

Guide for Teachers and Students – The second is a guide for teachers and students, to help them select different types and levels of opportunities for inquiry available through NASA. This guide would be web-based, and provide a way to search across missions based on the depth of inquiry experience and the type of inquiry skills. Students and teachers should be able to select opportunities based on a progression of skills, from entry-level guided inquiry to more open-ended explorations. This progression across missions enables students to develop and apply skills with greater depth and sophistication, even across multiple missions.

We recommend that this emphasis on inquiry-based learning not be limited to Space Science, but integrate across all NASA enterprises. If activities use a common framework, students could develop entry level image analysis skills in one project, then use those skills as a basis for more advanced image analysis with other missions. This would provide an opportunity to develop diverse skills at increasing levels of depth over time. This would not only help answer the national need to support inquiry-based learning for all students, it would also help develop a future cadre of NASA scientists, since these are the same inquiry skills that NASA needs for its long-term success.

We believe that engaging students in the thrill of exploration and discovery provides a powerful uniting theme across NASA E/PO efforts. It helps students achieve essential learning goals that are central to science education standards (and often missing in current practice). And it provides a clear and vital pathway for NASA to contribute to improving science education ... as only NASA can.

Yerkes Observatory Professional Development Programs

Vivian Hoette
Yerkes Observatory, 373 West Geneva Street, Williams Bay, WI 53191-0258, vhoette@hale.yerkes.uchicago.edu

Hands-On Universe
(http://hou.lbl.gov or http://handsonuniverse.org)

Major funding for Hands-On Universe (HOU) is provided by the National Science Foundation. Hundreds of teachers nationally and internationally participate in HOU. HOU provides astronomy curriculum for high school students based on analysis of images in fits format using HOU Image Processing software. Teachers may request new images from the HOU telescope network. Currently HOU has an NSF grant to compare the effectiveness of face-to-face workshops with moderated or independent on-line learning. Yerkes Observatory is involved in the following ways:

- Host of HOU Teacher Resource Agents (TRAs) Annual June Conference. TRAs lead workshops, provide teacher support year round, and lead research and observing projects.

- Telescope Support, Image Requests, Observing Projects, Explorations

- On site experiences with observing using the 24 inch Telescope and CCD imaging systems.

- Remote Internet Access to Rooftop Telescope Observatory at Yerkes for teachers who are part of the HOU, Near and Far Sciences for Illinois (NFSI), and Space Science for Illinois Teachers (SSIT) projects, and undergraduate students of University of Chicago campus classes for non-majors. Collaborative remote observing with KIT telescope at Tokyo Science Museum.

- UNIVERSE Liveshow – Observing with Yerkes Telescopes for museum audience at Science Museum in Tokyo. Friday night Yerkes skies for Saturday afternoon Liveshow.

- Center for Adaptive Optics (CfAO) Optical Powers curriculum in collaboration with Univ. of Ca, Santa Cruz, and Hands-On Solar System curriculum in collaboration with Univ. of Ca, Berkeley, Lawrence Hall of Science and TERC, will soon be released.

Near and Far Sciences for Illinois
(http://nfsi-server.yerkes.uchicago.edu)

Funded by the Illinois State Board of Education (ISBE) first through Adler Planetarium and then through DePaul NASA Center 1997-2001. (NSFI continues with locally initiated funding led by NFSI teachers or project leaders when requested.) Hundreds of Illinois teachers have participated in various opportunities provided by NFSI. Many teachers participate in multiyear programs. The NFSI Reunion 2002 was funded by the OSS Broker/Facilitator program at DePaul University. NFSI is designed with emphasis on regional implementation and formation of networks. Seven regions statewide were organized as subsets of a systemic statewide program of professional leadership for teachers. Regional teams include teachers, regional offices of education, professional scientists, experts from museums and other informal science centers, amateur and professional societies.

- NFSI I - Introduction to Meteorology, Astronomy, Geology, Action Research.

- NFSI II - Projects in Earth and Space Science; Groundwater

- NFSI III - SKYWatch: Sharing the Skies; Picture the Universe; Stars and Scopes. Each region shares a Meade LX200 telescope and CCD (seven sets), and a weather station.

- NFSI IV - Build Your Own Technology. Projects: Dobsonian 6" telescopes, tornado simulators, Radio JOVE receivers, Solar projects, etc.

Space Science for Illinois Teachers

Major funding by Eisenhower from ISBE through DePaul NASA. About 20 teachers spend two weeks at a conference in the summer. Week one is at Yerkes; week two is at DePaul. Astronomers at Yerkes work with the teachers on classroom hands-on activities that embody astronomy instrumentation and observing. Topics include connections to NASA's SOFIA with the HAWC instrument team PI, Al Harper; Sloan Digital Sky Survey with Rich Kron, and telescope projects with Kyle Cudworth, Yerkes Obs. Director.

Museum Partners Science Program

Variety of funding including NSF Urban Systemic Initiative, ISBE, and Chicago Public Schools. About 60 teachers participate each year. This program has been active for about six years. MPSP was founded to provide museum support for professional development of science teachers in Chicago. The program offers Saturday museum classes in Earth and Space, Physical, and Life Sciences with science graduate credit through Aurora University. Yerkes, DePaul NASA, and HOU support this program by offering one or more of the Saturday sessions each year. This session usually occurs at DePaul. SOFIA Outreach. Teachers participating in all of the above projects with Yerkes Observatory keep up to date with progress on NASA's SOFIA mission, and are involved in the SOFIA Outreach Network.

Astronomy in the Ice: Bringing Neutrino Astronomy to the Secondary Schools

James Madsen

University of Wisconsin-River Falls, 410 S. Third St., River Falls, WI 54022, james.madsen@uwrf.edu

Robert Atkins, Matthew Briggs, Jodi Cooley, Brenda Dingus, Francis Halzen, Susan Millar, Terrence Millar, Katherine Rawlins, and David Steele

University of Wisconsin-Madison, 1150 University Ave Madison WI 53711

Steven Stevenoski

Lincoln High School, 1801 16th Street So, Wisconsin Rapids, WI 54494

    Astronomy in the Ice introduces the science of the AMANDA (Antarctic Muon And Neutrino Detector Array) and IceCube projects into the secondary school curriculum. This 8-day course for secondary school teachers is designed to provide teachers enabling background in neutrino science and the classroom resources needed to understand the design, operation and science potential of AMANDA, IceCube and similar projects. The course is structured so that a new topic is presented in the morning, with an afternoon activity to illustrate the material and homework in the evening to reinforce the day's learning. Five activities are refined into useable classroom modules by the teachers in collaboration with the course instructors. The connection between the teachers and the scientist instructors is reinforced, and direct connections between the scientists and high school students are established through classroom visits during the school year.

    Astronomy in the Ice is a summer course for secondary teachers in the Master's of Science Education program at University of Wisconsin-River Falls. The course is taught by scientists in the AMANDA/IceCube collaboration (http://icecube.wisc.edu). Our broad goals are:

- help teachers learn more physics

- help teachers develop new or better activities they can use in class

- share the excitement of AMANDA and IceCube scientists with teachers, so that they can pass this excitement on to their students

- help AMANDA and IceCube scientists learn about middle and high school teachers and teaching.

    AMANDA is the first of a new generation of telescopes designed to map the universe with neutrinos rather than photons. We are about to begin constructing a much larger and more sophisticated neutrino telescope, called IceCube, that uses the same principle as AMANDA. These telescopes will help us answer fundamental questions in both physics and astrophysics pertaining to, among other things, the properties of neutrinos and the types of cosmic accelerators (such as black holes) that create the ultra-high energy particles that strike the earth.

Detecting neutrinos is difficult because they rarely interact with anything. They have no charge and almost no mass. For this reason, we need a very large, dark and transparent chunk of material in which to search for the telltale blue light flashes that come from a rare neutrino interaction. The place that best satisfies these requirements is the ice deep under the South Pole. Our AMANDA telescope has demonstrated that freezing an array of light sensors into the ice one mile below the surface at the South Pole creates a working neutrino telescope. IceCube will scale this design up to a size where we can create a new sky map of the universe using neutrinos.

The combination of this new, fundamental research with the rich history of exploration and the intriguing environment of Antarctica make AMANDA and IceCube very attractive topics for the secondary school curriculum. Antarctica attracts and holds the interest of the teachers and students. Evaluation findings show that the fundamental questions that neutrino scientists are asking interest teachers and students in physics and astrophysics.

Our approach to Astronomy in the Ice is guided by the belief that the teachers themselves are the best judges of what works in the classroom. Accordingly, we first help the teachers develop new scientific knowledge and insights, and then give them the lead in developing laboratory activities that will be effective in their classrooms. Challenges that the course instructors and teachers face in developing these activities include:

- creating activities that are accessible to secondary school students
- creating activities that capture some of the interest of these topics
- ensuring that the teachers acquire sufficient training to be comfortable with the new course materials
- connecting these activities to the standards applicable to the teachers' curricula.

The Astronomy in the Ice instructor team consists of a course leader on the UW-River Falls faculty who also is an AMANDA scientist, a high school "master" teacher with Antarctic and curriculum development experience, and 4-5 other scientists who are actively involved in astrophysics research, ranging from graduate students to the IceCube project's Principal Investigator. The format consists of approximately three hours of lecture/discussion each morning with an associated activity in the afternoon. Lunch together each day helps build relationships among all the participants. The teachers are offered the option of evening meetings with the master high-school teacher to go over daily homework assigned to solidify the ideas presented during the day.

The course instructors provide ideas, resources, and background for each afternoon's activity, which all the teachers perform. At the end of the activity session, the teachers and scientists reconvene to discuss the strengths, weaknesses and viability of the activity. On the second to the last afternoon, the class splits into groups of three, based on the teachers' interest in a particular activity, and each group refines their activity into a useable module, using a uniform template. Module development includes identifying the national standards applicable to the activity. On the last afternoon, each group presents their activity to the instructors and other teachers.

Throughout the course a team approach is stressed. The course instructors provide background and ensure the scientific integrity. The teachers maintain the classroom focus, refining the activities into modules appropriate for secondary schools. The course instructors make follow-up visits to the teacher's classes, and write a quarterly newsletter to build on the connections made during the two-week course.

In the summer of 2000, 18 teachers enrolled in the course and the scientists visited 12 schools, addressing more than 1400 students. The class of 2001 enrolled 22 teachers, including 2 who travelled to the South Pole station to work on the AMANDA project during the 2001-02 austral summer as part of the NSF sponsored Teachers Experiencing the Antarctic and Arctic program (TEA). The scientists visited 8 teachers and addressed more than 1000 students. The 2002 class enrolled 15 teachers, one of whom is a TEA teacher slated to work in Antarctica in the 2002-03 austral summer.

Astronomy in the Ice demonstrates that getting scientists, teachers and students together can bring the excitement of scientific research to the classroom and the excitement of teaching to the scientists. Evaluation data from both the Astronomy in the Ice course and the follow-up classroom visits demonstrate that outcomes include: (1) significant gains in teacher knowledge of astrophysics and the AMANDA/IceCube experiment, (2) the development of useful laboratory activities that help achieve the state science standards for the secondary schools, (3) valuable professional experience with teaching and outreach for the instructor team, particularly for the graduate students, and (4) development of new knowledge and interest in general science, astrophysics, particle physics, or the South Pole for high school students.

Acknowledgements

This course was supported by National Science Foundation NSF Grant DGE 139335 (Kindergarten Through Infinity – KTI http://www.wisc.edu/gspd/kti/kti.html) awarded to the UW-Madison, by the Teachers Experiencing the Antarctic and Arctic (TEA) program, administered through Rice University, by the AMANDA project, partially supported by NSF grant OPP-9980474, and by additional funds from the UW-Madison.

Research-Based Public Outreach and Education

Greg Bothun

University of Oregon, Department of Physics, Eugene, OR 97403, nuts@bigmoo.uoregon.edu

Astronomical public outreach is ubiquitously in the form of either informed press release or "glitzy" display. While this is a necessary starting point, it need not simultaneously be the termination point. Yet it seems that most resources are still being directed toward packaging and delivery of passive and static presentations of the science underlying either the space mission of the ground-based observation. While the use of the Web as a public education outlet has facilitated and expanded the kind of information that can be presented, in general, such material is again passive and static.

New Internet technologies, however, offer the possibility of true interactive public outreach and experimentation where the public can run or use a simulation of the mission/observation and/or gain access to the data in a form that allows them to do analysis. While this approach is more expensive to implement, the result is a layered outreach product that can be accessible to people with interests on different levels. Such a product will also help promote scientific literacy as the public can be engaged in an inquiry mode which is based on the actual mission data/images. For example, while the Hubble Deep Field was a spectacular scientific success, the publics' exposure to this science is largely in the form of a cool poster with lots of pretty colored galaxy images on it. The data behind the poster, of course, is digital data that can be measured and analyzed, but that part of the science, for some reason, is generally not delivered to the public (but for an outstanding exception see http://www.ifa.hawaii.edu/~cowie/tts/tts.html).

This panel presentation highlighted some examples of how the Web browsing environment can be used as an interface for data analysis by anyone. Such an environment has direct transfer to the K-12 community and therefore could also become part of the classroom curriculum.

NASA/JPL Solar System Educators Program

Eric Brunsell
*Space Explorers, Inc., 1825 Nimitz Dr., DePere WI 54115,
eric@spaceed.org*

Introduction

The NASA/JPL Solar System Educators Program (SSEP) is a professional development program with the goal of inspiring America's students, creating learning opportunities, and enlightening inquisitive minds by engaging them in the Solar System exploration efforts conducted by NASA. SSEP is a Jet Propulsion Laboratory program managed by Space Explorers, Inc. (Green Bay, WI) and the Virginia Space Grant Consortium (Hampton, VA). However, the heart of the program is a large nationwide network of highly motivated educators. These Solar System Educators lead workshops around the country that show teachers how to successfully incorporate JPL materials into their teaching.

The NASA missions and programs that have supported SSEP include the Cassini Mission to Saturn, the CONTOUR mission, the Deep Impact Mission to a comet, the Deep Space Network (DSN), the Galileo Mission to Jupiter, the JPL Space and Earth Science Directorate, the NASA Office of Space Science Solar System Exploration Education and Public Outreach Forum, the Mars Exploration Program, the Navigator/Planet Quest program, the Outer Planets/Solar Probe Program and the STARDUST Comet Sample Return Mission.

Description

SSEP is a growing community of educators that are passionate about Solar System exploration and its impact on science education, and motivated to share that passion with other educators. JPL, through its contract with Space Education Initiatives (SEI), provides these educators with the knowledge, skills, materials and support that they need. SSEP is a multi-year program. Solar System Educators attend a training institutes at the Jet Propulsion Laboratory, receive online training, support and invitations to special events throughout the year. Solar System Educators are then required to conduct workshops for at least 100 teachers each year.

Community

> SSEP has helped me personally to understand more about the science behind the various missions, and it helps the educators who attend my workshops to feel more comfortable with the subject of space exploration. As a direct result of SSEP, NASA is reaching many thousands of students who may not otherwise learn of these programs and missions.

–Mal Cameron, NH

Solar System Educators are master educators from both formal and informal educational settings. The Solar System Educators are all actively involved in leading professional development opportunities for classroom teachers. Solar System Educators are competitively selected based on their educational background, professional development experience, and knowledge of NASA missions and educational products. There are currently 50 active Solar System Educators representing 30 states. Solar System Educators have received many National awards including the Brennan Award by the Astronomical Society of the Pacific for Outstanding Contributions to the Teaching of High School Astronomy, USA Today All-American Teacher Team, Great Lakes Planetarium Association Fellow Award, two National Board Certified Teachers and two with applications pending, and 10 Presidential Awards for Science Teaching winners and finalists. At least five of the Solar System Educators have won or been finalists for their state's Teacher of the Year or Informal Science Educator of the Year awards. Besides SSEP, these educators are involved in leading a variety of professional development, mentoring, and consulting activities, including facilitating NASA Educational Workshops (NEW) and NASA Explorer School workshops at Goddard and Langley Research Centers and the Jet Propulsion Laboratory, acting as an American Astronomical Society Resource Agents, acting as an American Meteorological Society Resource Agents, acting as Special Term Appointees to the Decision and Information Sciences Division of the Argonne National Laboratory and mentoring and consulting for a variety of Universities and State Departments of Education. At least six Solar System Educators also work directly with pre-service teachers.

Partnerships

> *As a novice teacher, my concept of the nature of science was shaped mostly by textbook authors and other educators. Only the most fortunate of teachers ever enjoy the opportunity to learn directly from individuals who are actively engaged in the very fields of science which they teach. The Solar System Educators Program has provided me with this priceless opportunity. To learn about space exploration from mission scientists, to view the actual space hardware, and to feel the passion that the scientists have for their work, have all had a very positive impact on my teaching. I now feel a greater sense of ownership in our nation's space program, and I can better transfer that passion for exploration and discovery to those I teach, students and other professional educators.*

–Tim McCollum, IL

Solar System Educators come from a variety of K-12 institutions, universities, and the following science centers:

- Alamogordo Space Center, NM
- Ellison Onizuka Space Center, HI

- Carnegie Science Center, PA

- FlintMulticultural Center, MI

- Cernan Earth and Space Center, IL

- Maryland Science Center, MD

- Challenger Space and Science Center, TN

- East Kentucky Science Center, KY

- Christa McAuliffe Planetarium, NH

- Muncie Planetarium, IN

- Rain Water Observatory, MS

- Science Museum of Virginia, VA

SSEP works to foster partnerships between State Space Grant Consortia and Solar System Educators. This effort has lead to more than 50 events throughout the nation. One of the concerns discussed during the NASA OSS E/PO conference in Chicago, IL on June 13, 2002 was building partnerships between scientists, engineers and educators. SSEP includes these types of partnerships at a variety of levels.

First, all of the Solar System Educators have been trained at a minimum of 2 training institutes at JPL. These institutes include a variety of activities including content presentations by scientists and engineers and activity modeling by E/PO staff and participants. The summer 2002 institute focused on increasing the participants' ownership of the training. The participants were divided into teams of five educators, a mission E/PO lead, and at least one scientist. Before the institute, each team was assigned a mission and educational activity and instructed to develop a 2-hour session. Teams met via teleconferencing and in-person to prepare. During the institute, each team led their session for the rest of the institute participants. Participant reviews of this method of professional development were very positive. Additionally, Solar System Educators provided critical feedback on mission educational products. This feedback resulted in expansion and modification of some of these products before they were published. SSEP partners plan to increase their involvement with Solar System Educators for educational product needs and reviews.

Communication between scientists, engineers and educators is also facilitated through the use of periodic telecon-based trainings. The general format for these trainings include a content presentation by a scientist, a question and answer session and a discussion of the educational use of the content, facilitated by an E/PO lead. Anecdotal and qualitative data show that this is an effective means of continuing professional development and could serve as a valuable model for cost-effective and efficient partnerships between scientists, engineers, E/PO personnel and educators.

During FY2003, SSEP staff are expanding their efforts to build lasting partnerships between scientists, engineers and educators. This spring, the STARDUST mission is working directly with Solar System Educators to determine needs and opportunities for education and public engagement activities surrounding the Jan. 2004 encounter with comet Wild-2. Additionally, SSEP staff is collaborating with the Mid-Atlantic Region Broker/Facilitator at Wheeling Jesuit University, Space Grant Consortia located within that region, and Solar System Educators to implement a series of professional development opportunities. The Space Grant Consortia are identifying events that could be enhanced with a Solar System Educator workshop and the Broker/Facilitator is working to provide science content speakers for those events. The outcomes of our work with the STARDUST mission and in the Mid-Atlantic region could become models for future partnerships between NASA, scientists, engineers and educators. SSEP staff is interested in developing partnerships with additional missions and programs that are involved in the exploration of our Solar System. This will be a major focus in upcoming years.

Grassroots

> *The inservice I did at Neil Armstrong Elementary, early this year was because they NEVER taught space before in the 5th grade.*
>
> –Dan Malerbo, PA

> *The State of Maine does not have a Teacher Resource Center in our state. SSEP has been a wonderful opportunity to bring NASA to Maine teachers through workshops I've done for preservice teachers at the University of Maine at Orono and with the Gulf of Maine Aquarium to all teachers. I have spoken at state conferences and bringing the message to math teachers to show high alignment with mathematics has been a real plus!*
>
> – Marshalyn Baker, ME

The importance of enthusiasm and creativity expressed through grassroots efforts and the need to reach a larger base of educators were two themes that were brought out by the participants during the formal education strand of the NASA OSS E/PO conference. SSEP is a grassroots effort started initially by E/PO leads for the Galileo, Stardust and Cassini missions. Growth of the program has lead to a "bigger-picture" view of Solar System exploration, but the mission-focused flavor of the program is still intact. Additionally, by creating a large nationwide cohort of master educators with a variety of backgrounds, SSEP has the feel of a grassroots organization. Solar System Educators are often considered local experts and have earned respect within their communities. This has allowed them to reach educators that would not normally attend a NASA related workshop. Over the past 30 months, Solar System Educators have conducted more than 1100 workshops with a direct impact on more than 25,000 educators. These are meaningful interactions that range from a 1-hour workshop to sustained semester-long courses. Using a conservative estimate, the direct impact on these teachers translates to more than 750,000 students reached each year. Solar System Educators conduct workshops in high-need areas including rural regions in Arizona, Idaho, Maine, Mississippi, Texas, Utah, Washington, and Wisconsin and urban regions in Arizona, California, Connecticut, Illinois, Nevada, Pennsylvania, and Wisconsin. SSEP staff is working on a data collection plan that will improve our ability to measure program reach to underserved populations.

> *All of our students are from underserved and special education populations. I have given several workshops for the teachers in Los Angeles County ... the County covers 4000 square miles and there are numerous special programs. Teachers attending my workshops serve special education, adjudicated students, emotionally disturbed students, mentally and physically challenged, and the juvenile camp schools. In all of my workshops I address the fact that we need to meet the needs of these students and that if I could/can do it ... so can these teachers. I am currently supervising 50 speech therapists serving inner city students. They use my materials for speech lessons! Special education has traditionally been ignored, now that I am here, this will NOT HAPPEN again.*
>
> – Adair Teller, CA

Conclusion

> *As an experienced classroom teacher, I bring credibility to what kids can learn and do by utilizing NASA's discovery missions. I connect teachers to NASA websites, state resources, to lesson plans that they can take back to their classrooms. I integrate subject areas and share my enthusiasm about how teaching about space has made a difference in my classroom.*
>
> *My goal is to empower teachers to teach science using NASA's missions. The Solar System Educator Program recognizes that teachers are an important part of NASA's mission outreach.*
>
> – Kathy Chock, HI

SSEP is a successful education outreach effort that is able to reach a large and diverse audience of teachers at the "grassroots" level. Lessons learned over the past three years have helped SSEP evolve from a project where NASA was a "source" and the participants were the "user" to a truly collaborative program where scientists, E/PO leads and educators are able to work together to identify and address needs and opportunities to engage the education community in Solar System exploration.

Empowering Teachers to Address Space Science Content Standards in the Classroom

T. Gregory Guzik and Roger McNeil

Department of Physics and Astronomy, Louisiana State University, Baton Rouge, LA 70803, guzik@phunds.phys.lsu.edu

Erin Babin

Praireville Middle School, Ascension Parish, LA

Introduction

Many middle and secondary school science teachers do not have the background in astronomy and astrophysics needed for them to engage their students in learning about astronomy concepts. These concepts, nevertheless, constitute a significant fraction of the content standards found in the Louisiana Science Curriculum Framework (Louisiana Center for Educational Technology 2002). Further, according to the 1999 Louisiana School Staffing Survey, elementary certified teachers who are only required to have a total of three college credits in the physical sciences, teach the majority of middle school science classes which is usually the last time students are exposed to astronomy content. This lack of background directly affects how effective a teacher can be in addressing content standards in the classroom. A number of research studies have found that students come to class with a common set of naïve conceptions that must be directly addressed in the classroom by teachers who posses a good understanding of the nature of science (Aikenhead 1999a; Aikenhead 1999b; Lederman 1990). However, in 1998, the National Center of Education Statistics found that only 41% of classroom teachers felt very well prepared to implement new teaching methods, and only 36% felt well qualified to implement state or district curriculum or performance standards (Snyder & Wirt 1998). Without the appropriate background, teachers will either avoid the content or teach it poorly.

Seven years ago, the Louisiana State University (LSU) Space Science group began to delve into education and public outreach (E/PO) in an effort to help address the problems identified above. Over the years we tried several different approaches including classroom material development, teacher professional development workshops, informal education outreach, and student involvement in research programs. In the process, we learned much about how to develop and implement effective E/PO programs, what works, and what does not. In this paper I will discuss a few of our more successful programs as well as the lessons learned from both our successes and failures.

The Highland Road Park Observatory (HRPO)

In the mid-1990's there were few resources in our area that served teacher needs and public interest in astronomy. At the same time the LSU Department of Physics and Astronomy needed to upgrade its venerable 11" Clark refractor to a system that could be used to teach LSU students modern observational instrumentation and techniques. As a result, a partnership was formed between LSU, the Parks and Recreation Commission for the Parish of East Baton Rouge (BREC), and the Baton Rouge Astronomical Society (BRAS) to develop a public astronomical observatory (Guzik et al. 1998). BREC provided the park land site (about 10 miles from LSU campus), the 2,300 sq. ft. facility building (with 25 ft. diameter dome), utilities, and facility maintenance, while LSU obtained funding through the state for the 20" Ritchey-Chretien f8.4 telescope, back-illuminated CCD camera, filter wheel, computer control system, and Internet T1 link to the LSU campus. Amateur astronomer members of BRAS provided expertise and assistance during the observatory development and continue to support observatory public programs. Overcoming local seeing conditions, the system has greatly exceeded expectations, allowing many different kinds of student training, classroom activities, and amateur observation projects. In particular, BRAS members used the HRPO to discover 55 new asteroids and to perform photometry to an accuracy of 0.02 magnitudes. We have used the HRPO setting to support a weekly informal science lecture and hands-on activity series, as well as more formal activities in support of BREC summer camps and classroom field trips. Today, the HRPO plays a key role in our teacher professional development program and our effort to bring new astronomy resources to teachers in their classrooms.

Physics Learning and Astronomy Training Outreach (PLATO)

Most school systems address space science content in middle school, and these teachers usually have had little to no background in astronomy. As a consequence, only the most basic space science concepts are generally taught to students. Further, many high school science teachers who teach physics are not certified in this subject. As a result, we developed the PLATO program (Guzik et al. 2002) to provide professional development for middle- and high-school science teachers to improve their command of basic astronomy and physics concepts, and to acquaint them with current inquiry-based teaching methodology. The project was funded by the Louisiana Systemic Initiatives Program (LaSIP) from June 1999 through May 2001. Over the course of the project, 74 teachers in East Baton Rouge and surrounding parishes successfully participated in the program, receiving 6 hours of graduate credit at LSU in either Astronomy or Physics. PLATO incorporated two identical three-week workshops held during June and July, followed by a coordinated program of academic year activities. We capitalized on faculty expertise already in place at LSU, including faculty who previously taught in middle or high school. These faculty typically also had a background in education reform. Additionally, we included several scientists interested in E/PO.

We also hired a full time Site Coordinator from the middle school teaching ranks to provide an interface between scientists and educators, as well as to help manage daily activities. Each three-week workshop included 15 full days of activities, five days a week, plus a few evening sessions at the HRPO. Each day was split into two parts: a "common time" period and a "splinter group" period. The "common time" period, which all participants attended, covered the following areas: (a) Science Reform, (b) Technology, and (c) Science Content based on the theme "Waves, Light, and Optics." During the "splinter group" period, the teachers divided into two groups for an in-depth study of astronomy for middle school teachers, or physics for high school teachers.

For the astronomy splinter, we linked our discussions with the space science benchmarks, used exemplary materials such as those from "Project Astro" to guide the inquiry process, and provided sessions on telescope basics, including the assembly and operation of small telescopes from purchased kits. While assembling the telescopes, principles of reflection, refraction, absorption, and scattering were studied and, with working telescopes, they received practical experience in using the telescope for student projects in the classroom. This training provided the teachers with the confidence, knowledge, and practical experience necessary to maintain the instrument and to use it effectively in the classroom.

The physics splinter was organized in the same fashion with the exception that we provided in-depth investigation of wave characteristics, electromagnetic radiation, geometric optics, and wave interference. The academic year program included follow-up activities that allowed us to monitor the progress of the workshop participants and re-emphasize the workshop material. Three times during the year each participant was required to prepare and present a hands-on activity for the public at the HRPO. This activity provided teachers with practice in using the knowledge they gained during the summer, as well as allowing us to enhance the public outreach program at the HRPO. We also held several one-day long "Update Workshops" for participants to share their classroom experiences, showcase new classroom activities, and inform teachers about LSU research projects. Finally, the Site-Coordinator made at least two to three visits to each participant classroom to assist and advise teachers as needed.

Multiple, real assessments were built into the PLATO program to evaluate the effectiveness of the program. One component in our evaluation was to give the participants the Astronomy Diagnostic Test (Hufnagel 2002) prior to, and following, the summer workshop to assess how much they learned. In general, the teachers showed a 20 to 30 point improvement on the post-test and most participants scored above 70%. Toward the end of the workshop, participants were also required to develop a lesson plan incorporating content, inquiry-based methodology, and assessments. These lessons were submitted to the Louisiana Center for Education Technology (LCET) where they were evaluated and scored using the LCET rubric. Participants were also scored on their technical capability with the telescope including correct assembly, optical alignment, and ability to locate objects. Finally, during the academic year participants were observed and evaluated in the classroom and during their activity presentation at the HRPO.

Robots for Internet Experiences (ROBIE)

This project builds upon both HRPO and PLATO by developing the hardware, software, teacher guides, classroom lessons, and teacher training necessary to bring particular space science associated instruments into the classroom over the Internet. From the primary ROBIE website (http://www.bro.lsu.edu/), teachers can currently access flight data and videos from the NASA sponsored ATIC Cosmic Ray balloon-borne experiment, and track satellites in real time using a HAM radio station. We have also successfully tested the ability of a small group of teachers to obtain images of Messier objects using the HRPO telescope over the Internet. In the near future we will be adding two more optical telescopes plus a radio telescope to this collection of Internet accessible "robots." The teacher guide and lessons, also available through the website, are tied to the science content standards and serve as a framework for the use of the "robots" in the classroom. Over the past year, we held a pilot workshop for about eight teachers who previously successfully completed the PLATO program. During the workshop, we detailed the ROBIE instruments and materials and then followed these teachers over the academic year as they used the ROBIE "robots" and lessons in their classrooms. In general, teacher and student responses to ROBIE have been very favorable, indicating that advanced technology in the classroom, when accompanied by teacher training and appropriate guide materials, can be very successful.

Lessons Learned

A key element in the development of our current program was the availability of resources through NASA that support scientist involvement in public outreach. Often, obtaining "startup" funds through conventional education support agencies can be very difficult. NASA, however, encourages and facilitates scientist involvement and most grant managers we worked with were understanding and supportive. In one of our initial projects, we proposed to develop classroom materials that would utilize the HRPO. Along the way we discovered several problems. First, most teachers did not have the background in astronomy necessary to make effective use of the materials we were planning. Second, the materials were integrated with a classroom field trip to the HRPO site and, for most teachers in our area, field trips are severely limited by logistic, funding and/or administrative constraints. Third, merely providing new materials or resources falls short of meeting teachers' needs, since they rarely have the time during the academic year to even locate such materials, let alone integrate them into their curriculum. These problems led us to conclude that we should focus on teacher professional development, provide materials based upon content standards, train the teachers to use these materials, show them how the materials fit within existing curriculum, and provide resources that can be used over the Internet.

These conclusions provided the basis for our successful PLATO and ROBIE programs. We have also found that it is very important and satisfying to involve teachers in all phases of an E/PO project. While the scientist has command over the content, teachers contribute knowledge of what is feasible and appropriate in a classroom, plus are highly motivated to produce a high-quality product. Further, we can not over emphasize the importance of having a full time E/PO project Site Coordinator recruited from the ranks of in-service teachers. In choosing such a coordinator, enthusiasm and dedication are much more important than the initial level of content knowledge. Since the cultures of science and education can be quite different, excellent communication and facilitation skills are essential qualities in a coordinator of an E/PO program. A good Site Coordinator significantly increases project efficiency, enabling scientists to focus on content and teachers on pedagogy. For the future our group is currently developing an expanded version of PLATO that would integrate elements of ROBIE. In this program we would provide teachers with content and pedagogical knowledge, provide opportunities for teachers to use standards-based materials aligned with exiting curriculum, and provide them with professional development in the use of resources that can be accessed over the Internet.

Acknowledgements

These projects are supported by NASA, Louisiana State University, the Louisiana Systemic Initiatives Program, and the Louisiana

References

Louisiana Department of Education, Louisiana Center for Educational Technology 2002
  http://www.doe.state.la.us/conn/index.php
Aikenhead, G. S. 1999a, Research Matters – to the Science Teacher
  http://www.educ.sfu.ca/narstsite/publications/research/authentic.htm
Aikenhead, G. S. 1999b, Research Matters – to the Science Teacher
  http://www.educ.sfu.ca/narstsite/publications/research/authentic2.htm
Lederman, N. 1990, Research Matters – to the Science Teacher
  http://www2.educ.sfu.ca/narstsite/publications/research/nature.htm
Snyder, T., & Wirt, J. 1998, National Center for Education Statistics Electronic Catalog
  http://nces.ed.gov/pubsearch/pubsinfo.asp?pubid=98013
Guzik, T. G., Motl, P. M., Burks, G., Fisher, P., Giammanco, J., Landolt, A. U., Stacy, J. G., Tohline, J. E., & Wefel, K. 1998, in Proceedings of SPIE, 3351, 13
  http://www.bro.lsu.edu/hrpo/3351-3.pdf
Guzik, T. G., Babin, E., & McNeil, R. 2002, in Louisiana and LaSIP Best Practices: Innovative Standards-Based Approaches to Teaching Mathematics, Science, and Technology, ed. Reuben Farley and Cathy Seeley (Baton Rouge, LA: Louisiana Systemic Initiatives Program)
Hufnagel, B. 2002, Astronomy Education Review, 1, 1

A Research Based Approach to Developing Engaging Classroom Ready Curriculum

Edward E. Prather and Tim F. Slater

University of Arizona Steward Observatory, Tucson, AZ 85721, eprather@as.arizona.edu

Every year, tens of thousands of non-science major undergraduate students enroll in introductory science courses. For many of these students, these courses represent the last formal science education experience they will ever have. Even though students may patiently listen to lectures, copy detailed notes, observe demonstrations, carry out the necessary experimental exercises, perform the required calculations with formulae, and reach predetermined conclusions, they all too often emerge from their general education, elective science courses without becoming intellectually engaged at a level sufficient to obtain a fundamental understanding of the phenomena under investigation (McDermott & Redish 1999). If we want our students to develop an interest in science and to emerge with a fundamental understanding of science concepts, it is necessary to treat the teaching and learning of science as a complex and interconnected process rather than a one-way transfer between the curriculum and the student. In particular, we must work to create more effective instructional environments that take into account student needs including their pre-instruction conceptual and reasoning difficulties.

The remainder of this paper is partitioned into two threads of discussion. First we provide an example result of an investigation into student's pre-instructional ideas about the topic of the Big Bang in an effort to illustrate how research results are used to inform instruction. From a constructivist perspective, it is vital to understand the prior knowledge that students bring to the learning endeavor.

In the second portion of the paper we highlight the results of an investigation into the effectiveness of student-centered instruction over the more often practiced conventional teacher-centered form of instruction.

Part I – An Example of Research into Pre-Instructional Student Beliefs

Introductory science students enter the classroom with many beliefs about the nature of the world around them that differ from scientifically accepted beliefs. To explore the frequency and range of student ideas regarding the Big Bang, nearly 1000 students from middle school, secondary school, and college were surveyed and asked if they had heard of the Big Bang and, if so, to describe it (Prather, Offerdahl & Slater 2002). An overwhelming 94% of the 177 college students surveyed reported that they had heard of the Big Bang. Of these students who reported having heard of the Big Bang, one-quarter gave responses suggesting that it was a theory describing the creation of stars, planetary systems, solar systems, or Earth, whereas more than half stated that it was a theory describing the creation of the universe. A full 80% of those students stating that the Big Bang is a theory describing the creation of the universe gave statements clearly indicating that the Big Bang was an explosion of some form of pre-existing arrangement of matter. The results for the middle and high school students were overall very similar to this data for the non-science major college students. These results suggest that the majority of students from all three populations have pre-existing beliefs about the Big Bang that differ from the contemporary cosmology model of the origin of the universe. It seems that many students of all ages, and likely the general public, carry with them the mistaken idea that the Big Bang was an event that organized a pre-existing arrangement of matter. These inaccurate ideas are well positioned to interfere with any instructional interventions designed to help students adopt a scientifically accurate view of the Big Bang.

To better inform instruction we felt it was necessary to look deeper into what students believed was occurring during the Big Bang. We administered a second survey to a different group of 133 college non-science majors who had not yet received instruction on the Big Bang. This follow-up survey more closely targeted student's beliefs by stating and asking students to provide a detailed, written description of what they believe existed or was occurring (i) just before, (ii) during, and (iii) just after the Big Bang. As with the initial survey, the majority of these students (nearly 70%) provided a written response clearly indicating that matter existed in some form prior to the Big Bang. Their ideas most often include atoms, molecules, and gas particles existing within an otherwise empty space or the existence of a massive object such as a star or planet. A description involving an explosion that either distributed matter throughout an already existing universe, and/or formed planets, stars or galaxies was given by 49% of students, while 17% described a scenario in which matter combined or came together. Overall the results from this second survey further illustrate that students hold scientifically inaccurate ideas about the modern astronomy topic of the Big Bang when they enter the classroom. One interpretation of these results is that many students are consistently using a mental rule to infer the behavior the natural world-the idea of "you can't make something from nothing"-despite targeted instruction intended to help them think otherwise (di Sessa 1993). An instructor who follows the tenets of constructivism will need to alter the conventional textbook-based, teacher-centered, approach to instruction if they wish to help students who appear to think about the topic of the Big Bang in this manner.

Part II – The Shift From Teacher-Centered to Learner-Centered Instruction

There is a growing body of research that illustrates how teaching methods that actively engage students in the process of scientific inquiry are more effective than conventional lecture-based instruction for helping students to develop a powerful conceptual understanding of science (Hake 1998). The focus of an inquiry based teaching method is away from a passive, teacher-centered classroom towards a student or learner-centered environment promoting the active and intellectual engagement of students. In contrast to training students to perform specific algorithms and deduce results from a set of well-defined general principles, students encounter tasks that emphasize inductive reasoning and require them to apply and extend their understanding to conceptually rich situations.

Recent research is revealing that students emerge from these courses with a lower level of conceptual understanding of fundamental concepts in astronomy than faculty would hope for (Deming 2002; Bailey & Slater 2003). One interpretation is that the most commonly found form of instruction – teacher-centered lectures – is insufficient to bring about significant conceptual gains. A rapidly growing number of university-based scientists are beginning to understand that we must treat the teaching and learning of astronomy for this population as a complex problem that requires a scholarly approach if we are to be successful.

To promote the active and intellectual engagement of students we have designed a suite of activities (known as Lecture Tutorials) that are focused on guided inquiry in which students work in small collaborative groups within the large-lecture classroom (Adams, Prather & Slater 2004). Two overarching research questions guided the evaluation of the materials we developed

1. What is the effectiveness of a conventional lecture on student understanding?

2. What is the effectiveness (both cognitive and affective) of the materials?

This study used a mixed-methods, one-group, multiple measures design. Students enrolled in "Introductory Astronomy" for non-science majors at major southwest Research Level-1, doctoral granting institution served as the primary data source with supplementary data coming from five additional field-test sites at varying size institutions. Initial baseline data was collected using a pre-course, post-lecture, and post-Lecture-Tutorial instrument, which included a 68-item conceptual inventory. A pre-course/post-course Likert-scale style attitude survey was also administered. For the pre-course data, the conceptual inventory was split into two forms with approximately one-half of the questions on each form. The division of the questions onto two test forms was done to reduce the survey administration time to about 15 minutes. This is important so as not to infringe on the survey's construct validity which can be breached if students are given too long of a test or survey to complete, in which case they cease to respond thoughtfully to items near the end of the test.

To collect the post-lecture and post-Lecture-Tutorial data, students responded to two to three closely related multiple-choice items immediately following a lecture related to that particular days topic and then again later once they had completed the corresponding Lecture-Tutorial. These were the same multiple-choice items that were given on the pre-course conceptual inventory. This approach to data collection was relatively straightforward in design, albeit highly complex to carry out in the day-to-day context of a large enrollment course.

At the end of the course, a two component qualitative study was conducted to determine student impressions and beliefs about their learning using the materials. One component was a large-group format focus-group interview, which was conducted by the University Teaching Effectiveness Center. The other component was an inductive analysis of 3 clinical interviews of student-participants conducted by a trained evaluator.

The first research question pursued the effectiveness on student learning of a conventional lecture. We, the authors, served as both researchers and lectures. Based on the rhetoric of the ineffectiveness of instructor-centered lectures, we had anticipated that there would be only modest gains in student scores from pre-test to post-lecture test. The average score on the 68 item conceptual inventory, deigned specifically for this project, administered as a pre-test, was 30% correct and the post-lecture average score was 52% correct. Although there is a statistically significant increase in scores ($p < .05$), we are wholly unsatisfied when our students are only able to answer half of these questions correctly. To illustrate the types of questions used and their corresponding pre-test and post-lecture results, several of the items and corresponding scores are provided in the appendix. These results confirm that instructor-centered strategies are largely ineffective at promoting meaningful conceptual gains on traditional astronomy topics presented to non-science majors.

The second research question focused on student cognitive and affective gains after using the materials. Students in all courses observed made statistically significant cognitive gains insofar as the 68 item conceptual inventory could measure. Because of highly varying student attendance and requirements for anonymity surrounding human subjects research, we chose to aggregate all student results and use unpaired pre-post data. Although students were able to provide correct responses to about one-half of the items after lecture, we were not sure how much improvement in scores would actually result from only investing 15 minutes in student-centered instruction by doing the materials. Surprisingly, the mean percentage correct increased from 52% immediately following lecture to 72% after completing the materials. This 20% gain is judged to be quite high considering the small amount of class time invested. Because of the conceptual difficulty and attractive nature of the distractors in the questions used we do not believe that this dramatic increase is simply a result of students memorizing the answers.

The mean scores on the Likert-scale affective survey were nearly identical pre-course to post-course. In almost every individual item, students showed no statistically significant gain insofar as the instrument could measure. These results are consistent with other studies using the same instrument (Zeilik & Bisard 2000) and lend support to the idea that student attitudes are difficult to impact even when conceptual knowledge increases. This is consistent with further interview results.

From these collective results, we conclude that implementing materials in this setting, which somewhat reduced the amount of time available for lecturing, made dramatic and significant positive impacts on students' achievement in learning astronomy. Furthermore, rigorous evaluation of how materials are being used is an essential and interconnected component of the curriculum development process. It does little good to create scientifically accurate materials that cannot be implemented into the classroom because they are either not cognitively appropriate for the students or because they make use of instructional strategies that are not pedagogically appropriate for the skills of most teachers.

Finally, effective learner-centered scientific inquiry can only be achieved by student led investigations. Today more than ever, students have easy access to powerful computer simulations that can serve as the basis for hands-on activities. With careful implementation and evaluation (in order to ensure students' intellectual engagement over their using simulations as video games), curriculum that incorporates computer-mediated activities which make use of existing online data bases can provide students with authentic inquiry experiences consistent with the goals set forth by the National Science Education Standards.

References

Adams J. P., Prather E. E., & Slater T. F. 2004, Lecture-Tutorials for Introductory Astronomy (Upper Saddle River, NJ: Prentice Hall)

Bailey J. M. & Slater T. F. 2003, The Astronomy Education Review, 2(2)

Deming G. 2002, Astronomy Education Review, 1(1)

di Sessa A. 1993, Cognition and Instruction, 10 (2-3), 105

Hake R. R. 1998, American Journal of Physics, 66(1), 64

McDermott L. C. & Redish E. F. 1999, American Journal of Physics, Physics Education Research supplement, 67(9), 757

Prather E. E., Offerdahl E. G., & Slater T. F. 2002, Astronomy Education Review, 1(2)

Zeilik, M. 2000, Conceptual Astronomy: A Tale of Two Classes, Talk presented at the American Association of Physics Teachers national meeting, Guelph, Ontario, Canada, August 2000

Additional Information

For information on the development of curriculum and the research conducted by the Conceptual Astronomy and Physics Education Research (CAPER) team at the University of Arizona please contact the first author or visit the group www site at URL: http://shiraz.as.arizona.edu

Exploring the Solar System: A Science Enrichment Course for Gifted Elementary School Students

Walter S. Kiefer, Robert R. Herrick, Allan H. Treiman, and Pamela B. Thompson

Lunar and Planetary Institute, 3600 Bay Area Blvd., Houston, TX 77058, kiefer@lpi.usra.edu

"Exploring the Solar System" is a science enrichment program taught at the Lunar and Planetary Institute for gifted fifth grade students from the local Clear Creek Independent School District. The school district selects which students are qualified to participate in the gifted enrichment program. In fifth grade, each student is allowed to choose one enrichment course to attend during the fall semester. "Exploring the Solar System" has been taught 16 times since 1992. The course is taught during the normal school day. Approximately 20 students meet at LPI for three hours on each of 12 consecutive weeks. We use a pre-test and post-test to assess what the students have learned, but no formal grades are assigned. In teaching elementary and middle school students, it is essential to have good activities. Slide presentations are sometimes necessary to convey basic facts, but they work best if they are kept short and are interspersed with related activities. It is also essential to make lectures interactive, by posing questions to the students.

Setting the Stage

The focus of "Exploring the Solar System" is not simply on what we know, but on how we know it (or at least think that we know it). The first two weeks are used to build a basic foundation. Astronomers tend to use words like million and billion a lot. To build a conceptual understanding of how large these numbers are, we start with some simple counting exercises. Students typically have a good sense of how long 100 seconds are. We ask them what they were doing 1000 seconds ago? 10,000 seconds? 1 million seconds? What were their parents doing 1 billion seconds ago? We use several scale models to illustrate the size of the Solar System (emphasizing that it is mostly empty space!) and the relative sizes of the planets. We illustrate the time history of the Solar System with our Wall of Time, a 46-foot long poster in which each foot represents 100 million years of history.

Week 2 includes a preliminary photographic tour of the Solar System. In this week, and indeed throughout most of the course, we focus primarily on the inner, rocky planets of the Solar System. We consider several key concepts, such as how the density of craters on a planet's surface provides clues to the age of that surface. We also consider the relationship between a planet's size and the duration of its volcanic and tectonic activity. These quantities are related to the rate at which planets have cooled. A simple experiment measuring the rate at which hot water cools in large and small containers illustrates this well.

Geologic Mapping and Processes

Subsequent weeks focus on two major themes. The first involves understanding basic geologic processes. We introduce some simple geologic mapping principles such as the law of superposition. This concept is easily understood if it is first presented in the context of the everyday lives of the students. For example, in their laundry piles at home, clothes on the bottom of the pile have been there the longest, and the clothes on the top of the pile were placed there most recently. The same basic concept can be used to infer the order of geologic events on the surface of planets. This concept is reinforced several times, as the students map the Apollo 15 landing site on the Moon and landslides and flood channels on Mars (LPI 2003). We perform several experiments that illustrate how geologic processes have shaped the surfaces of the terrestrial planets. An impact cratering lab shows how the mass and velocity of the impactor affects the size and morphology of the resulting crater (LPI 2000). A volcanism lab using molten wax as "lava" illustrates several basic styles of volcanic eruptions. This lab also emphasizes the role of volcanism in resurfacing planetary surfaces. By combining the impact and volcanism labs, we illustrate how complex surfaces can be developed. This reinforces our basic theme of superposition and the sequence of events. In conducting these labs, good safety practices and adequate supervision are essential. Other labs examine analogs of lunar rocks and of martian soil and show how rock samples provide important information about a planet's history.

Remote Sensing of Planetary Surfaces

The second major theme is remote measurements of the properties of planetary surfaces. We conduct several lab sessions on computer image processing (Kiefer & Leung 2004). We use an inexpensive spectrometer to measure the visible and near-infrared spectra of soil samples and discuss how this provides information about the chemical composition of distant planets (LPI 2001). We also make use of low-tech experiments. For example, we use styrofoam inserts in shoe boxes to create "mystery planets." The students measure the heights of the features through small holes in the box top, thus creating a simple topographic map (LPI 2000). This simulates how NASA missions such as Magellan and Mars Global Surveyor have mapped the topography of Venus and Mars using radar and lasers.

Mars Mission Planning Activity

The capstone event for the course is a Mars mission planning activity. This occupies an entire 3 hour class session, usually on the next-to-last week. Teams of 3 or 4 students work together to define an objective for a Mars exploration program. They then plan a sequence of robotic and human missions that are intended to meet this objective. We provide the students with a list of possible robotic missions such as orbital cameras and radars, simple landers, and rovers. Similarly, we list various components for a human mission to Mars. Each piece has a cost, and the teams must plan their exploration programs within a fixed budget. At the mid-point of the session, we impose a 33% budget cut into the process. This forces the teams to carefully consider which missions they deem most important. At the end of the session, each team makes a 5 minute presentation of their plan to the class.

Helpful Hints for Getting Started

The most direct way to begin a similar course is to approach the principal or a teacher at your local school. This may be easier than working with a school district's central administration. In designing a course, it is not necessary to develop the curriculum from scratch. Many excellent activities already exist (NASA Education Enterprise 2004) and can be adapted to a variety of circumstances. Keep your plans flexible – classroom activities rarely go precisely as planned the first time through. Teaming with another scientist or with a classroom teacher can be very helpful. This provides an extra set of hands during activities as well as a source of feedback on your presentation and classroom management styles.

Acknowledgements

This work was supported by NASA Contract NASW 4574. Lunar and Planetary Institute Contribution 1098.

References

Kiefer, W. S., & Leung, K. 2004, page 318 of this volume
Lunar and Planetary Institute 2000
   http://www.lpi.usra.edu/education/EPO/fun_w_sci.html
Lunar and Planetary Institute 2001
   http://www.lpi.usra.edu/education/products/spectro.html
Lunar and Planetary Institute 2003
   http://www.lpi.usra.edu/expmars/expmars.html
NASA Education Enterprise 2004
   http://spacelink.nasa.gov/

Issues in Formal Education as Systemic Programs

Gwen Pollock
Illinois State Board of Education, 100 N. First Street, C-277, Springfield, IL 62777, gpollock@isbe.net

With an audience of enthusiastic scientists, experienced teachers and dedicated program providers, a lone state education agency voice offered broad ideas from the formal education perspective. The nation-wide emphasis on the role of standards-led decision-making permeates accountability issues from range of the classroom teacher's daily curriculum planning and implementation to the documentation of the students' mastery levels. It goes beyond to the development of commercial textbooks and passionate special programs marketed to students and teachers. Still farther, it enters the evaluation of teachers and administrators and unfortunately, real estate values. This accountability extends into the decisions of legislative bodies at the states and federal levels and the responding agencies with resource arms dedicated to educational opportunities.

This venue focuses on the interests of NASA's Office of Space Sciences toward meeting the systemic needs of the partnerships between scientists and educators. Opportunities to broadly describe specific programs with specific audiences and share the common lessons learned (including the successes and frustrations) were set for willing conversation. However, it is strenuous to think more generally in settings of specific activities with localized accomplishments and individualized circumstances laced with personal pride. Systemic perspectives seem to be relegated to others, sometimes without apparent names or faces. For the sake of this difficulty, I tried to share some generalities.

The role of systemic impact, action and involvement for the success of programs in the formal education sector is absolutely vital. The issues of best use of resources, including time, expertise, funding and materials, as well as sustainability, "scaling up" and responding to actual and perceived needs, must all be a part of systemic decision-making process at all levels. As the needs for standards-led professional development and curricular materials development at the formal and informal levels rise, those who are affected, either by the process of receiving or producing or deciding, must accept these parameters as paramount for actualizing personal, professional and programmatic goals.

The recent passage and implementation of the federal legislation for "No Child Left Behind" has pressed new and stringent directives and mandates into state-to-classroom dimensions for reading and mathematics. The loss of directed professional development funds at the state and local levels for science and mathematics from the Elementary and Secondary Education Act (Title II, commonly known as Eisenhower funds) was diverted into highly competitive (and under funded) grants from the federal level. Economic downturns have forced state agencies into cost-cutting measures that have reduced services significantly. In light of such policy and budgetary measures, the emphasis on stretching and multiplying science education resources based on national and state standards for learning and teaching and accountability has intensified significantly.

The primary role of the nation's states' department of education science consultants may be explained as coordinating, collaborating, facilitating and catalyzing resources for systemic impact. From the Illinois setting: Very successful partnerships have been developed between state agencies (Natural Resources, Agriculture, Commerce and Community Affairs, Environmental Protection and Public Health), county offices (for University of Illinois Extension Services, Soil and Water Conservation Districts, etc.) and federal agencies (US Fish and Wildlife, Environmental Protection, etc.) and local entities (park districts, nature centers, zoos, aquariums, planetariums, botanical gardens, etc.). These partnerships provide a multiplying effect for implementation of the Illinois Learning Standards for classroom teachers from the perspective of informal service providers.

In each of the collegial partnerships, the catalytic effect from the state education agency and its resources have yielded marketability to classroom teachers in the definition and meaning of systemic impact of our teaching and learning standards. The dimensions for these responsibilities include our states' classrooms' students which directly connect to the potential opportunities for our states' science teachers and teachers-to-be. The alignment of college and university courses of study to the Illinois Content Area Standards along with the legislated Certificate Renewal Process for current teaching certificate holders has created significant prospects for sharing standards-led professional development opportunities for teachers. This application of standards-alignment is happening nationwide.

Foreknowledge about the development and access to the progress and research associated with state, regional, national and federal resources of programs can become other factors that affect our roles as catalysts. I cannot stress strongly enough the impact potential in this collaborative foreknowledge and planning. State data resources can be offered to help direct and substantiate choices for program providers and developers; anticipated directives and directions can be shared for potential partners. The commonality of state-level programs nationally is phenomenally striking. Perhaps more pointedly, we contend with translating legislative and regulatory policies and standards-led research findings and applications into practices that are manageable and useful for our districts.

As an example of the prospects of systemic reform, the Council of State Science Supervisors (CSSS) has worked successfully with NASA's Education Office since 1999, under the direction of Frank Owens and Larry Bilbrough, to define Science as Inquiry in the NLIST project (Networking for Leadership, Inquiry and Systemic Thinking). Sue Darnell-Ellis, AES Curriculum and Staff Development Specialist and former Kentucky science supervisor, has helped in the coordination of the project designed to facilitate systemic reform around this definition. The elements addressed from this definition, so far, include instructional resource alignment, professional development and administrative leadership. This effort uses the basic definition of the common denominators of Science as Inquiry displayed from the perspectives of each of the elements. The work can be used by accessing: http://www.inquiryscience.com/.

From the mandates for standards-led classrooms, professional development, assessments and the national concentration on reading and mathematics, we are stretching resources to their maximum values, in ways that certainly stretch traditional imaginations. Our teachers have precious little time to devote to the processes of curricular evaluation, modification and use, much less for actual development of the materials. Data-driven decisions about research-established impact, analysis of assessments results to establish needs for the students and valid teacher surveys, as well as projected budget allocations, must be prime factors in that decision-makers matrix.

The following remarks cannot be emphasized enough. The degree of difficulty in addressing most or all of these "wheel spokes" is extreme. Deciding what to focus and what to minimize is complicated. In the times of budgetary cut-backs, programs must be able to demonstrate effectiveness through multiple facets and greater, more fundamental conceptual ideas. Very specific programs about very specific sub-concepts may not be desirable. Documented scaling-up dimensions will be very marketable. The often-used cliche associated with inventing wheels may be the link necessary in this consideration. While we must be on the watch for inventors at all times, we must be vigilant to "wheels" that really work. These "wheels" are successful because in some combination, they:

- work in the challenging situations for students, teachers and classrooms in innovative ways; they work with less costly outlays for dollars and time out of the classroom,

- work with more direct connections to experts,

- work with developmentally appropriate strategies for learners with the right depth and breadth,

- work systemically overtime,

- work to provide continued access to resources,

- work in remote, urban, suburban, impoverished and wealthy districts, as they fill a real need,

- work because they are not personality-driven, and

- work because they really work!

Less discussion was offered for this specific Illinois program. The presentation closed with the reference to the common activity for students during which SLIME is made. Students really enjoy making slime in classroom settings (and then hiding it in innumerable places in the school!). The realization that SLIME is not standards-based (and probably not even standards-coincidental!), is undeniable, unless the factors of non-Newtonian fluids are described. While teachers and students enjoy this activity, it is an example of how programs, courses, activities, materials, etc, may need to "let go of the slime!!"

In Illinois, a program associated with the Earth and Space Sciences has emerged for K-12 teachers can serve as an example to incorporate many of these criteria. The Near and Far Sciences Program, funded through the Scientific Literacy appropriation to the Illinois State Board of Education, has served approximately 500 teachers in its five-year heritage. It is specifically mentioned at this time to bring to mind certain characteristics that may be contributing to its systemic impact. The program was designed to incorporate:

- a focus on the Illinois Learning Standards, specifically the two standards on earth and space sciences, where the number of certified teachers is among the lowest of all science teachers and in which lower student scores prevail on the state assessment,

- graduate course credit option for teachers as learners who choose to formalize the opportunity, both in content courses and pedagogy strategies, (including 2 hours in each of the sciences and 2 education hours for a total of 8 graduate credit hours),

- national grant program curricular resources and strategies for magnified classroom applications and materials (such as DataStreme, A-Astra, etc.),

- scientific expertise from US Weather Service, the Illinois State Geologic Survey, various university and planetarium astronomers, DePaul's NASA Office of Space Science broker, amateur associations and field-based local programs,

- the educational strategy of Action Research for classroom application and professional reflection and teacher networking electronically and personally,

- job-shadowing requirements for teachers to see the sciences of astronomy, meteorology and geology in action and to build personal and professional networks and relationships with these professionals, and

- three two-day regional sessions culminating with a closure showcase for all participants at the state's Capitol.

The success has been personal to the participating teachers. They have built relationships with professionals beyond their classrooms, including colleagues throughout the state and nation, and within scientific circles that were uncommonly open to them. They have a newfound confidence in the sciences for themselves and their classrooms. Many have built their networks for parent and community involvement. The costs were slight, in comparison to the systemic impact. We have modelled other programs from this template.

The table discussions were fruitful and intriguing. Many of the factors that were identified on the synthesis notes were discussed among our small group. The ideas of passion, enthusiasm, creativity, leverage, sustainability, coordination, coherence, and communication were agreed to be absolutely necessary for success. The opportunity to share was extraordinary.

Issues in Informal Science Education

Issues in Informal Science Education

Edited transcription of the keynote presentation given at the beginning of the session on issues in informal science education

Paul Knappenberger

Adler Planetarium and Astronomy Museum, 1300 S. Lake Shore Drive, Chicago, Illinois 60605, paul@adlernet.org

I will identify issues in informal science education to help stimulate discussions in the Conference sessions that follow. I will not present an exhaustive list of issues, but will group related issues in ways that may lead to creating strategies for solutions. My focus will be on astronomy and space science rather than all of informal science education, and I'll use examples from the Adler Planetarium to illustrate certain points. Several people provided suggestions and guidance in the preparation of this Plenary Address, including Naomi Knappenberger, Bonnie Van Dorn at the Association of Science-Technology Centers (ASTC), Alan Friedman at the New York Hall of Science, and several Conference participants. It's a pleasure to acknowledge their assistance.

Differences between Formal and Informal Science Education

Reviewing some of the differences between formal and informal science education will assist in understanding some of the issues we want to address. Formal education is characterized by a structured environment. There's usually a curriculum that guides the scope and sequence of content. Classes or meetings occur daily or at some regular interval over a sustained period of time. There is some homogeneity of students, such as groupings by age or ability level. There's formal assessment of progress, like tests leading to a grade and then a diploma.

Informal education frequently takes place in a more casual environment. There's usually not an established curriculum determining the sequence of informal learning experiences. These educational encounters are often one-time occurrences, not linked sequentially to others, but forming a series of lifelong learning experiences. They may include visits to museums, watching science education programs on TV, or attending club activities and lectures on science. There's usually not a formal assessment - no test score for a museum visit or planetarium show. There may be very little homogeneity of students, but likely a great diversity of learners sharing a common experience.

These and other characteristics distinguish formal from informal education. But both formal and informal experiences help individuals learn science, as well as other topics. Learning occurs when new neuronal connections form in an individual's brain, and a concept can be "learned" through experiences in formal or informal environments. There have been some studies that suggest that most learning takes place outside the formal school environment.

Some institutions are involved in both formal and informal science education. Museums, science centers, planetaria, and community organizations, which are cited as major providers of informal education experiences, also conduct formal education programs, frequently in partnership with schools and universities. The Adler Planetarium's professional staff spends about 50% of its time partnering with the formal education systems in Chicago, Illinois and the midwest. These cooperative programs include professional development workshops for teachers and classroom activities for students that are part of the school's curriculum. The Adler also offers formal courses for the public that are taught by area astronomers in the evenings throughout the year. Some of these earn academic credit at area universities.

Any successful long-term strategy for improving math, science and technology literacies must include both formal and informal education, and their interrelationships.

Issues

1. The nature of astronomy and space science may account for some of the issues encountered in teaching these subjects. Both are highly visual areas of study involving objects beyond our physical reach and exploring environments quite different from our everyday experience. Stunningly beautiful, exciting and compelling images of exotic celestial objects result from much of the research. The words 'inspirational' and 'motivational' have been used quite a lot at this conference to describe the nature of astronomy and space science. There is an on-going series of conferences called INSAP (Inspirations From Astronomical Phenomena) that explores the impact that astronomical phenomena have had on cultures throughout history. These conferences occur every several years and include representatives from literature, art, philosophy, theology, history and science – areas of human endeavor that have been significantly influenced by astronomical phenomena. Comets, eclipses of the sun and moon, meteor showers, and other celestial events and objects have helped shape beliefs and understandings in many areas of earlier cultures. The highly visual results from the exploration of space and the contemporary theories based on vast amounts of new data are impacting various aspects of today's cultures. Ed Weiler stated in his talk here earlier this week "The most important result of the NASA science program is the sense of wonder and imagination it inspires in America's youth."

These characteristics of astronomy and space science fall largely in the affective domain. They provide gateways to interdisciplinary studies and ways of viewing science as a highly creative human endeavor. But they present some challenges as well. It's hard to find examples of concrete experiences in astronomy. As one speaker pointed out yesterday, there are few hands-on lab experiments in space science that can be done in schools. Many of astronomy's concepts require formal reasoning skills, and a teacher has to get students to that level before these topics can be explored in a meaningful way.

2. A concern that many people express about teaching astronomy and space science is the perceived lack of relevancy to everyday life. One of the panelists yesterday suggested that what happens at home - the direction the parents encourage the student to take - may be a large factor in determining whether a student studies topics in space science. Many parents encourage their offspring to take only the subjects that are relevant to careers that provide future financial

success. So, the question of relevancy of astronomy for today's students must be effectively addressed. The answer will be quite different than it was for earlier civilizations. The Adler recently opened an exhibit on cultural astronomy, exploring the varied roles of astronomy in different cultures around the world over time. In earlier times astronomy was more directly involved in the daily lives of people than it is today. Survival sometimes depended on a knowledge of the sky. Crossing deserts or oceans in search of food and shelter was a daunting experience, and knowing how to navigate by the stars made the difference between life and death. The critical time to plant crops on which survival depended was determined by astronomy-based calendars and observations of celestial objects. The design and orientation of temples, pyramids and other significant structures, as well as cities and roads, depended on knowledge of the sun, moon, planets and stars. Social hierarchies and the divine right-to-rule often stemmed from the spiritual influences of celestial bodies. While the heavens have less of a direct role in day-to-day matters in the 21st Century, they still have a large effect on our curiosity and imagination, and provide powerful inspirations for human endeavors.

3. One of the main issues in informal education is a *lack of a sufficient research base to assess and guide our efforts.* While studies have been conducted to determine the effectiveness of individual exhibits or programs, an ample base of substantive research has not been created such as exists for formal education (For a list of such studies, see http://www.astc.org/resource/case/index.htm. For a recent study of Space Science Media Needs of Museum Professionals, see http://cse.ssl.berkeley.edu/spacescience.pdf). Because of the limited number of studies in this field, there's no overarching framework of theory to inform the practice of informal education. Yesterday during the science education research panel, the basic ingredients of doing education research were enumerated - the philosophical underpinnings, the theoretical framework, the review of prior findings, the methodologies for conducting the research, and the interpretation of results. All of these things have been in place for many years in formal education and have greatly benefited that profession. But aside from a few efforts in some of these areas, a comprehensive and systematic approach to research has not been established for informal science education – maybe the time has come to do this.

4. Informal science education also *lacks the well-developed structure that characterizes the formal education field.* In his presentation this morning, Pinky Nelson articulated the very detailed structure of formal science education and the great consideration that has gone into establishing the goals for that endeavor. Are there well-articulated and accepted goals for informal education? Is there a well-developed structure for attaining these goals? There's been a lot of discussion about what these goals should be. For instance, there is a great range in content knowledge and in the skill-sets of museum visitors. So, many museums have set a goal of moving each person a little bit higher on the learning curve then they were before they visited. Is that a valid goal and how does it fit within a larger context? Should goals in informal education be aligned with the national standards? Should the standards drive lifelong learning experiences? Should the goals involve depth and richness in content and/or be compelling entertainment? A lot of efforts in the informal arena depend on their ability to attract sizeable paying audiences to fund the program expenses. Yesterday

during one of the discussion sessions, a producer of educational TV programs said that his primary goal is to make a show that's compellingly entertaining. If it's not that, nobody's going to watch it and he's not going to get a contract to do the next show. What should the goals for informal education be and how do we develop an organized structure to attain them?

5. The *lack of effective communication among the practitioners of informal education* is a critical issue. There are a growing number of people working in this area and many projects going on, but very few formal lines of communication exist to document, track and critique these efforts. The on-line *Astronomy Education Review* that Sydney Wolff and Andy Fraknoi told us about is a big step in the right direction. Everyone in the informal science education community should read it and give feedback to the authors and editors. Articles should be submitted for publication presenting educational research results that can benefit the whole community. Museum studies programs (in informal science education) at universities are needed to develop professional staff for science museums, planetaria and the many other places now employing E/PO specialists. Most museum studies programs in existence today are focused on art or art history. There are not many professional development or training opportunities for people doing informal science education.

6. *Informal science education lacks the substantial tax-based support that sustains formal education.* Federal, state, and local taxes are collected to provide on-going support for formal education. That's not the case for informal education programs. I don't know if it should be, but it might be worth discussing this at various levels. Yesterday, Cary Sneider pointed out the lack of recognition by national organizations and federal agencies about the power of informal education. Since that recognition isn't there at the National Research Council, as well as other key places, the understanding which translates into support on an ongoing basis is missing. That's a major issue that needs to be addressed.

7. *Intelligent use of communications technologies* in informal science education is another issue. These technologies recently burst onto the scene and they're evolving very rapidly. They hold great potential to bring about significant changes in the way education is done, both in the formal and informal arenas. These technologies are now being used in a variety of ways in museums, planetaria and other informal education organizations. But as with any new tool, we have to be careful about how we use this technology. The objectives for its use must be clear and tested for effectiveness. If not used wisely, technology could lead to loss of personal contact and a decline in visitors that see actual artifacts or experience shows in the planetarium theater or other immersive environments that facilitate learning in unique ways. A systematic approach to planning and assessing the use of computer/telecommunications technologies is needed.

8. *The relatively small level of minority participation in astronomy and space science* is an issue that could be addressed in informal science education. That's a major challenge facing our nation. It could be very beneficial to offer programs that provide encouragement and build confidence among minority participants in the informal education area. Sometimes language can be a barrier. Our experience at the Adler in trying to attract more members of the Latino

community into our programs has met with some problems because English may not be the first language for a great number of people in that community. As a result one of our astronomers, Jose Francisco Salgado, has begun offering astronomy courses in Spanish, working with the Mexican Fine Arts Center in Chicago, as a way to get over this hurdle. Panelists later this afternoon are going to discuss both the positive and the negative experiences they've had with similar programs. There may be cultural barriers that stand in the way of encouraging young students to get involved in careers in math, science, engineering, and technology-related areas. We need to learn more about the needs of communities who are underrepresented in these professional careers and work with them in overcoming the obstacles.

9. Another issue is the *accessibility of astronomy and space science experiences to people with special needs*. We saw a wonderful example in Ed Weiler's talk yesterday of a project that addressed the needs of the visually impaired. There are several other projects that have had some positive results in serving people with special needs that could also serve as models. There are many needs that aren't being addressed and the challenge is how best to expand initial efforts to reach larger numbers of people (Please see papers by N. Grice et al. on pages 185 and 301, and the description of SERCH Broker/Facilitator programs in the Appendix on page 457)

10. There are some issues that relate to interfaces. The *interface between the science research community and the education community is an ongoing challenge*. Some recent progress has been made as different approaches are being tried that could be topics for discussion in the sessions that follow. There's a model at the Adler that's working reasonably well. We have arranged for our astronomers to have faculty appointments at the University of Chicago and Northwestern University. The majority of our ten astronomers pursue research with colleagues at the universities, but instead of teaching undergraduate or graduate level courses or advising graduate students, they spend time at the Adler working on K-12 and public education programs, materials, exhibits, sky shows, and doing public talks, activities and events. This experience enables them to provide a conduit for members of the larger research community that want to get involved in K-12 education. *The interface between large and small institutions and organizations is not as productive as it might be.* How large programs that are national in scope or institutions that serve a nationwide audience or a very large audience can interface with smaller community-based institutions and organizations needs further attention. Whether it's a community organization or a planetarium in a small community with very limited resources, we need to further explore how to better interface with larger planetariums like the Hayden in New York or Adler. There has been recent progress here thanks to some cooperative programs between OSS E/PO and the international planetarium society, as well as regional planetarium societies. *How can NASA and OSS in particular best interface with the informal science education community?* The 1996 OSS E/PO implementation plan addresses this question and presents strategies, some of which have been tried over the last 3 or 4 years. There are other opportunities that could be explored that would be mutually beneficial. I'm sure many of you have ideas and I encourage you to share them in the following discussion sessions.

As Terry Teays mentioned when he introduced me, NASA's Space Science Advisory Committee (SScAC), which is the committee that advises Ed Weiler, has established a Task Force at Jeff Rosendhal's request, to assess the progress of the OSS E/PO program and to make recommendations for its future evolution. That Task Force started its work this spring and wants your input. Members of the Task Force are at this meeting and I introduced them yesterday. They would like to find out about your experiences with the E/PO program by talking with you either here at this meeting, or receiving communications from you over the next couple of months. It is important for us to get your insights, perspectives and suggestions. Your input will be considered along with other information that the Task Force gathers and will be incorporated into a report to SScAC that can help shape the future of the E/PO program (Please see the recommendations of the SScAC Task Force in the Appendix).

I have tried to identify and organize into categories some of the most important issues facing informal science education today, in the hope of providing some direction for the discussion sessions that follow. Now I'll turn briefly to another topic.

Adler Event Information

I've gotten a lot of questions about the visit out to the Adler Planetarium this evening, so I'll take a few minutes to address some of those questions. The Museum Campus consists of the Field Museum of Natural History, the Shedd Aquarium, and the Adler Planetarium. We share a beautiful park along the lakefront that includes parking lots and Soldier Field. This summer Soldier Field is undergoing a major reconstruction.

The reception for our group will be in the Sky Theater, which was the first planetarium in the United States. It opened May 12, 1930 on Max Adler's birthday. Mr. Adler was an officer at Sears Roebuck & Co., which is headquartered here in Chicago, and did very well on the company grow during the early years of the 20th century. He retired in the 1920s and desired to give something back to the community that supported him. He visited Germany and saw the first planetarium at the Deutsches Museum, then decided to build one in Chicago. So, he paid for it and donated it to the city Park District. It's now a non-profit museum on park district property, governed by a private board. The reception will be in the original night sky theater, and there will be a show in our newer Star Rider Theater. The 7:00 show is called "Journey to Infinity." The 7:45 show is "Solar Storms," which is our Sun-Earth Connection show. Our theme for this year is the sun-earth connection. The CyberSpace gallery will be open with demonstrations taking place throughout the evening. It is our newest gallery, a high-tech gallery, with three three parts to it. One is a public walk-through area, which has 19 plasma screens mounted on the wall and several vision stations. All the information that's conveyed to the visitor here is done so electronically. So it can be changed easily, literally at a keystroke. Topics can be updated without changing the hardware or furniture in the gallery. There's a second section that is used for training teachers, students, and the public on using the Internet. Visitors can link up to the Internet to probe in more depth any of the topics they may have come across in the museum or in our shows.

There are 10 different galleries at the museum that you can visit. There is also a special event called Luna Cabana, which is a Latino music and dance celebration that is happening this evening. We do this every Thursday in the summer to attract people that ordinarily wouldn't visit the planetarium. We keep the galleries open and run shows in the theater.

Discussion

Question: You mentioned that there isn't enough conscious structure in informal education like there is in formal and there isn't a goal and a road map of where you want to go. However, I just wondered from your perspective in informal education, is there a concern that too much structure might not be beneficial ultimately? Where is that balance?

Answer: That's a valid question and I don't know the answer to it. I think there should be more formal structure than presently exists. There are many good projects that are going on and I think they could benefit from shared information. I wouldn't want to see a structure that would limit creativity, but rather one that would enable thoughtful exchange of information.

Question: I don't know if you're familiar with the Park Service paradigm for informal education. It's also been carried over to a group called the National Association for Interpretation and there is actually a framework for defining informal education. The Park Service has developed an educational tool book that includes blueprints for identifying whether something falls within what they call interpretation or outside of it. I'd be happy to share that full suite of things with anybody who's interested. It falls between formal education and what some people call informal education in that it does set goals and uses criteria like: there is some learning that occurs, it has to engage the visitor and so forth. It answers some of the questions that your issues raised. The National Association of Interpretation is now embracing some of that framework. A couple of other land management agencies are also beginning to use this approach, including the Forest Service, so it's not just something that's been squarely within the Park Service.

Answer: I'm glad you're here. I had no knowledge of this and I think this would be a good topic to pursue further in our discussion sessions. This is an excellent example of how the community can benefit by communicating and sharing experiences with each other.

Question: What about the role of the International Planetarium Society and other large associations having to do with informal education? Do you see upper level interactions between these organizations and NASA? Not at a grassroots level or individual partnerships but on a more systemic, higher level?

Answer: That's a good question. Last night we had 3 officers here from the International Planetarium Society that gathered to talk about the Society's interfacing with the OSS E/PO program, as well as other issues. There are probably a number of other organizations similar to IPS that should be involved in this type of discussion.

Biography

PAUL H. KNAPPENBERGER Jr., Ph.D., is President of Adler Planetarium & Astronomy Museum in Chicago. He is actively involved with Chicago's Museums in the Park, a consortium of the nine major museums on Chicago Park District property; and with Museum Campus Chicago, which incorporates the Adler and the nearby John G. Shedd Aquarium and the Field Museum of Natural History. He is also Chairman of the Museum Partners group of the Chicago Systemic Initiative, part of a nationwide effort focusing on systemic improvement of math and science instruction in schools. Before coming to the Adler in 1991, he served as Director of the Science Museum of Virginia in Richmond, Virginia since its founding in 1973. Over the past 25 years he has led efforts to develop interactive exhibits on astronomy and to create educational activities for elementary and secondary schools. Currently he is participating in nationwide efforts to reform math and science education, both formal (classroom) and informal. He is an active member of the International Planetarium Society and regional planetarium associations, and was a reviewer on the National Science Foundation's Panel for the Public Understanding of Science. He has served as President of the Association of Science-Technology Centers and on the Council of the American Association of Museums. He has taught introductory astronomy at Emory University, Georgia State University, University of Richmond, University of Virginia, and Virginia Commonwealth University. His research efforts include work in optical interferometry at the University of Virginia.

Summary of Informal Education Discussions

Isabel Hawkins

Center for Science Education, Space Sciences Laboratory, MC 7450, University of California, Berkeley, Berkeley, CA 94720, isabelh@ssl.berkeley.edu

This is a summary of the discussions that were held on 2002 June 13, following presentations and discussion on informal education topics by panelists, and includes references to some of the papers submitted by the panelists and printed in this section of the proceedings. It represents a distillation of the common themes and issues that surfaced at many of the small group discussion tables. The information was obtained from the facilitators at each table and the flip chart notes that each table generated.

Characteristics of Informal Education

Panelists and small group participants pointed out key characteristics that set informal education apart from classroom education. The following scenario covers many of the parameters that were identified: When we think of informal science education, particularly when dealing with space science, often what comes to mind is the type of experience that we get when going to a museum, science center, or planetarium. When we visit these types of places, we are usually accompanied by friends and family of varying ages and backgrounds, we expect to spend a few hours browsing through the exhibits, playing with the hands-on equipment, viewing a show, and perhaps also interacting with museum staff asking them questions, directions, etc. A stop by the cafeteria and the gift shop are also typical of such visits. We expect the experience to be fun, engaging, and educational. We expect to see and experience new, exciting, and cutting-edge science. We expect to be "wow'ed" and even inspired.

Informal education goes beyond the realm of science museums, encompassing a variety of programs (such as camps, guided tours, informational talks, star parties, etc.) at national parks, community based organizations, youth group organizations, libraries, and even malls and other settings that we may connect more naturally in our minds with the word 'entertainment' than with the word 'education.' Common threads that run through all these experiences include the fact that informal education is voluntary, is typically offered at a cost that can range from moderate to expensive, occurs in a variety of non-traditional settings, covers a wide range of topics through unstructured or semi-structured activities, and includes offerings appealing to a wide range of ages and backgrounds, as well as different learning styles. Even though many informal education institutions have more flexibility in the type of content and experiences they provide their audiences, it is also true that informal education institutions that partner closely with school districts or the government sector are accountable and responsive to state or federal educational mandates.

Programs and activities often provide opportunities for exploration, hands-on inquiry, and interaction. The general public constitutes a primary audience for informal education, but these institutions also serve as a key science resource to K-12 educators, their students, and parents. Words to characterize informal science education that surfaced repeatedly in the notes from the small group discussions included: Inspiration, passion, play, engagement, inquiry, exploration, unconstrained, excitement, fun, educational, interdisciplinary, discoveries, "aha!" moment.

In the context of informal education, the word "Edutainment" was often mentioned, emphasizing the fact that informal education is voluntary, and that museums, parks, and other venues must compete for audiences with other media, particularly television. An important concern was the need for safeguarding scientific accuracy and currency, and not 'watering down' the message and content for the sake of entertainment. This tension is often seen in interactions between informal science educators and the space science community, since it is very challenging to share with the public technical jargon and space science concepts in a way that is neither diluted nor intimidating.

Defining and Intersecting the Goals of NASA and Informal Education

To enhance the impact of partnerships and collaborations, participants identified the need for defining and intersecting the goals of NASA Office of Space Science (and NASA in general) with those of informal education communities. Coherence and alignment between the needs of the informal education community and what NASA and space scientists can provide will result in less duplication of effort and increased benefit to target audiences. Participants offered the following possible goals for NASA and Informal Education:

NASA:

- Highlight NASA missions & discoveries, and the people that make them possible

- Provide opportunities for students and the public to engage in critical thinking

- Enhance the scientific literacy of the general public

- Contribute to the improvement of pre-college science education

- Help develop the next generation of informed adults, scientists, and engineers

Informal Education:

- Help develop a better-informed citizenry, with critical thinking skills, who care about and understand the importance and relevance of science to our daily lives; strengthen the democratic process by improving the thinking ability of voters

- Provide an environment for life-long learning to improve our quality of life

- Highlight and interpret the latest discoveries in science
- Demystify science, making it accessible to diverse audiences

There is a great deal of overlap between the two sets of goals stated above. Participants felt that NASA and the various constituents of informal science education can form natural alliances and partnerships. There was clear agreement among participants that NASA discoveries are tremendously appealing and exciting to the public. The NASA "brand" can play a key role in attracting audiences to informal education venues. For example, travelling exhibits at museums generate good public relations from the local media, attract large numbers of the general public, school field trips, and provide special opportunities for museum members. NASA can play a key role in keeping science fresh and exciting. Through its missions, discoveries, and the people that make them possible, NASA provides a unique inspirational value that catches people's attention. NASA's presence can offer insights unlike any other endeavor-the combination of scientific accuracy and the excitement of mission discoveries is inherently interesting to the public. NASA Office of Space Science Education and Public Outreach effort also provides an opportunity to involve the space science community at large in informal education. The use of unexpected NASA role models, particularly highlighting women and minorities, at informal education events and programs can serve as a powerful tool for encouraging young people to become interested in science.

The Need for Educational Coherence

While NASA can effortlessly add excitement to informal education offerings, there needs to be coordination and coherence of NASA programs that can smooth out the multi-faceted nature of the agency, the demands of finite mission cycles, and the limited time and resources from the space science community. Since the public is most interested in the 'big ideas' of science and how the latest discoveries help answer age-old questions about the nature of the Universe and our place within it, there is also a need to align mission science goals along more broad and fundamental themes, concepts, and strands.

The Importance of Partnerships

A common theme that emerged from the documentation was the importance of establishing sustained partnerships between NASA and informal education institutions. At the institutional level, understanding and addressing the different needs of various partners (e.g., large vs. small institutions, formal vs. informal, scientists vs. amateurs, scientists vs. educators, etc.) is key to the success and long-term sustainability of a partnership (e.g. the paper by Semper on page 163 emphasizes the role of the science museum as a "transformer" agency between the space science community and the public, enabling efficient two-way communication).

At the individual level, scientists and informal science educators need to be informed and sensitive to the needs and constraints of the other community, as well as validating of each other's strengths. Informal education professionals have expertise in demystifying science, revealing scientific phenomena in interdisciplinary and more familiar terms, such as the physics of sports, the science of cooking, the chemistry of winemaking, or investigating light through rainbows. Missions and scientists add unique elements of excitement, discovery and currency. Scientists can provide input on scientific accuracy, help interpret press releases and technical jargon, and share the process of scientific discovery.

Partnerships are also critically important for the goal of engaging diverse user communities in space science. Informal educators are often more knowledgeable about the needs of diverse communities than space scientists, and can help serve those needs with space science resources. Particularly when reaching out to people with disabilities, it is essential for the space science community to work with experts in providing access to exceptional audiences through closed captioning systems, sign language interpretation, tactile resources, etc. (e.g. see papers by Grice et al. on pages 185 and 301, and the description of the SERCH Broker/Facilitator group in Appendix 2, page 457).

The work and ongoing relationships between scientists and informal educators could be improved by: (a) increased communication; (b) more preparation to understand each other's needs, assets, and common ground; (c) access to existing informal education research results from surveys and audience studies; and (d) professional development for all partners.

Increasing the Reach of Space Science Through "Stealthy Science"

Another common theme that emerged from the discussions dealt with increasing our reach by showcasing space science where it is not expected (e.g. national parks, malls, even airport shuttles) and by utilizing multi-disciplinary and cross-cultural approaches that combine and integrate space science with other subjects such as sports, history, theater, art, literacy, etc. (Johnson et. al. on page 178 describe the benefits of cross-disciplinary integration, a strategy that has been used successfully in the Windows to the Universe project since 1994. Bishop on page 196 describes how planetarium programs for students often integrate astronomy with other subjects, such as social science, history, mythology, reading, writing, and literature.) An integrative approach enhances the potential for engaging sectors of the general population beyond the traditional informal and formal space science audiences.

The strategy of reaching out to audiences through non-traditional means included working directly with the target audiences, e.g. through programs at playgrounds, airports, gardens of science, malls, etc., as well as indirectly through groups such as Migrant Corps, church groups, Rotary Club and other adult civic organizations, the Analemma Society, amateur astronomers, national park interpreters, scout and 4-H leaders, etc. For example, amateur astronomers often conduct astronomy outreach activities with schools, community and youth groups, and at national parks. Interdisciplinary themes that integrate space science with historical events and cultural values give additional opportunities to reach out to underserved audiences. At times, controversial topics like evolution, the face on Mars, or the landing on the Moon "hoax," can provide 'teaching moments' and hooks to teach evidence gathering and critical thinking skills. *Star of Bethlehem* planetarium shows can provide a cultural and historical context. In these situations, however, it is important to understand your audience.

Community "buy-in" and cultivating long-term relationships are essential when working with groups that have been underserved by science outreach programs, since parents, other volunteers, and community leaders, can help facilitate informal activities. The local community can also help us understand the needs of audiences from diverse cultures and backgrounds. Community leaders can rally their peers to become volunteers and mentors, reaching out to non-English speaking parents and their children, especially girls and others who are not typically encouraged to pursue careers in science. Other important issues that came up in the context of reaching out to underserved communities through informal education included the high cost of some informal education venues and offerings. NASA could help through mobile program alternatives to give access to rural and other areas.

NASA can provide "staying power" to the science message through exciting discoveries, making it easier for informal education institutions to get the word out and engage the community. Scientists can serve as powerful role models, especially women and minorities, and those scientists who speak Spanish and other languages including sign language. Informal education institutions value the access to experts who can provide the latest content, passion for research and discovery, and a view into the workings of professional laboratories, observatories, and space-based telescopes.

Internet and Media as Tools for Outreach, Dissemination, and Visibility

Informal education space science programs that are broadcast over the Internet and the media can provide an efficient way of reaching out to large numbers of remote audiences. The Internet can provide access to NASA data, simulations, animations, movies, and scientist interviews in a high-leverage mode. Many of the potential audiences that NASA wants to reach lack adequate access to the Internet (e.g. the paper by Thomas, Carruthers, & Takamura on page 187 identifies several barriers that make it difficult for underserved communities to access space science resources through the Internet). Thus, partnerships with broadcast media such as the Discovery Channel, PBS, talk radio, etc. can broaden the effectiveness of our efforts. Remote broadcasts through live webcast or video that capture unique natural events (e.g. total solar eclipses) or that provide a window into the way scientists do cutting-edge research (e.g. from laboratories, observatories, and satellite operations centers) give informal education institutions the opportunity to share the most current scientific endeavors being carried out by NASA today (e.g. the paper by Semper on page 163 discusses innovative uses of webcasting to make the process of current scientific research accessible to large public audiences.) The public is fascinated with the most current and fast-changing images from NASA satellites, and technology gives the opportunity to share them with large audiences. The latest images and data are inspirational and generate continued interest.

Computer games and engaging educational software can also provide an appealing hook for audiences with access to computers. NASA and informal education institutions could investigate partnering with software developers to adapt popular existing games to space science. For example, the game SimCity could be adapted into a "SimUniverse" in which a user could change system variables, e.g., gravity, ratio of baryonic or regular matter to dark matter, etc. or create a space colony with various resources for survival. Such games could be used in extended versions at home, and shortened versions in museums or other venues.

There needs to be a balance between high tech and low tech, however, and clear pathways for local access through community centers, libraries, museums, etc. (e.g. Thomas, Carruthers, & Takamura on page 187). These strategies can make live Internet-based events accessible to underserved communities. High tech offerings need to be combined with low tech approaches based on cultural backgrounds to provide relevance. For example, storytelling is way to expand our reach in an engaging and relevant way. The paper by Moroney on page 160 makes the case for combining Native folktales with "fact tales" as a means to both enchant and educate audiences in informal education venues.

High visibility space science discoveries and exciting astronomical events have also demonstrated great success in garnering extensive national and international media attention. Using such events and other live opportunities opens avenues for collaboration between television and museums (e.g. the paper by Salgado on page 191) discusses the benefits and challenges of using the Spanish language media to reach the Hispanic community in the Chicago area, with the goal of increasing participation of this community in Spanish language courses at the Adler Planetarium.

TV programming like *NASA Connect* produced by NASA Langley Research Center already reaches thousands of classroom teachers nation-wide, and could be tailored to informal education audiences. Working with television can be challenging, however, because of the different format needed for TV, and the limited duration of public television programming.

Informal Education Institutions as Bridges to Formal Education

Informal education institutions can serve as venues to reach out to a varied audience, bridging formal and informal education, fostering parent involvement, and extending the science learning opportunities of school children and their teachers through camps, field trips, professional development, and access to science Internet resources from home and libraries after school (e.g. the paper by Morris et. al. on page 175 discusses a partnership that combines in-service training and community events for families to engage the underserved Hispanic population of an urban charter school). Bishop, on page 196, makes the case that most small planetaria are located at middle schools and high schools, particularly in the Eastern part of the U.S. These facilities serve mostly classroom students in grades 1 through 6, serving as a resource for school districts.

The amateur astronomy community can also help to bridge the focus on space science from informal to formal education settings. Many amateur astronomers nation-wide are already active in education and public outreach activities. In the paper by Bennett on page 246, the Night Sky Network is described, as well as kit resources that are used with the public, teachers, and students.

Questions pertinent to this topic that arose from participants included: (a) To what degree is there a need to align informal education products and programs to Science Education Standards? (b) How can NASA and informal education institutions provide access to underserved populations that have limited resources? (c) In view of the emphasis on literacy and math, how can we develop offerings for young students to spark their interest in science early on? (d) Can museums help align NASA educational materials with science education standards mandated by States? and (e) Can informal venues provide a forum for scientists and teachers to meet as colleagues?

Examples of How NASA Can Help Informal Education Institutions

Participants consistently identified the following informal education needs that NASA could help address:

Measuring the impact of informal education programs on the target audiences:
The evaluation of impact in the realm of informal education is equally important to the accountability and types of outcome measures expected from formal education efforts. Existing methodologies and research on effective forms of evaluation need to be shared among the informal education community and NASA. As discussed in the paper by Wyatt on page 169, it is critically important to pay attention to the learning process occurring in informal education settings such as planetaria. Asking planetarium audiences whether or not they liked the show is not a sufficient measure of impact. NASA E/PO components of missions and research programs could require that a set percentage of the budget be dedicated to both formal and informal education products and program evaluation. For example, NSF and the US Department of Education require that at least 10% of the budget be used to involve independent, professional evaluation groups.

Professional development for informal educators as well as NASA E/PO personnel and space scientists:
Informal educators need professional development as much as formal educators do. Informal education institutions could also serve as appropriate venues for training E/PO personnel and scientists on education and public outreach strategies. A variety of opportunities for all involved could include mentoring programs, internships, focus groups, and advisory working groups. Networks involving informal educators, NASA E/PO personnel, scientists, etc. could range from one-to-one partnerships to a cadre of trained "NASA Associates" at informal education institutions and trained "Informal Education Associates" working with NASA missions, at NASA Centers, etc. A network of well-informed and trained "associates" would facilitate the transfer of timely and up-to-date information and opportunities for engaging the public. This network could serve as a resource to research programs eligible for the NASA Research Announcement (NRA) small E/PO grants, providing a menu of high-leverage options. An example of an emerging network involving informal science institutions and NASA is the collaboration between the International Planetarium Society (IPS) and OSS. NASA OSS is participating in IPS's strategic planning, which resulted in a nascent memorandum of understanding between the two organizations (e.g. the paper by Ratcliffe on page 182 identifies ways in which NASA and IPS are

already working effectively as a team to meet the needs of the varied planetarium community in terms of resources and professional development.)

An efficient pipeline of NASA information and resources, tailored to the needs of all types of informal institutions, large and small:
NASA has a responsibility for doing outreach to, and catering to the needs of, a wide range of informal education institutions. Programs from small museums or libraries, youth groups, and community-based organizations have the potential to reach thousands of interested youth and members of the public, and can address needs that cannot be filled by products and programs more appropriate for large-scale venues (e.g Bishop's paper on page 196 discusses the fact that the approximately 1000 small planetaria in the U.S. serve millions of visitors, albeit with small budgets of $500 per year or less.) Specific suggestions for meeting the needs of a wide range of institutions included:

- Flexibility of formats for shows and exhibits: Small institutions cannot easily use 'canned' shows, and need more audience interactivity, slides, etc. NASA could help provide resources to "translate" big shows and exhibits for smaller venues that do not have IMAX theaters, high-end technology capabilities, large amounts of space, or the funds to attract a keynote speaker (e.g. the papers by Sumners & Reiff on page 155, Wyatt on page 169, and Sweitzer on page 165, identify new digital technologies that provide immersive experiences for the public in planetarium domes large and small. These authors highlight that NASA should provide images, data, visualizations, etc. in a variety of formats to help simulate a digital planetarium experience in smaller venues).

- Modular resources for various types of infrastructure: Not all science museums have planetariums, so there is a need for more modular resources that can be shared with the public in a variety of formats.

- Speaker's Bureau: A speakers' bureau would provide small and large institutions with access to scientists who can engage an audience and share the latest NASA discoveries. Funding for the speaker's travel and stipend is typically not available, particularly at small institutions. NASA could provide small grants to scientists who are members of the speakers' bureau.

- Press Release Information: Participants also expressed the need to have access to press release information before the discoveries are announced by the media, since informal education institutions are often the first to receive phone calls and inquiries from the public regarding interpretation of NASA research results.

Conclusion

It is clear that NASA can play a unique role in informal science education. The varied nature of the informal education arena requires sustained partnerships and networks where information, resources, and expertise can be easily shared and enhanced. The needs identified by the participants, and the suggestions offered for next steps can provide the beginning of a roadmap for action.

NASA Office of Space Science Education and Public Outreach Conference 2002
ASP Conference Series, Vol. 319, 2004
Narasimhan, Beck-Winchatz, Hawkins & Runyon

Creating Full-Dome Experiences in the New Digital Planetarium

Carolyn Sumners

Houston Museum of Natural Science, One Hermann Circle Drive, Houston, TX 77030, csumners@hmns.org

Patricia Reiff

Rice Space Institute, 6100 Main St. Houston, TX 77005

Content Challenge

The last decade has brought an explosion of image and other databases describing Earth and space systems. Digital libraries archive and provide structure for the information, yet the connection with the public and schools remains fragmentary. False color images and media sound bites have little impact on an audience that remains unsure of even what causes the seasons (Schneps & Sadler 1988). Standard textbooks and even well designed videos have been shown ineffective in changing deeply rooted misconceptions (Anderson 1994, Finney 2002).

Meanwhile dynamic processes continually reshape the universe, but at scales that are challenging for humans to visualize. Either the change occurs too slowly, too infrequently, or its effect is too extensive for direct observation. Many changes are beyond the scope of direct experience because they occurred long ago, will happen in the distant future, or are only observable outside the visible spectrum. Finally there are other changes that are best understood when the situation is viewed from a vantage point that is not available to most humans – from inside a hurricane or in space watching sunlight angles change during the year. Traditionally learners have been forced to create their own mental images to understand situations that cannot be viewed directly. In many instances the result has been a misconception, which takes on a reality of its own. Baseline research described below indicates that learners who are given the opportunity to view changes and to conceptualize cause and effect from their own direct observations are more likely to form correct conceptions as is documented by the advantages attributed to hands-on instruction and laboratory experiences.

It will require more than captions on photographs to make Earth and space science discoveries meaningful and engaging. Audiences need to "experience the data," seeing with their own eyes the effects of gradual and dramatic changes. Creating these experiences requires a new kind of meta data – volumetric image bases in which the public can explore and experience the universe in motion – and a large-format playback system – a domed digital theater.

155

Opportunity

In the last five years, a revolution has occurred as planetariums transform into these digital theaters. No longer are planetariums confined to a starfield and pointer, plus obsolete slide projectors and laser disk players. Modern planetariums are filling their domes with pixels – creating full-dome still and moving images that put audiences inside the action – from nebulas and solar storms to changes on Earth ranging from seasonal variations to plate tectonic motions. Hubble photos are no longer flat images hanging in front of a fixed starfield. Now audiences fly through stellar birth clouds like the Orion Nebula and along the galaxy's spiral arms, journeys enabled by computer modeling of these three-dimensional structures. As an Earth science example, in a show called *Force 5*, audiences are transported inside a hurricane, tornado and, as an example of space weather, a coronal mass ejection from the Sun – with the action happening all around them.

This full-dome color moving digital imagery is created in a three-dimensional rendering package (such as 3D Studio Max, Maya, or Lightwave). Five virtual cameras (facing ahead, up, left, right and behind) capture 30 frames each second. Frames are combined using proprietary software to create "dome masters" – hemispheric polar views. In a small theater, a single fisheye lens projects the dome masters on the dome. For larger theaters, multiple projectors are required and the dome masters are sliced and edge-blended with a separate computer feeding each video projector.

Market Reach

As of this publication, there are over 25 large format digital theaters (most in planetariums) in the United States and more than 10 in other countries. Over 30 shows have been produced, either largely or totally full-dome high-resolution digital video. Creating content is expensive – perhaps a third to half-million dollars per 20-25 minute show. However, this is a tenth of the cost of an IMAX production. Several producers of large-format volumetric material are now actively seeking both content partners and funding partners. For example, the *Force 5* show was funded 67% by Earth Science and 33% by the Office of Space Science with in-kind matching funds. A commercial movie distributor is now making the show available to the domed digital theater community. As more shows are being produced, the challenge is to make the shows entertaining, engaging, and educational. Audiences of all ages will learn to expect new, exciting, and cutting-edge science – to be wowed and even inspired by their digital theater experience.

The number of digital theaters is also expanding significantly with last year's introduction of small digital theater projectors by three different companies and with this year's introduction of the portable digital theater by the authors. With these new systems, digital theaters can be installed in any small fixed dome and even in inflatable portable planetariums. Portable dome theaters can reach populations far from urban areas where IMAX theaters and major planetariums are found. The most expensive large-format digital theater production can now

Educational Merit

In collaboration with the Houston Independent School District, Dr. Will Weber of the University of Houston undertook an independent formal research study of the fourth grade planetarium experience since the addition of the SkyVision immersive digital projection system at the Houston Museum of Natural Science. One week was selected at random in March 1999, and all HISD schools scheduled for the planetarium that week became participants in the study. The eight schools represented a population that was 85% economically disadvantaged, 66% African-American, 33% Hispanic, and 1% white. From these schools 438 students completed pre and post written multiple choice assessments. For the treatment, students received an astronomy lesson from a visiting teacher and a 45-minute planetarium program. Most classroom teachers did not provide additional astronomy instruction in the period between the classroom visit and the planetarium experience. The t-test for paired samples yielded a t (19.39) that was statistically significant (p <0.001). In addition, the analysis yielded an effect size (+1.27) that suggests that the gains made by students were both statistically significant and educationally meaningful. In a post hoc item analysis of the Planetarium instrument, researchers identified three subscales of seven items, each focusing on the educational experience of the students. Scale A represents concepts presented through words and non-moving flat projections. Scale B represents concepts presented primarily through moving videos, and scale C represents concepts introduced through immersive video experiences. Table 1 shows the results.

Scale	Presentation	Percent of Questions Answered Correctly		
		Pretest	Posttest	Gain
Scale A	words and slides	43.07%	54.60%	11.53%
Scale B	standard video	31.28%	49.02%	17.74%
Scale C	immersive full-dome video	33.22%	57.21%	23.99%

Table 1. Comparison of student gains

All three presentation techniques were mixed throughout the planetarium program and used by the same presenter. The only differences were the difficulty of the concepts, the students' initial knowledge, and the presentation format. These results strongly suggest that students achieve mastery of more difficult concepts when they can discover behaviors and relationships by observing the concept in action. Standard television video seems to be more effective than still images and immersive video is more effective that standard video. These results may also speak to the level of student attention created by immersive experiences.

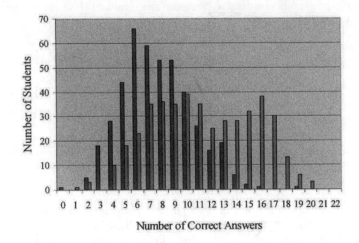

Figure 1. Comparison of correct student answers before and after the planetarium experience

Further analysis addressed information gathered through the instrument's demographic and attitudinal questions. Achievement gains were not related to the sex of the students, their interest in science or their prior experiences in the Museum. Students who indicated that they were good in science did perform slightly better than those who did not. It was also determined that the planetarium experience causes an increase in the number of students who expressed an interest in a science career and who wanted to read books about space.

In Summary

Immersive Earth and space science experiences are now possible with volumetric rendering of dynamic situations and projecting of scenes in immersive digital theaters. These new digital theaters are showing products more immersive than IMAX experiences, but for less than a tenth of the production costs and with less physical distortion of the images. These digital theaters are also well equipped to handle digital databases and can modify programs to include dynamic real-time imagery in a way that is not possible in a linear IMAX theater format. These large-format digital theaters can also deliver a range of content products, not just the astronomy topics normally associated with a planetarium. The theaters are also self-supporting, but, like IMAX theaters, need financial assistance in the development of content. Models established in the last two years indicate that once shows are developed, licensing fees and spin-off products will generate funds for the production of additional shows or clips. In summary, these theaters have the public and school audiences and scientists have the volumetric models and databases. It's an ideal recipe for public outreach with a proven educational effect.

References

Anderson, R. C. 1994, in Theoretical models and process of Reading, 4th ed., pp. 469-495, ed. R. B. Ruddell, M. R. Ruddell, & H. Singer (Newark, DE: International Reading Association)

Finney, M. J. 2002, Electronic Journal of Literacy through Science, 1(2) http://sweeneyhall.sjsu.edu/ejlts/archives/language_development/finney.pdf

Schneps, M. H. & Sadler, P. M. 1988, A Private Universe (Santa Monica, CA: Pyramid Films)

Coyote, NASA, and other Informal Educators

Lynn Moroney

1944 NW 20th St., Oklahoma City, OK 73106, skyteller@aol.com

With many years of designing and contributing to education projects for museums, science and nature centers, and other community agencies, along with working on my own, I have long known that I am about "informal education" but not until today had I ever looked up the word, "informal." I'm glad I did.

In·for·mal *adj.* 1. Not formal or ceremonious; casual. 2. Not being in accord with prescribed regulations or forms; unofficial. 3. Suited for everyday wear or use. 4. Being more appropriate for use in the spoken language than in the written language.

When I read number four, I lifted out of my chair. I thought, "I knew that!?" I remembered the neighbor who asked me to imagine a bowl the size of a football stadium and fill it with grains of sand. He then told me that there were still more stars in the universe than grains of sand! No way! He even told me that one day, we would go to the moon, walk around on it, and return to earth. Yeah, right. It was he who told me the "truth" about falling stars.

Because of these and other stories he told, it was my neighbor, the most informal of informal science educators, who introduced me to the sky world. More than that, it was he who instilled a life-long enchantment with the sky and its mysteries. Informal education is not new. Long ago people entered the sky's mysteries with the help of Storytellers who enchanted them with myths, legends, and folktales about the comings and goings of the sun and moon, the stars, the ongoing seasons, and more. Built into the storyteller's "explanations" of how the world worked were values, traditions, beliefs, and tribal wisdom. The stories invited listeners to create their own images, thus making the world's secrets more personal. When early peoples turned to the storytellers to learn about the world and how it worked, they learned something about themselves as well.

Today we turn to scientists to help us enter into these same mysteries. Their stories, i.e., theories, discoveries, and explanations, dispel confusions so that we may better 'stand-under' the universe and be led more fully into its mysteries.

The problem is that all too often science educators (formal and informal) begin the learning process with investigation and its resulting discoveries. We do not first engage or motivate students and audiences to want to know about the universe. We do not encourage wonder or curiosity. Instead, we give answers to questions no one has asked. We too often bombard the public with seemingly unrelated facts, detailed in a vocabulary for which they are ill-prepared, with the result that they not only miss, but are often led away from the real wonders of, and curiosity about, the universe. Informal and formal educators alike are dismayed at the lack of science literacy in our country, but perhaps, at least in some measure, we ourselves are to blame.

In an attempt to not fall into the trap of listing numbers and formulas, we too often err on the other end of the spectrum. We offer programs that are all bells and whistles, slides, special effects, whiz and bang, with a little or no science teaching. We leave the science museum or planetarium program exhausted, having experienced wheels, flashing lights, and things that "whirrrr..." but are not one bit closer to the reality of the workings of the world, much less our place in that world.

We can learn something from the ancient storytellers whose stories sparked imagination and invited people to create models of the universe. This imaging helped solve earthly puzzles such as the best planting cycles or predicting seasonal changes. Even today these same stories do not fail to establish a sense of wonder as they once again engage our minds, spark our curiosity, and sense of possibilities.

Most of us have an impoverished view of the universe. We think the universe is somewhere 'out there' and in no way connected to us and our daily lives. We need to find ways to reconnect with the universe, to once again become enchanted and delighted with the world and ourselves. To that end, educators are constantly being encouraged to seek new teaching approaches.

Responding to this need, The Lunar and Planetary Institute is now in the process of producing a learning resource package for small planetaria. *SkyTellers: Storytellers & Astronomers* is made possible with a grant awarded by the National Science Foundation. Along with planetaria, this resource will be of use to museum educators, classroom teachers, libraries, and other informal as well as formal education venues.

Short, ten to twelve minute, "mini" programs are being produced to offer twelve astronomy subjects, i.e., sun, moon, stars, day and night, Milky Way, seasons, etc. Each topic will be introduced by a traditional story gathered from the oral literature of Native Peoples. Following the "AncientStory" will be today's "ScienceStory" on the same subject. A Resource Booklet will be included in the package. The tale of Coyote randomly tossing stones into the sky to create stars, paired with the story of how stars are formed in giant clouds of gas and dust, allow the storyteller and the astronomer to work together to help the listener learn "how our world came to be." The folktale and the "fact tale" each have the ability to charm and enchant the listener.

Free of jargon, subjects will be explained and explored in a vocabulary "suited for everyday wear or use" and "more appropriate for use in the spoken language than in the written language."

StarTellers will reveal that science is not a dead body of knowledge to be mastered but a continuing voyage of discovery motivated by wonder and a playful curiosity. The storyteller and scientists will work together to excite the listener about science subjects and the place of humankind in science endeavors. And just as storytellers end their tales with the suggestion that there is more to know, but "that is a story for another day," the scienceteller will tell that there is more to the science story for another time. Is not science a story that is refined if not changed with each new science discovery?

In our society the official task of education has clearly been given to formal educators. It is the schools, not the science museums, planetaria, and nature centers, whose feet are held to the fire and scorned if Johnny can't read, Jenny scores poorly on her math test, and Harvard graduates can't explain the seasons.

It is true that as informal educators, we are free from standard testing, but we are not free from a commitment to our community. We do have a mandate. We are ourselves a community of individuals, museums, planetaria, science and/or nature centers, who by our own choice are committed, or if you please, whose by-laws clearly pledge and commit, to science education (pre-school through adult and life long learning) and to the promotion of science literacy.

Creating Authentic and Compelling Web-Based Science Experiences for the Public: Examples from *Live @ The Exploratorium* Series

Rob Semper

Exploratorium, 3601 Lyon Street, San Francisco, CA 94123, robs@exploratorium.edu

For most of the public, the activity of science and scientists is a great mystery. Not only does the public not understand the nature of science, for many it does not seem to be a very inviting endeavor in the first place. This is not so surprising when one considers that the last experience most people have had "doing science" was in fact in high school biology or chemistry laboratory. These experiences have as much to do with real scientific activity as piano lessons have to do with the life of a concert pianist.

The *Live @ the Exploratorium* project was developed to provide an opportunity for the public to participate in authentic experiences of scientific discovery in an exciting way. It was designed to provide a window onto the everyday activities of scientists for a public visiting a museum as well as for an on-line audience. Through a combination of remote media production, a robust Website and a series of Webcasts using the Exploratorium Webcast Studio as a hub, the *Live @* series provides an audience with the experience of current science activity by providing a virtual field trip to locations which for most of the audience are places where they will not be able to visit.

There are a number of projects designed to bring science events and scientists directly to students using video and the Web including *Passport to Knowledge* and the *Jason Project*. The *Live @ the Exploratorium* project makes use of a museum's facility and sensibility to create a program connecting scientists with a public audience. The museum operates as a mediator between the world of science and the world of the public.

The origins of *Live @* were the early planetary flyby public events held at the Exploratorium beginning in 1974 with the Pioneer flyby of Jupiter and continuing with the planetary encounters of Saturn. These early events demonstrated the public interest in real time scientific exploration. With the advent of the World Wide Web and Webcasting, the Exploratorium has recently produced a number of *Live @* programs including the Kuiper Astronomical Observatory, the Hubble Space Telescope Servicing missions, the Total Solar Eclipse from Aruba (1998), Turkey (1999) and Zambia (2001). For the Zambia Eclipse, a network of 70 museums worldwide was created with the support of the Sun Earth Connection Education Forum to offer the programming to a wide public audience. This network allowed the distribution of the program to many smaller museums that would not have had the resources to develop this kind of programmatic offering on their own.

Recently through the *Live @ Origins* project the public has been able to visit CERN, the particle accelerator in Geneva, the Space Telescope Science Institute, and McMurdo Station in Antarctica. In each of these cases a visiting public and an on-line public have been able to connect live with researchers in the field, ask questions and explore an on-line environment. These programs are archived on a Website that provides contextual information for later viewing. The project also has had a continual assessment program in place to study the impact on the local and remote audiences.

Over the past five years of work we have learned a number of important lessons. First, it is clear that through the Web and Webcasting it is possible to make use of exciting real-time events (stealthy science activities) to discuss scientific issues and to present the work of scientists to a broad public. There is an interested and attentive public who find this access meaningful. This audience is made up of students and teachers as well as members of the general public. Second, the use of museum development expertise has permitted the creation of programs and Websites that are accessible and compelling. The museum's strengths in design, education and science have proven to be valuable in connecting the world of science to the public. Third, marketing support and making use of existing channels of distribution are key features that support the goal of wider impact and this reinforces the rationale of a partnership between NASA and the science museum community. And finally, a long-term relationship between NASA and the science museum community is critical to the development of the trust and infrastructure needed to create good educational programming. The information flow from NASA to the museum world was critical to the creation of authentic programs.

The conclusion of this work is that if one looks at the question of how best to facilitate the interaction between the world of science and the world of the public from a systemic point of view, the interface becomes critical. A good way to improve the interface is through the use of a transformer agency. A transformer takes on the task of creating an efficient transfer of energy between systems by matching impedances between one system and another. The key idea is that each side affects the other rather than it being a one-way transmission. An educational transformer, in essence, allows each side to talk to the other in a way that the other can understand. To do this, a transformer agency must be credible to both audiences, the scientists on the one hand and the public on the other. Science museums, with their staff of scientists, educators, and designers, and their developed audience channels offer a unique function. Our experience with *Live* @ has shown that science museums can be critical transformer agencies in the presentation of science to the public.

Understanding Planetariums as Media

Jim Sweitzer[1]

Rose Center for Earth and Space, American Museum of Natural History, Central Park West at 79th St., New York, NY 10024, jsweitze@amnh.org

This short paper outlines the impact planetariums have on their audiences by investigating them as distinct educational media. Understanding planetariums in this broader context can help shape their development and use in space science partnerships between scientists and educators.

I will attempt to answers three questions: What is special about planetariums? How are planetariums different than other media? How can NASA scientists work most effectively with planetariums? Planetariums are a medium with both a rich history and tremendous potential as they now adopt digital technologies. But, they will be successful in the future only if they enjoy active participation by space scientists and rapid access to the latest, large-format data sets from the frontiers of space exploration.

What is special about planetariums?

Planetariums immerse their audiences in panoramic views of outer space using dome-shaped theaters. As such, the experience is relatively unique, no matter what the size of the planetarium or projection technology. Their traditional trump card for over seventy years has been simply showing the starry sky. All planetariums can do this and most audience members will say that this is one of the primary reasons they're there. Since 98% of Americans live in light polluted night skies, this experience is now becoming one that only planetariums can deliver. The night sky is our primordial contact with deep space. Without it, the reality of the wider universe would soon be eclipsed by science fiction films.

For generations planetariums have been places that offer a cosmological perspective. Planetariums model the scientific universe. Ironically, before digital projection systems, planetariums could only accurately depict the geocentric model of the universe. But that has all changed with shows like those at the Rose Center's Hayden Planetarium, which fly people through a multi-scale, continuous data set from the earth to billions of light years in space. When shown the scope and scale of the universe in this way, audiences have a deep and meaningful experience. Planetariums are missing their primary function when they do not offer such a mind-changing perspectives, and choose instead to unsuccessfully imitate seemingly competitive media like PCs and television.

---

[1]Current address: DePaul University, Space Science Center for Education and Outreach, 990 W Fullerton, Suite 4400, Chicago, IL 60614, jsweitze@depaul.edu

Finally, planetariums are a form of what is called the technological sublime. This term refers to new large-scale technologies that are appreciated or viewed in their own right as demonstrations of the power of science. The launch of Apollo XI was an important example of the technologically sublime. The awe that many often feel for a sophisticated mechanical planetarium projector is a minor example of the technologically sublime. Technically sublime experiences change with improvements in the hardware, so that what was technically sublime for our parents may not be so for us. Since planetariums deal primarily with a science that is future oriented, to remain technologically sublime, they must continue to remain at the cutting edge of technology.

How are planetariums different than other media?

Planetariums are obviously distinct from other media like television and the Internet. Each has its own strengths and weaknesses at conveying different types of experiences. To understand how planetariums are different than these media, it helps to compare them. I will review both traditional and digital planetariums as panoramic or 3D media, then consider 2D media like television and the Internet.

Traditional planetariums like the sky theater with the Zeiss MkVI projector at the Adler Planetarium are descended from 18th century orrerys and 19th century magic lantern shows. The significant experience is being immersed among the stars. The stories the projectors illustrate best are ones of cyclic star patterns seen from different locations on earth. Any other story must be told using auxiliary projectors, employing the planetarium projector as simply an outer space backdrop. The primary meaning viewers draw from such a spectacle is the vastnesses of space. They also enjoy an opportunity to re-connect to the night sky from which modern light pollution has deprived them.

Digital planetariums like the Rose Center's Hayden Planetarium and the Adler's Star Rider Theater use blended images from six or more video projectors to immerse audiences in 3D data sets. These data can be displayed in real time in both of these planetariums using high-end graphics computers. The progenitors of these planetariums are flight simulators and the virtual reality CAVE at the University of Illinois. The Rose Center can also play back stored, highly-rendered images that require massive amounts of computing. The scene of Orion in the initial space show, *Passport to the Universe*, required 1,700 processor-hours of computing to create an ultra-realistic 90 second segment. Just as with the traditional planetarium, the audience experience here is one of immersion and flight, this time in rendered 3D data bases. Digital planetariums allow audiences to fly through deep space and they need not be limited to astronomical data sets. Viewers respond to flight through the digital stellar data bases with the same awe with which they respond to the stars of a traditional planetarium. The digital planetarium can reveal, however, an accurate model of the modern universe. Audience reaction is one of wonder at the immensity of the universe and a better understanding of their place in it than the more traditional planetarium could give.

Planetariums are fundamentally 3D environments because they are immersive. Properly choreographed true 3D digital data induce motion cueing that make audiences believe they are actually out in deep space. It's valuable to contrast this experience with media that have been equally successful at teaching space science: video and the Internet. In the case of documentary videos, the progenitors have been stereoscopic slides, travel log films and human documentaries. The experience is confined to the flat screen, but the video medium can take audiences almost anywhere and give them a stunning sense of immediacy when the programs are live. Like the cinema, video is a medium best suited to linear story lines and stories involving living people. The best space stories told by this medium in recent decades have been the moon race and Voyagers' encounters with the giant planets. Audiences draw their deepest meaning from these programs when they contemplate how their senses have been extended to the depths of space or when they've felt part of the human adventure of the moon landings. Television has been the medium most successful at bringing this chapter of the human exploration of the world to billions of people.

The Internet is the ascendant popular medium for delivering the news from space. When a PC is connected to a web site like those for the Hubble Space Telescope, one has, in effect, a virtual space observatory. This technology is directly descended from naked-eye, ground-based telescopes and the appeal is similar. Now, however, one may select from a menu of the latest, best images of objects from many realms of the universe. Since the HST has a very narrow field of view, it's best suited to telling stories about individual objects. The virtual observatory is like the photo album of our evolving universe.

On the horizon is a new variant on the planetarium experience. It's the Small Digital Planetarium (SDP) - designed for individuals or small groups. It consists of a 3D astronomical data set running on a PC and projected using a single video projector. To be immersive it needs to be projected onto a dome or small curved screen using a fish-eye projection lens. Currently, there are beta versions of software that can duplicate much of what the large digital planetariums can project, but operating off of a single PC. SDPs can be fully interactive in ways that voting systems in large planetariums can never be. Another advantage is that immersive SDPs can cost hundreds of times less than current large digital systems.

How can NASA scientists work most effectively with planetariums?

NASA science missions are expeditions to the frontiers of outer space. The images and data from these missions need to be made available and in the proper format for planetarium use in a way as timely as press releases. Digital planetariums require dramatic 3D data bases. These are extremely difficult to construct, and, if they are to be perceived of as real, must be fashioned directly from the scientific data. NASA is one of the only organizations with the resources and scientists able to deliver these data to planetariums. The visualizations needn't all be immense 3D worlds, but can also be small-scale visualizations that may be inserted in the digital frameworks of digital planetariums.

NASA should help to stimulate the migration of the digital planetarium experience to commodity computer platforms, thus insuring the success of SDPs. This will allow more educators to participate. A useful analog is the small inflatable planetariums that were so popular in the 1980's and 90's. Soon, a digital version will be competitively priced. This is a perfect "vehicle" for teaching space science on a broad scale. Furthermore, SDPs, because of their data flexibility, will actually allow students to "ride along" with spacecraft in ways 2D media cannot.

NASA can help by integrating the efforts of its scientists, education partners and planetariums into a meaningful whole. This can best be done by understanding how the different educational media I've analyzed complement one another. Although planetariums may not be the best medium for telling personal stories, they are the best for giving people the big picture and inspiring them with the vastness of space.

Scientists and educators need to appreciate the scientific modeling role of planetariums to generate meaningful experiences. Planetariums should employ their new digital visualization powers to serve space missions. Digital planetariums now have the ability to present the modern cosmological worldview to the general public in a way they can readily understand. Developing data so that future planetariums can keep up with the science of the day is crucial. This challenge has been with planetarium professionals for hundreds of years.

When successful, the digital planetarium experiences of the 21st century will continue to echo the words London journalist Sir Richard Steele penned in 1713. Reporting on his encounter with Lord Orrery's new mechanical device designed to demonstrate the Copernican worldview, Steele gushed:

> *It is like receiving a new Sense, to admit into one's Imagination all that this Invention presents to it with so much quickness and Ease. That which would have taken up to a Year of Study to come at a familiar Apprehension of it, is communicated in an Hour.*

The Big Picture
Planetariums, Education and Space Science

Ryan Wyatt

Rose Center for Earth and Space, American Museum of Natural History, Central Park West at 79th St., New York, NY 10024, jsweitze@amnh.org

A planetarium truly presents a "big picture," with images that immerse an audience in science stories. Although planetarium stories have typically revolved around the night sky, planetarium technology today can represent the discoveries of space science better than ever before. With immersive video technology, domes can be filled with computer-generated visuals that depict current astronomical discoveries with unprecedented fidelity.

In the current Rose Center Space Show, *"The Search for Life,"* each image (out of more than 42,000) covers about four million square inches of dome surface. Audience members view a show that fills almost half their field of view, and at a rate of 30 big pictures per second, which visually approximates an alternate reality – corresponding not to an experience under a dome, but an experience inside an environment.

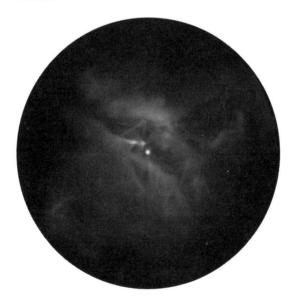

Figure 1. One frame (of approximately 42,000) from the American Museum of Natural History's space show, *"The Search for Life,"* depicting a stage in the formation of a HII region. The circular "polar fisheye" view represents what is projected over the entire planetarium dome.

At its best, immersive video allows audiences to experience a virtual environment in an exceedingly visceral way. An "immersed" audience member becomes part of the action – and part of the science! Award-winning large-format-film director Ben Shedd's article, "Exploding the Frame," describes an approach to large-format cinema that seeks a new cinematic language to work in this medium. He writes, "The whole group of giant screen film formats have one thing in common: the gigantic images extend the edges of the projected film image to the edge of our peripheral vision or even beyond it. I believe we are not just talking about bigger films here, but a new cinematic world. It is a frameless view, an unframed moving image medium" (Shedd 1997). With computer-generated, geometrically-correct imagery, immersive (sometimes called "fulldome") video continues the trend established by large-format film over the last several decades.

What does this mean for those who come to see a contemporary planetarium show? Because the emphasis shifts from story to environment, a modern planetarium show is more about taking a journey than watching a narrative. At the end of a trip, fellow travellers may compare notes and find they have gleaned very different experiences from the same itinerary. Likewise, at the end of a planetarium journey, every audience member takes home something unique to him or her. (In the sense that museums allow for travel without leaving a building, or science centers offer opportunities for exploration, the planetarium "journey" mirrors other paradigms in informal education.)

The individuality of the experience presents challenges to those of us who would like to evaluate the quality and effectiveness of planetarium programs. (Again, a challenge throughout the informal education realm.) Somehow, one would like to account for the matrix of reactions from the cognitive to the aesthetic to the visceral, while probing further than, "So, did you like it?"

To that end, the American Museum of Natural History (AMNH) conducted pre- and post-viewing surveys of audiences who attended the Rose Center's debut space show, *"Passport to the Universe."* Those surveyed responded positively to the show and showed significant gains in comprehending many of the show's underlying concepts: an understanding of humanity's "cosmic address," the relative size and location of stars, the structure of the Milky Way Galaxy, and the origin of heavy elements through nucleosynthesis (AMNH 2002). Further surveys of audiences who saw *"The Search for Life"* indicated that the immersive feel of the show had broad appeal, from eight-year-olds to adults. As one teenager commented, "It was much better than seeing it in a movie theater. The special effects were like actually being there" (Insight Research Group 2002).

Every survey helps, but overall, greater attention needs to be paid to the learning process that occurs under the planetarium dome. Increased evaluation can help pinpoint what works and what does not – an especially important step as the technology driving the shift in planetariums reaches an increasing number of theaters and the audience for immersive video widens. Implementation of the technology in new theaters should take advantage of what their predecessors have taught.

Right now, most immersive video productions are the purvey of only a small number of sizable venues associated with fairly large-scale institutions: only a few dozen theaters are in operation around the world. But as the medium evolves, smaller theaters will have access to similar technology, and the variety of presentations (from pre-recorded to real-time, fairly passive to highly interactive) will increase dramatically.

For example, Small Digital Planetariums (affectionately called "SDPs") will soon offer unprecedented interactivity with the cosmos, in a format that permits each participant to control their own experience. In the spring of 2001, AMNH rolled out its astronomy-oriented Moveable Museum, featuring a 1.5-meter-diameter vertically-positioned dome running software that allows students to pilot around the solar system. Although similar opportunities for one-on-one interaction may be rare, the same technology supports single-lens projectors in school planetariums: both in terms of the hardware to project higher and higher resolution images and the software to navigate through space and time.

Figure 2. The Small Digital Planetarium (SDP) in the American Museum of Natural History's Moveable Museum, shown running a 3-D tour of the solar system.

Particularly as the medium continues to evolve, the quality of tools and access to supporting media need to improve. With an increasingly large audience of planetarians (with varying technical expertise) interested in incorporating immersive video in their presentations, hardware and software tools need to support easy acquisition and inclusion of materials into fulldome programs.

Many fulldome systems include real-time displays – of traditional planetarium functions such as sidereal motion and orrery simulation as well as 3-D data and virtual spaces. Because the field is new and the market remains relatively small, an overall improvement in 3-D capabilities and investment in user-friendly interfaces seems both necessary and likely in the coming years. Real-time solutions gain particular importance in light of the fact that pre-rendered, high-resolution fulldome video will remain expensive to produce for the foreseeable future.

Indeed, a further barrier to successful implementation of new technology into the planetarium sector is, quite simply, economics. Although costs continue to drop, fulldome video remains expensive, especially relative to the budgets traditionally allocated for planetarium production. The American Museum of Natural History invests large-format-film-sized budgets in its space shows, which represent substantially larger budgets than most productions, and a significant investment to recoup.

Pre-produced video sequences, especially those created in high-resolution fulldome format (i.e, those with at least ten-megapixel resolution and a two-pisteradian field of view), will be of tremendous use to the planetarium community. (Lower-resolution SDPs currently make use of megapixel resolutions, but technology should rapidly bring even these systems into the three- to four-megapixel realms.) One can imagine the need for a library of standard video segments – depicting astronomical processes, displaying data on objects, or illustrating basic astronomical concepts (particularly those in 3-D). Ideally, these segments will combine scientific accuracy with visual engagement, capitalizing on the new cinematic language of which Ben Shedd writes.

In fact, the Space Telescope Science Institute has already taken an unprecedented step in providing pre-rendered fulldome material to planetariums free of charge (STScI 2004). Several planetarium programs have already incorporated these segments into storylines, and the imagery has received positive response.

Another challenge planetariums face is a variety of audience expectations that range from sitting under the stars with a lecturer to watching slide shows with pre-recorded narration, from listening to rock music accompanied by laser projections to (perhaps) a large-format-film-style immersive production. Audiences do not understand the diversity of experiences that take place under planetarium domes, let alone the changing nature of the medium, and most people's expectations are defined by the trips that they took to planetariums as elementary-school students. The typical planetarium-as-experience (as opposed to planetarium-as-venue, where a changing slate of programs might be more expected) places most visitors in an "oh, I've done that before" mode of thinking that curtails return visits to a facility. According to a frequently-quoted planetarium adage, the typical person visits a planetarium three times in their life: as a child, with their children, and with their grandchildren.

Unfortunately, because most data about planetariums are approximately as anecdotal as the child-to-grandchildren adage, it is difficult to identify means by which planetariums can help define expectations and attract a wider audience. With any luck, immersive video will help attract more people into planetariums and perhaps increase the visibility of the field in general.

Our culture is immersed in science – science inextricably linked to people's everyday lives. Astronomy and space science have proven to be an appealing and effective in-road to science education, and planetariums are part of that success. As planetariums continue to immerse audiences in increasingly realistic scientific visualizations and narratives, they can give people big pictures that contextualize complex science stories. Each person can take away their own "big picture" and a unique experience of their place in it.

References

American Museum of Natural History 2002, In-house survey

Insight Research Group 2002, Evaluation for the American Museum of Natural History

Shedd, B. 1997, Exploding the Frame
http://www.cs.princeton.edu/ benshedd/ExplodingtheFrame.htm

Space Telescope Science Institute 2004, Informal Education web site
http://informal-sci.stsci.edu/sources/video/dome/

Resources

American Museum of Natural History (AMNH) and Hayden Planetarium's Digital Universe

Even without a dome, you can experience AMNH's Digital Universe! Via a relatively simple interface through the National Center for Supercomputing Application's Partiview software. Users can introduce themselves to the 3-D universe by navigating through a model of the Milky Way Galaxy developed at AMNH. Similar software runs in the Hayden Planetarium dome, used for live lectures and monthly programs. Future fulldome venues can potentially offer immersive versions of this type of experience – either for a group viewing a guided tour, or for an individual exploring alone. Pilot activities for grades 4-6, 7-9, and 10-12 are also available on the site. Download Partiview and the Digital Universe from http://www.haydenplanetarium.org/.

The Fulldome Video Mailing List

The fulldome video mailing list brings together an interdisciplinary cast of characters from around the world with a common interest in the technologies that make immersive planetariums possible. Discussions range from the mundane to the philosophical, and all are welcome to join in. Join the fulldome video mailing list at http://groups.yahoo.com/group/fulldome/ or by sending a message to fulldome-subscribe@yahoogroups.com.

Planetariums in K-12 Science Education

Shawn Laatsch[1]

Gheens Science Hall & Rauch Planetarium, University of Louisville, 108 West Brandeis Avenue, Louisville, KY 40292, 102424.1032@compuserve.com

Planetariums in the K-12 schools systems and Universities are often the main resource in bringing astronomy and space science to students. In this panel discussion the author will discuss the role of smaller planetaria and the unique opportunities they can provide for students. These include individualized attention, live interactive programming, and on-site visits to schools and other educational groups.

In the last few years the importance of astronomy in K-12 curricula has been challenged. School systems often rely only on state standards and teach very minimal astronomy to their students. These standards are often based on Project 2061: Benchmarks for Science Literacy (AAAS 1993) and the National Science Education Standards (NRC 1996). Both of these publications, while well intended, cut short the importance of fundamental astronomical concepts in the classroom. They often give only one of two minor indicators per level, and these in many cases miss the mark. For example, the grades 3-5 Benchmarks state "learning constellations is not important in itself" and "Once students have looked at the stars, moon, and planets ... no particular educational value comes from memorizing their names or counting them." These statements emphasize process over content. Good science education needs both. What should be taught to students at the K-12 level in regards to astronomy? Is process more important than content? We will examine these questions in detail during the panel discussions.

References

American Association for the Advancement of Science Project 2061 1993, Benchmarks for Science Literacy (New York: Oxford University Press)

National Research Council 1996, National Science Education Standards (Washington: National Academies Press)

---

[1] Current address: East Carolina University, Department of Math and Science Education, 323B Austin Hall Greenville, NC 27858

An Urban Partnership: Inservice and Science Enrichment Programs for a Hispanic Serving Charter School

Penny A. Morris[1], Olivia G. Garza[2], Marilyn Lindstrom[3], Jackie Allen[4], James Wooten[5], and Victor D. Obot[6]

[1] *University of Houston Downtown, 1 Main St., Houston, TX 77002, smithp@zeuss.dt.uh.edu*

[2] *Raul Yzaguirre School for Success, 2950 Broadway Blvd Houston, TX 77017*

[3] *NASA/Johnson Space Center, 2101 NASA Rd. 1, Houston, TX 77058*

[4] *Lockheed Martin/JSC, NASA Johnson Space Center, C23, Houston, TX 77058*

[5] *Houston Museum Of Natural Science, One Hermann Circle Drive, Houston, TX 77030*

[6] *Texas Southern University, 3100 Cleburne Houston, TX 77004*

Introduction

The Houston area has minority populations with significant school dropout rates. This is similar to other major cities in the United States and elsewhere in the world where there are significant minority populations that recently moved in from rural areas. The student dropout rates are associated in many instances with the absence of educational support opportunities either from the school and/or from the family. This is exacerbated if the student has poor English language skills. To address this issue in the greater Houston area, a NASA minority university initiative enabled us to develop a partnership between a Hispanic serving institution (University of Houston Downtown-UHD), a historically Black institution (Texas Southern University-TSU), an urban museum (Houston Museum of Natural Science-HMNS), NASA/Johnson Space Center (JSC), and a predominantly Hispanic charter school (Raul Yzaguirre School for Success-RYSS). Our main goals in the program are to reach out to minority and other underrepresented groups, encouraging an ongoing interest in education, especially in the areas of science and space science. We are achieving this through community events and professional development. In the following paragraphs we will discuss the RYSS program as it includes in-service programs for teachers, workshops for parents and their children, and summer enrichment programs led by high school and college Student Ambassadors for RYSS students.

Results

In-Service Teacher Programs

The twelve hour in-service program is centered on developing computer skills, i.e., using technology to enhance teacher science skills. The course includes the following components: HTML programming, Web Quest, Search engines.

HTML

The purpose of training RYSS teachers is to use a very basic HTML, enhance their computer skills, and teach problem solving skills.

Web Quest

This is an inquiry-oriented, computer based, learning activity (Dodge 2001). It uses a variety of Internet resources to develop a project, in this instance, a space science project. The program is designed for increasing the teachers' analysis, synthesis, reasoning, and evaluation skills. The teachers, after completing the project, involve parents and their children.

Search Engines

Participants research numerous sites, learning to compare and evaluate the sites. Teachers learn to be observant of the investigated sites and learn to evaluate their effectiveness and validity.

Parent and Student Workshops

After completing the training sessions described above, teachers conduct parent-student workshops composed of six to eight participants. Each workshop, led by a teacher, is different, but all follow the Web Quest components. Projects included a solar system travel brochure, why black holes exist, creation of a star, colonization of planets within our solar system, and using mathematics to navigate the solar system. The participating students range from pre-K to grade 5. The parents, mostly mothers, have a variety of educational backgrounds, including some with fifth grade educations. All participants learn how to use computers, including e-mail as a learning tool. The youngest students (pre-K) and their mothers studied black holes, creating black hole paintings that closely resembled images downloaded from the web. The group decided that a space ship journal would enable them to study the black hole more closely. The parents designed their children's space suits and avidly discussed their design.

Summer Enrichment Programs

At RYSS there is a six-week space science enrichment camp for children ranging in age from 6-15. The children, including a disabled child, vary in their English skills. The program is taught by our space science Student Ambassadors who are recruited from TSU, UHD, RYSS and Houston area high schools and are trained by NASA/JSC and HMNS. The activities included Mars soil sleuth, volcanoes, edible rocks, search for a habitable planet, impact cratering, strange new planet, rockets, space station, chemistry day, sun dial, solar systems, etc. The Student Ambassadors developed a space science jeopardy game covering the summer's topics. Teams had a flag and a cheer; the team with the most points received a prize. Parents are invited into the classroom and are encouraged to ask questions. Both the children and their parents were enthusiastic about the class. The summer program added a significant level of space science to the school's curriculum.

Discussion

Outreach programs to minority and underrepresented groups are important. The school dropout rates for Latino youths are higher than African Americans or whites (Klinenberg 1996). Their educational attainment at every level is lower than other groups (Veciana-Suarez 2002). Latino children are less likely to be enrolled in head start or pre-kindergarten programs and they are less likely to take rigorous academic classes in high school. The dropout rate does not help society as we may deprive ourselves of space scientists, technologists, and perhaps a future Einstein.

This type of program can be applied to other minority or underrepresented groups in the United States and other countries with high dropout rates. Most of the groups falling into this category are immigrants from rural areas. Space science can be an important tool for increasing self-confidence. The program will enhance school access and encourage the parents to be part of the educational process. These steps will help ensure that these children complete their schooling.

References

Dodge, B. 2001, Learning & Leading with Technology, 28(8), 6

Klineberg, S. L. 1996, Houston's Ethnic Communities, Third Edition: Updated and Expanded to Include the First-Ever Survey of Houston's Asian Communities (Houston: Rice University Publication)

Veciana-Suarez, A. 2002, Some kids won't wear caps, gowns, Houston Chronicle, June 2

Windows to the Universe – A Web Resource Spanning Formal and Informal Science Education

R. M. Johnson[1], C. J. Alexander[2], J. J. Bergman[1], C. R. Deardorff[1], L. Gardiner[1], J. Genyuk[1], S. Henderson[1], M. LaGrave[1], D. Mastie[3], and R. Russell[1]

[1] *University Corporation for Atmospheric Research, Office of Education and Outreach, P.O. Box 3000, Boulder, CO 80307-3000, rmjohnsn@ucar.edu*

[2] *Jet Propulsion Laboratory, Mail-Stop 169-237, Pasadena, CA 91109*

[3] *Pioneer High School, Ann Arbor, Michigan 48104*

Introduction

The *Windows to the Universe* web site (http://www.windows.ucar.edu) was started in 1995, with funding from the NASA High Performance Computing and Communications Public Uses of Remote Sensing Databases Program. The project was originally developed as a resource for informal science education, with a focus on museums and libraries. However, from the beginning, we recognized the potential of web-based information resources to support the needs of students and teachers in the K-12 classroom. Through our partnerships with educators and outreach specialists, we worked to develop a resource that could function as a small, no-cost computer-based museum exhibit (assuming computer and internet access), as well as a supporting information resource for research and exploration in libraries and the classroom. Now in our ninth year of development, our audience of over ~4 million users per year includes large numbers of students accessing the site from schools, from home, as well as from other community centers providing web access. Our approach of providing interdisciplinary science and humanities content in an intensively hyperlinked format is appreciated by our users, and facilitates both K-12 education as well as informal, curiosity-driven science exploration.

Project Description

In order to serve our diverse user population, our focus has been to provide an attractive, interdisciplinary web site that engages users and invites them to exploration of the geosciences by providing scientific and cultural information in context. *Windows to the Universe* therefore spans the Earth and space sciences, ranging from the Earth as a planet to astrophysics, and including interdisciplinary arts and humanities content. In addition, the site provides information on new research discoveries, current events and space missions, links to databases and other "safe" sites of interest on the World Wide Web. To date, our site consists of 30 major sections, together composed of ~6000 html files, ~6000 images and ~690 image maps. The site is continually revised and expanded in support of Earth and space science research.

Key popular aspects of the project include three levels of content, representing the upper elementary, middle school, and high school reading levels. By providing three levels of content for our users, we provide a remediation opportunity for users that are having difficulty understanding content at a higher level, allowing them move to a lower level. Similarly, we provide younger people the chance to challenge themselves by reaching to a higher level of content (see comment below).

Site Use

The *Windows to the Universe* website has served ~4 million users annually over the past several years. Analysis of responses to user surveys over the past three years indicates that 74% of respondents are students. 78% of these students are in K-12, with the remainder at the undergraduate, graduate, or continuing education levels. Furthermore, 46% of survey respondents use the site once per week or more frequently, with users accessing the site most frequently from home and school (accounting for over 70% of responses).

Major Activities

The major focus of activity in the project is development of new content supporting our collaborations (see below) and our users. Our major thrusts for content development include the Earth sciences, space science, and space weather. A major enhancement underway in our space weather area is development of multiple levels of space weather content (a portion of which was formerly separate from the main body of the website and available only at one level), and the development of an undergraduate-level survey of space weather content.

Maintaining and updating the existing content on the website is also a vitally important and challenging task, given the large size of the website and the speed of research progress in the Earth and space sciences. In addition to our content efforts, we regularly present workshops, demonstrations, and short courses and participate in Share-a-Thons and in the exhibit area at regional and National NSTA conventions since 1996. As a component of this activity, we provide training to over 800 educators per year (at NSTAs and other venues) on standards-based classroom activities that can be included in curricula linked to the National Science Education Standards and associated state and local standards.

A new major initiative begun in January 2003 is translation of the entire website into Spanish, with funding from the NSF Geoscience Education program. The first sections of the translated website will be released in fall 2003, and will provide Spanish-speaking audiences around the world access to our content at all three levels. Users will have the ability to switch between English and Spanish versions of each page (at each level) at the click of a mouse, facilitating both science and language education. The translation of the website is planned for completion in fall 2004. As new content is added in the future, it will be made available in both English and Spanish, with an on-going translation effort.

Another focus of our recent activities has been the development and improvement of tools for users as well as tools to facilitate website development, tracking, and maintenance. We have restructured our search tools to provide more information about requested content in priority order, with the ability of selecting or deselecting different aspects of the site such as content, images, Ask-A-Scientist questions, and news. We have also developed interfaces for content development and Spanish translation that facilitate and help manage the work of content developers and translators on the site. We are also developing an image database for the website that includes metadata about the images on the website (sources, captions, credit, copyright information, size, etc). This database is integrated into the content development interface, allowing developers to select images from the database and automatically transfer the needed credit and copyright information to newly developed pages or pages in revision.

Finally, we also strive to be responsive to our large user population, which communicates with us through multiple venues in the website. In addition to our user surveys, which provide us information about our user population, we also receive comments, suggestions, compliments, criticisms, requests, and questions from our users.

Collaboration

Major support for the *Windows to the Universe* project is provided by the NASA Earth Science Enterprise as well as the Science Information Systems Program of the NASA Office of Space Science. The project is now being used in collaboration with missions and research efforts to leverage web-based investments in education and outreach, based on our extensive existing content and large user base. We have developed (or are currently developing) web content, interactives, classroom activities, news reports, and dissemination materials for numerous NASA missions and research projects including: the SWICS instruments aboard the Ulysses, ACE, and WIND missions, the AIM mission, the Ulysses mission, the ISTP Solarmax program, the JPL Mars Program, the Galileo mission, and the Rosetta mission. In addition to its support of our Spanish translation effort, the National Science Foundation has also provided targeted development support through the NSF-supported Space Physics and Aeronomy Research Collaboratory (SPARC), Comprehensive Space Environment Model, and Boston University Center for Integrated Space Weather Modeling. Now maintained at the University Corporation for Atmospheric Research, the project is also expanding in support of atmospheric and related sciences with support from the National Center for Atmospheric Research.

Quotes from Users

"This is a fantastic site that we are now utilizing at our middle school. We are beginning an integrated approach to teaching science & your site is a perfect teaching tool."

–Female Middle School Educator

"Dear *Windows to the Universe* 'Creators',
I really think your web page information is excellent and helped me a lot in science class. Your graphics are wonderful and really gave me an idea of how the planets look. Also, I compliment you on your, 'BEGINNER INTERMIEDIATE AND ADVANCED' program. You can learn info and terms in simple and more challenging layouts. Once more, thank you for such a great website."

–Female Middle School Student

The International Planetarium Society

Martin Ratcliffe

The International Planetarium Society, c/o Exploration Place, 300 N. McLean Blvd, Wichita, KS 67203, mratcliffe@exploration.org

The International Planetarium Society (IPS) represents planetarium theaters around the world. With over 90 million visitors per year to planetariums, these facilities are often the place where the public is exposed to space and astronomy for the first time. Planetariums come in almost as many flavors as particles in the baryonic zoo, and no two are alike. Some of them interact with each other, and some don't.

Working Relationships

Developing effective working relationships with the planetarium community requires an investment of time and willingness to learn about the internal culture of the planetarium world to better understand how things work. Do you want to make a planetarium show? It is not simple, and rarely should it be a taped-slide presentation converted from lecture room to domed theater.

Each institution operates differently, prioritizing their reach to diverse segments of the community (schools, teachers, university students, general public, etc.) in a unique way.

Likewise, the planetarium community values opportunities to understand the NASA culture. There is much for the community to learn about the NASA Office of Space Science and its Education and Public Outreach infrastructure and program, about the new NASA Education Enterprise, about the thematic organization of Forums and the role of regional Broker/Facilitators. Developing such an understanding of how each community works is a prerequisite to long-standing collaborations. Long-term relationship building is far better than aiming for short-term goals. Some major steps towards relationship building have begun.

Planetarium Diversity

Planetarium theaters come in a wide range of sizes and incorporate a wide range of technologies in addition to catering for different audiences. Slide projectors are going the way of the dinosaurs, but you can't count them out yet, since such technology is going to last some time in the majority of smaller theaters.

Newer planetarium theaters utilize multiple video projectors and seamlessly overlap adjacent projection areas on the dome to create entire scenes. This can be useful for creating unique views, in three-dimensions, of astrophysical processes: star formation, quasar jet formation, galaxy formation and interactions, supernovae, etc. In addition, these techniques can help rid the planetarium theater of a pre-Copernican view of the sky/solar-system/universe that traditional electromechanical star projectors inherently reinforce; i.e. that of an Earth-centered view of the sky.

Some theaters focus on inspiring a new generation of young minds and hoping that spin-off will focus minds in the classroom, providing science classes with greater meaning. Other planetarium theaters 'are' the classroom. Some theaters are both of these, in some cases all the time, and in other cases some of the time, depending on the application for different audiences.

Through the Eyes of Hubble

As an example of a planetarium show that resulted from a collaboration with the Associated Universities for Research in Astronomy (AURA) and the NASA Hubble Space Telescope, "Through the Eyes of Hubble," resulted in the following statistics:

- 110 shows distributed in 14 countries
- Translation into 12 languages
- Show production underwritten by AURA, paid back in full through show sales

The show involved a collaboration between the Carnegie Science Center and the Space Telescope Science Institute Office of Public Outreach, with scientific leadership provided by Dr. Anne Kinney.

The goal of the show was to bring the story of the technological success of the first servicing mission and the ensuing brilliant science. In the planetarium world, this was considered an outstanding success, in spite of the challenges that naturally occurred when bringing two different cultures together.

Challenges of Show Production

For large theaters that cater to a general public audience, a natural tension develops between scientists and show producers for a variety of reasons. Such pitfalls can be defined most simply as follows:

- The scientists goal is content
- The producers want an appealing story with good content.

This view is admittedly oversimplified, but this is done to convey the point, often what has to be done in a show. It is also true to say that the planetarium community themselves don't agree on the division of content versus story. It is a dynamic thing that must be worked through. Consequently, a clear scope for scientist and producer is worth developing up front.

Cross-Cultural Efforts

Collaborative efforts by IPS and JPL's OSS E/PO team have resulted in at least four meetings involving planetarium staff and NASA E/PO personnel. These direct meetings generate many ideas, some easy to implement, others will take time to mature.

Example of Short-Term Benefit

The Jet Propulsion Laboratory's (JPL) OSS E/PO team distributed a Mars mission compilation video that IPS could economically reproduce on videodisks for sale at cost price to a number of planetarium facilities. Over 60 have been distributed. The disk was about to be mastered when the launch video from the rocket cam of the launch became available, was sent overnight to IPS, and was being mastered on a video disk within a week of launch. It was very current and good material to get out to the attentive planetarium audiences quickly. Fortunately the JPL E/PO team understood the urgency of the IPS production timeline and was able to respond accordingly.

Longer-Term Planning

Working with the International Planetarium community involves getting to know the community better in all its diversity. In July 2002 (a month after this OSS conference) the IPS biennial conference was held in Wichita, Kansas, and NASA E/PO community took an active role in that meeting. Teacher training workshops and a panel session devoted to discussing planetarium needs took place. It is all part of a continuing dialogue that will build stronger relationships and lead to better products in the long run. It is hoped that more opportunities for closer ties can be developed (a new example is the Mars Visualization Alliance), so that Planetariums can do their job better, and NASA can leverage their products through a medium that can display them in a way no other medium can.

The Accessible Universe: Making Space Science Accessible to People with Special Needs

Noreen Grice

Charles Hayden Planetarium, Boston Museum of Science, Science Park, Boston, MA 02114, ngrice@mos.org

A Planetarium is a virtual reality simulator that can take visitors on a journey to view the sky from any place on Earth and from any time in history. Images of stars and other celestial objects are projected on a curved dome ceiling. A narrator identifies constellations and other celestial wonders. But what if you could not see the stars? What if you could not hear or understand the narration? What if you were denied access into the planetarium or observatory because you used a wheelchair? These are obstacles faced by thousands of visitors each year.

Disabilities can create barriers to information and experience. The simple act of answering a telephone, mailing a letter or opening a door may be impossible for someone without personal assistance. Many people with disabilities have difficulty finding employers willing to accommodate their special needs. Others do not have access to computers, the internet, and other important technology.

In a museum or planetarium setting, accommodating visitors with special needs requires an environment that is both stimulating and accessible. The following resources are available at the Charles Hayden Planetarium in Boston.

- Tactile/Braille astronomy illustrations for visually-impaired visitors

- Modular closed captioning system for hearing-impaired visitors

- Booth and lamp for American Sign Language (ASL) interpretation

- Volume-adjustable assistive listening devices for hearing-impaired visitors

- Removable chairs and transfer seats for visitors in wheelchairs

To help other planetariums meet the needs of their disabled visitors, I have edited a book entitled *How to Make Planetariums More Accommodating and Accessible to Visitors with Disabilities* (GLPA 1996). This book, which was published by the Great Lakes Planetarium Association, offers suggestions to make astronomy more accessible for students with hearing, learning and visual impairments in a planetarium setting.

I have also worked on projects that make science accessible to disabled audiences outside of the museum and planetarium environment, for example through three tactile/Braille books entitled *Touch The Stars* (Grice 1998; Grice 2002b), *Touch The Dinosaurs* (Morgan 1999) and *Touch The Universe: A NASA Braille Book of Astronomy* (Grice 2002a). The latter was developed under a Cycle E/PO grant from Space Telescope Science Institute in collaboration with astronomer Bernhard Beck-Winchatz (DePaul University) and Ben Wentworth (Colorado School for the Deaf and the Blind). It makes images taken with the

Hubble Space Telescope accessible to blind readers (Grice, Beck-Winchatz, & Wentworth 2004).

Making space science accessible to *everyone* can be a challenge. Is it really important?

Consider these statements:

- It costs too much money to print our brochure/handouts in Braille.
- I've never had a deaf student, so why should I worry?
- Thank you for calling about our open nights. Sorry, our observatory is not wheelchair accessible.
- All students will be required to complete the test in 60 minutes. You'd better write fast – no exceptions.
- As you can see... this is the Ring Nebula.
- Everyone... point to Orion.
- This is the North star... and over here is the Big Dipper.
- Our planetarium has captioning units for deaf visitors... If you need to borrow one, please notify the attendant.
- The sign language interpreter will be here next month... maybe you should come back then.
- Sorry... I can't hold the class back just because you don't understand.

... yes, it really is important.

References

Great Lakes Planetarium Association 1996, How to Make Planetariums More Accommodating and Accessible to Visitors with Disabilities, ed. Noreen Grice
http://glpaweb.org/tips.htm

Grice, N. 1998, Touch the Stars (Boston: Museum of Science)

Grice, N. 2002a, Touch the Universe: A NASA Braille Book of Astronomy (Washington: National Academies Press)

Grice, N. 2002b, Touch the Stars II (Boston: National Braille Press)

Grice, N., Beck-Winchatz, B., & Wentworth, B. 2004, page 185 of this volume

Morgan, A. 1999, Touch the Dinosaurs (Boston: National Braille Press)

Collaborative Support for Solar Eclipse 2001 Activities

Valerie L. Thomas

The LaVal Corporation, 2004 Clearwood Dr., Mitchellville, MD 20721, vthomas@erols.com

George R. Carruthers

Naval Research Laboratory, Code 7609, Washington, DC 20375-5320

Eduardo Takamura

Minority University – Space Interdisciplinary Network (MU-SPIN), Goddard Space Flight Center, Greenbelt, MD 20771

Introduction

The Science Mathematics Aerospace Research and Technology (S.M.A.R.T.) Technology Learning Center hosted solar eclipse activities for the Solar Eclipse 2001 Webcast, and our experiences provide an excellent illustration of issues associated with informal education. The webcast, provided by NASA and the Exploratorium, allowed museum and community-based participants in the Northern Hemisphere to experience the event in real-time, on June 21, 2001 at 8:30 a.m. EDT, while the totality was only visible in the southern part of Africa. S.M.A.R.T. is a community-based organization, focused on preparing people of African descent in science and technology. The informal education issues discussed in this paper include: technology support, access to the community, charging versus free classes and workshops, and limited funding. This paper will discuss these issues in the context of the solar eclipse activities and how NASA and the National Society of Black Physicists (NSBP) helped in resolving the issues.

Background

The S.M.A.R.T. Technology Learning Center (TLC), in Lanham, Maryland, had planned to host two distinct solar eclipse activities: a S.M.A.R.T. Solar Eclipse 2001 Family Night on June 18, and a S.M.A.R.T. Solar Eclipse Workshop and Solar Eclipse 2001 Webcast: A View From Zambia on June 21. The S.M.A.R.T. TLC was located in Washington, DC in 1999 for the last solar eclipse of the millennium and recently relocated to Lanham, MD; therefore, arrangements had to be made for technology support and high speed Internet access.

The Family Night activity was held at the S.M.A.R.T. TLC. Presentations and demonstrations were given by scientists from NASA and the National Society of Black Physicists. There were two presentations on solar eclipses, including a scale model demonstration, two others on the sun, and a solar cookie construction activity.

Fewer people than expected attended the Family Night activity; however, there were some very young, well behaved, and highly motivated children in attendance. During the slide presentation of an actual total solar eclipse that the presenter had experienced, the images and descriptions made the audience want to have a similar experience. The presenter pointed out that the next total solar eclipse, on the East Coast of the U.S., will be visible in the year 2017; and that a 5-year-old girl in the audience would then be 21 years old. The other children were overheard calculating their ages, while the next two scientists were setting up for their presentations.

The children were obviously paying close attention to the presentations on the sun because they were able to answer questions about features on the surface and inside the Sun when asked during the solar cookie construction demonstration. Their answers surprised the adults who were present. This demonstration reinforced the information that the children had learned and each of them had a chance to construct their own solar cookie, which illustrated the surface characteristics, sun spots, magnetic prominences, and the interior layers.

High speed Internet access (i.e., DSL) had to be installed at the S.M.A.R.T. TLC to support the June 21 webcast. Due to technical difficulties, the DSL provider informed S.M.A.R.T. on June 20 that the installation was rescheduled for the following month and S.M.A.R.T. had less than 24 hours to relocate the webcast activities and notify the people who planned to attend. The webcast activities were held at the NASA/GSFC, with a full house and two attendees in wheelchairs (accommodations were also available for them at the S.M.A.R.T. TLC). The audience consisted of mostly GSFC summer students, along with the young children, professionals, and other adults. Three scientists from the National Society of Black Physicists joined the NASA scientists in informal presentations that accompanied the solar eclipse webcast. There were also demonstrations and hands-on activities: solar eclipse, UV beads, and solar telescopes.

Issue 1 – Technology Support

The members of S.M.A.R.T. are typically volunteers who hold daytime jobs and, therefore, are not available to provide the technology support for the special activities such as the solar eclipse webcast. To address this issue, S.M.A.R.T. partnered with the Minority University – Space Interdisciplinary Network (MU-SPIN). MU-SPIN provides education and research outreach support to minority institutions (K-12, 2- and 4-year colleges, and universities) through its Network Resources and Training Sites (NRTS). The MU-SPIN Project staff provided the support needed for the DSL access within the S.M.A.R.T. TLC. With the last minute relocation of the webcast activities, they (and MU-SPIN's summer students) also provided invaluable support for setting up the equipment and venue and establishing the connection to the webcast that was transmitted from Zambia, Africa.

Issue 2 – Access to the Community

This issue has two parts: the inability to determine the potential turnout and transportation. Since it was difficult to ascertain how many people would attend the two solar eclipse activities, plans were made to have concurrent sessions. Three concurrent sessions (i.e., the solar eclipse presentations, sun presentations, and the solar cookie construction demonstration) were scheduled for the Family Night, with the attendees rotating through each session. Because there were fewer people than expected at the Family Night activities, concurrent sessions were not necessary. The two concurrent sessions planned for the webcast activities would have been necessary if they had been held at the S.M.A.R.T. TLC. However, since the webcast activities were relocated to GSFC, concurrent sessions were not needed. If it had been determined that more people were going to attend than the S.M.A.R.T. TLC could accommodate with the concurrent sessions, then using GSFC as a relocation option may still have been needed even if DSL was installed.

Because of the early time for the solar eclipse webcast (8:30 a.m. EDT) and the desire to attract minority students, arrangements were made with GSFC to provide a bus to pick up their minority summer students and transport them to the S.M.A.R.T. TLC. This transportation need would be of general concern for minority students' access to the TLC during the day because of its location in a commercial area.

Issue 3 – Charging Versus Offering Free Workshops, Classes, and Special Events

This issue is the most critical with respect to the TLC's long term sustainability and is somewhat paradoxical. There needs to be a dependable income stream in order to pay the operational expenses (especially the rent) in order to ensure the continued existence of a physical location for the TLC, and charging a fee could provide that. However, the fee could become a deterrent to obtaining access to the target community for many reasons. One example is a family with several children. The young children who attended the Family Night activities were all from the same family. A total of eight children are in that family and they were all in attendance. If a fee had been charged, some of the children may not have been able to attend.

Issue 4 – Limited Funding

S.M.A.R.T.'s limited funding issue has an impact on the other issues. With limited funding, S.M.A.R.T. must make a choice between paying the rent and buying equipment, services, and supplies versus paying for staff support (for technology and facilitating classes and workshops). S.M.A.R.T. must also find a way to ensure continued funding to pay the rent to house the informal education activities. And since the target audience is not in close proximity to the TLC, arrangements must be made to pick up students, if necessary. This could be done with schools in the surrounding communities.

For the solar eclipse webcast activities, this issue was not critical. The Sun Earth Connection (SEC) Education Forum provided a grant to support the solar eclipse webcast activities; MU-SPIN Project provided the technology support staff; the SEC, Exploratorium, and National Society of Black Physicists provided the scientists and content; and NASA provided a bus to pick up the summer students. Partnerships such as these are critical for the success of informal education Centers like S.M.A.R.T.

Conclusion

The solar eclipse webcast and Family Night activities hosted by S.M.A.R.T. serve as a microcosm for illustrating the informal education issues experienced by small community-based organizations that host outreach programs for minority underserved audiences. These issues, although viewed through the solar eclipse lens, are generic issues that apply to all of the S.M.A.R.T. TLC activities.

Reaching the Spanish-speaking Community through the Adler Course Program

José Francisco Salgado

Adler Planetarium & Astronomy Museum, 1300 S. Lake Shore Drive, Chicago, Illinois 60605-2403, salgado@adlernet.org

Abstract. The Adler Planetarium & Astronomy Museum offered its first Spanish language course in 2001 as part of its efforts to reach the Spanish-speaking community in the city of Chicago. The naked eye astronomy course, entitled Espectáculo Celeste: Astronomía a Simple Vista, was designed for a general audience. It examined celestial phenomena using Adler's historic Sky Theater and its Zeiss projector. Challenges in informing this audience about museum programs were identified during the development of special programs and currently are being addressed. Given the strong positive response to this course, the Adler Planetarium is planning to offer courses in Spanish on a regular basis, as well as explore other opportunities to interact with the Spanish-speaking community in Chicago.

Introduction

Adler's course program serves adults of all ages, but has recently started serving families as well. The program brings in experts from all fields within astronomy, from history to cosmology to special interest topics. Approximately 1,300 people took advantage of the course offerings in 2001, and in 2003, with the success of family programs, that number should increase. The courses are fee-based and provide a formal learning experience for Adler members and non-members as well. The courses are held in the Adler classrooms, the Sky Theater, the Doane Observatory, and/or in CyberSpace, Adler's electronic gallery.

The Adler's prior efforts to reach the Spanish-speaking community had been limited to the presentation of *Estrella de Maravilla*, the Spanish version of *Star of Wonder*, a planetarium show about possible scientific explanations for the star of Bethlehem. With the addition of José Francisco Salgado to Adler's Astronomy Department in 2000, planning began for astronomy courses in Spanish.

Challenges

The Adler's main way of advertising its course program is through literature sent to its members (e.g, course program brochure, members magazine, electronic newsletter). Unfortunately, the Spanish-speaking community is under-represented in Adler's membership. Members' data do not include information about members' native language, therefore we counted the number of members with Spanish last names to approximate the number of Spanish-speaking members. This number accounts for less than 3% of the Adler membership.

To advertise our first Spanish language course, *Espectáculo Celeste*: Astronomía a Simple Vista, we had to seek new ways of promoting the course. We placed announcements in several electronic bulletins, including one accessible to Chicago Public Schools teachers. We also approached local television and radio stations. Salgado participated in several programs including a live TV broadcast for a morning show from the Adler Planetarium. The broadcast with the largest impact was a 2.5-minute news segment produced by Noticias Univision. It exclusively featured the course as an activity for families to enjoy. After this broadcast the number of course participants registered quickly rose from two to 22.

The Course

Espectáculo Celeste: Astronomía a Simple Vista, a naked eye astronomy course designed for the general audience, examined celestial phenomena using Adler's historic Sky Theater and its Zeiss projector. It was offered in 8 two-hour sessions on March 2001 for a cost of $45 for Adler members and $50 for non-members. This was the first Spanish language course ever offered by a Chicago museum other than the Mexican Fine Arts Center Museum (MFACM). The students not only enjoyed learning about celestial phenomena, but also showed a deep gratitude towards the Adler and its efforts in developing a course program in Spanish.

Other Education Activities in Spanish

Adler outreach activities in Spanish have had markedly varying levels of success correlating to the amount of media coverage (television, in particular) each activity has received. Planet Week, a series of observing nights in Spring 2001, was very successful in bringing Spanish-speaking visitors to the Adler. Many of these visitors learned about the event by watching segments of a video news release (VNR) about Planet Week in Noticiero Telemundo. The VNR was produced by the Adler and featured Salgado explaining where and when to observe the five planets observable to the naked eye. It also invited them to join the Adler during the observing nights.

Other events that news producers decided not to cover and were eventually cancelled included a second course, El Sol, la Tierra y Luna, and a lecture by Daniel Altschuler based on his book, *Hijos de las Estrellas*. The English version of the latter event took place where the author presented the English edition of his book, *Children of the Stars*.

Future Plans and Concluding Remarks

The success of Adler educational events in Spanish has so far depended on the coverage received by local television stations. Since this coverage has been very inconsistent we have started to develop ways of increasing the representation of the local Spanish-speaking community in our membership. That would make the current advertising of educational programs effective in reaching our target audience.

We are currently developing a partnership with the MFACM centered around collaborations in member services, including reciprocal member benefits for admission, gift shop discounts, special programs, etc. Through such a partnership the Adler would build its audience and reinforce its position as the Chicago science museum most connected to the Latino community.

NASA Office of Space Science Education and Public Outreach Conference 2002
ASP Conference Series, Vol. 319, 2004
Narasimhan, Beck-Winchatz, Hawkins & Runyon

Spacelink: Providing Updates in Space Science and Astronomy

Flavio J. Mendez and Jim P. O'Leary

Maryland Science Center, 601 Light Street, Baltimore, MD 21230, mendez@mdsci.org

Introduction

How do science centers, which have traditionally presented basic scientific information to visitors, stay current and interpret "cutting-edge" research and explorations? This is a challenge faced by the Maryland Science Center – indeed many science centers – to allow visitors to experience the immediacy and excitement of scientific discovery. Because of the enormous rate of creation of new knowledge, science centers can no longer just present basic textbook science. New faster technologies and digital data-driven visualizations are helping to drive these discoveries and breakthroughs. Part media center, part discovery room, and part newsroom, *SpaceLink* brings its visitors the "latest and greatest" in space science and astronomy.

Description

SpaceLink is a 1,500 square feet high-tech update center, which is part of our spectacular *OuterSpacePlace* exhibit that includes the Hubble Space Telescope National Visitor Center. Called by some a "Sports Bar for Space Science and Astronomy," *SpaceLink* is an event rich environment, offering hands-on activities, and "live" events on a regular basis.

SpaceLink is a science update center that explores the "latest and greatest" breakthroughs in solar system exploration, the search for origins, the sun-Earth connections, and the structure and evolution of the universe, among other themes. For general visitors, such as families, the exhibit provides experiences in robotics, remotely operated vehicles, the International Space Station (ISS), space-rich web sites, live NASA TV and space-related careers. Credited professional development for teachers is provided in *SpaceLink* through monthly *Teachers' Thursdays*, and as a professional development site for pre-service teachers. During school group sessions, *SpaceLink* becomes an incredible "classroom of the future" with expert facilitators using resources beyond that of any school. On special mission milestones, *SpaceLink* offers *Live Events* highlighting a specific space science mission or anniversary. The common denominator for all *Live Events* is the presence of a local scientist or expert on the topic celebrated.

Special programs developed in collaboration with Baltimore's FOX-45 TV and designed to educate the non-visiting public include *Mars Weather*, *Space Weather Center* and *Faces of Hubble*, among others. Each TV story included multiple segments, each reaching an estimated 100,000 viewers.

Since opening in April 1999, over 250,000 people have visited *SpaceLink*. Eighteen *Live Events* have been celebrated, such as the total solar eclipses in 1999 and 2001 and a live conversation with ISS astronauts in 2001. Over 400 teachers have participated in the *Teachers' Tuesdays* program. All these activities include the volunteer participation of over 86 scientists from institutions such as the Johns Hopkins University Applied Physics Laboratory, Goddard Space Flight Center, Space Telescope Science Institute, NASA Headquarters, Johnson Space Center, University of Maryland, Towson University, Jet Propulsion Laboratory, and ILC Dover, Inc.

Our two-year experience has proven hugely successful and has provided a national model for other museums now planning similar experiences. Large video screens, Internet-capable computers, exhibitry themed to current issues in space and astronomy, and experienced staff combine to interpret the latest findings. Because of the success of *SpaceLink*, other "*Links*" are now being developed at the Maryland Science Center. *BodyLink*, which will explore current health and human body news, will debut in Spring 2002, and has been funded by the National Institutes of Health under a $1.6 million Science Education Partnership Award. *TerraLink*, opening sometime in 2003, will interpret the latest news and discoveries in Earth sciences and paleontology for our visitors, students, and teachers.

The Small Planetarium: Its Nature, Mission, and Needs

Jeanne E. Bishop

Westlake Schools Planetarium, 24525 Hilliard Road, Westlake, OH 44145, JeanBishop@aol.com

Small planetariums appeared in the 1940's. These were made by individuals, including Richard Emmons of Canton and William Schultz of the Cranbrook Science Center, and those built and then manufactured by Armand Spitz and the Spitz corporation, beginning with the Model A in 1946. About 50 Model A's were manufactured prior to 1951, with Spitz Model A-1's and A-2's developed shortly thereafter.

With the advent of NASA, the shock to the US created by the Soviet Sputnik, and subsequent events leading to a series of manned lunar trips in the Apollo Space Program, the small planetarium's popularity grew rapidly. Spitz, homebuilt, and toy planetarium projectors could be found in many places.

These smaller planetariums, along with the major planetariums, brought excitement and knowledge to a public that did not know very much about astronomy. About 25 years ago the author concluded that much of the fault of decline in astronomy taught in the schools in the first half of the twentieth century was due to the failings of a very influential committee that met in 1892, the Committee of Ten (Bishop 1979a).

Today, we find a variety of manufactured small projectors in facilities with 20-60 seats, in the United States and in other countries. Major producers and models include Spitz Space Systems (A3P, A4, and Nova), Viewlex/(Apollo), Goto (Mercury, Eros) and Learning Technologies (Starlab portable). The portable planetarium has become a major type of small planetarium in the last twenty years (IPS 2000).

At least 1000 small planetariums, both fixed and portable, currently are operating in the US (IPS 2000). A variety of surveys have shown that many of these planetariums each present programs and lessons to thousands of people yearly. It appears likely that between one and ten million children and adults view the artificial sky in small planetariums each year, an opportunity for considerable impact.

All states have small planetariums, although more are found in the Eastern and Midwestern states than elsewhere. Planetariums are organized by US region, with planetarians of both small and large facilities meeting together to address common problems and share ideas. In the Great Lakes Planetarium Association (GLPA), the home organization of the author, the majority of planetariums are small permanent facilities. No differentiation for planetarium size is made in the International Planetarium Society, although some concurrent sessions treat different production issues for large and small facilities.

Most small planetariums are found in public schools; more in high schools and middle schools than elementary schools. Universities, museums, and libraries also operate many of the small planetariums. Some museums and large school districts and school service networks (e.g., the BOCES in New York State) own more than one portable planetarium (IPS 2000).

Equipment in small planetariums varies greatly. In some facilities only the planetarium projector is available, while a minority of others have computer-controlled large-screen video, a laser disk player, and many special effect projectors. Most permanent facilities have 1-3 slide projectors, some type of sound system, and several purchased auxiliary projectors. Talented directors and teachers in small facilities frequently make their own special-effect projectors at small cost. Background preparation of those operating small planetariums is varied, as in large planetariums. However, in public schools all teacher-directors necessarily have a minimum of a B.A. and many of them have advanced degrees. Also, many have learned a great deal through self-study and conferences. Just because equipment is less than that found in large planetariums, one should not conclude that knowledge and ability of the planetarian in the small facility is less.

The educational value of all planetariums is extensive (Bishop 1979b). The specific goals of small planetariums are determined by the setting. In K-12 schools, museums, and universities, the majority of planetarium visits are by students in grades 1-6. But a particular library or school may concentrate on preschool visits, and a college or university may present programs only for its students. A number of small school and university planetariums have special programs for the public, beyond its school curriculum. With its public programs, the small planetarium serves some of the same population as the major planetariums. Presentations have attractive topic names. (Likewise, the major planetariums usually have programs for schools.) Since most small planetariums do not have a lot of special equipment, the public programs they offer must be prepared to make best use of minimal equipment.

An example of the 2001-02 curriculum in a small planetarium can be seen from that found in the Westlake Public Schools, the author's facility. The planetarium is located in an intermediate school (grades 5-6). Programs were developed with teachers to go with classroom work. Many of the lessons require activity by the students. The author's high school astronomy classes are bussed twice a week to the planetarium for lessons about stars and constellations, earth motions, the celestial sphere, the sky from different latitudes, lunar motions, planet motions, stars and the Milky Way, and the sky as seen by different cultures. Kindergarten children attend for a special introduction-to-the-sky program. First graders participate in interactive programs, *Seasons and Weather*, *Indian Sky Ideas*, and *The Big Dipper Sky Clock*. Second graders participate in a program about use of the sky to escape from slavery, *Follow the Drinking Gourd*, and they write creatively in a rainforest night environment or a Canadian woods night environment (using appropriate background sounds). Third graders study the planets and their motions and positions and participate in simulated problems they would encounter on space trips. Fourth graders learn about the earth as a planet and observe motions in the sky that result from earth motions.

Fifth and sixth grade students, who are in the building where the Westlake planetarium is located, can attend many different programs. Most fifth grade programs are correlated with a very extensive astronomy unit in science: *The*

Northern Stars, The Autumn Evening Sky, The Winter Evening Sky, The Spring Evening Sky, Moon Phases, Eclipses and Tides, Seeing the International Space Station, and *Native American Uses of the Sky: Cycles and Ceremonies.* Social studies-related programs predominate for sixth grade: *Use of the Sky in the Age of Exploration, The Egyptian Sky, The Babylonian Sky, The Greek-Roman Sky, The Medieval Sky, Light and Color Activities, Astronomy Review for Sixth Grade Proficiency Tests, Writing Poetry Beneath the Stars, Music Appreciation: Music About the Sky Heard Beneath the Sky.*

The author teaches astronomy and geology for half a day at Westlake High School besides presenting all planetarium lessons. Also, the author gives topical programs on two weekday evenings each month, and trains Science Olympiad teams who are preparing for astronomy competitions at all levels in the planetarium. Occasional special programs are given after school for scouts and other groups.

Budgets for school and small museum planetariums are handled differently from those in most large facilities. Staff usually consists of one person, one part-time person, or a couple of part-time persons. Small planetariums normally do not need to "pay their own way," because professional salary is listed under other headings. But the amount of money budgeted for operation often is painfully small. The author has an annual budget of $500/year.

Two years ago a group of directors of GLPA from small planetariums were invited by Bernhard Beck-Winchatz of the DePaul University NASA/OSS Broker/Facilitator to begin offering suggestions about how NASA/OSS might serve the small planetarium population. There were fruitful teleconferences, which let staff from NASA/OSS education forums and Broker/Faciltators understand the nature of the needs of small planetariums. Dr. Beck-Winchatz spoke at GLPA conferences and provided opportunities for all conference participants to learn about programs. A grant program, the PLATO grant project, was organized, which gave $1000 for worthy astronomy/planetarium projects by GLPA members in small planetariums. Some results of the PLATO grants, as well as Chuck Bueter's IDEAS grant, can be found on web sites hosted by DePaul University's Broker/Facilitator (Bueter 2001; Bishop 2001).

Many directors of small planetariums have voiced the desirability of materials from which they can construct their own programs. Some very much appreciate complete interactive programs geared to the limitations in equipment in the small planetarium, like *The Explorers,* produced by the Bishop Planetarium in Honolulu. A NASA IDEAS grant funded the production of a planetarium show by GLPA members for GLPA members. Titled *The Stargazer,* it features astronomer Dr. James Kahler sharing his passion for astronomy and events which influenced him. Great care was made to make the program appropriate for use in small planetariums; one writer-producer is from a small planetarium.

This fall GLPA will continue its liaison with NASA/OSS by administering a survey designed by a GLPA committee on how NASA/OSS might best serve the planetarium community. Since a majority of GLPA planetariums are small planetariums, it is expected that the results of this survey will reveal much about what directors of small planetariums would most value from NASA/OSS. Those who would like to know the results of this survey should contact the author, who is editor of the survey. Results should be available by December 1, 2002.

References

Bishop, J. E. 1979a, S&T, 57, 212
Bishop, J. E. 1979b, IPS Planetarian, 8-1, 7
Bishop, J. E. 2001
    http://analyzer.depaul.edu/jbishop/
Bueter, C. 2001
    http://analyzer.depaul.edu/paperplate/
International Planetarium Society 2000, The IPS Directory of the World's Planetariums 1999/2000 Edition (Prince Frederick, MD: International Planetarium Society)

Journey through the Universe – Taking Underserved Communities to the Frontier

J. Goldstein, M. Bobrowsky, T. Livengood, S. Smith, and B. Riddle

Challenger Center for Space Science Education, 1250 North Pitt Street, Alexandria, VA 22314, jgoldstein@challenger.org

In 1999 Challenger Center for Space Science Education launched *Journey through the Universe* – an initiative to establish a national network of underserved communities committed to sustainable, community-wide science, mathematics and technology education. Funded by grants from NASA's Human Exploration and Development of Space Enterprise and Office of Space Science, *Journey through the Universe* uses human space flight and space sciences to engage entire communities within a multidisciplinary context.

The Partnership

Challenger Center provides diverse national resources to *Journey* sites, including: local programming for students and families, K-12 curriculum support materials, K-12 educator training, and ongoing support for Challenger Center staff scientists and educators in both content and pedagogical approaches in the classroom. The communities integrate these resources into their existing science, math and technology education programming in both formal and informal science education venues. The programming resulting from this partnership is meant to reflect the strengths and capabilities of the community, and provide access to resources that would otherwise be unavailable.

Philosophy of Approach

The program is meant to gradually build a local infrastructure – a Local Team – that can provide self-sustaining programming using national resources regularly piped in from afar. This team reflects the diversity of the community, including K-12 school districts, informal science education organizations, civic and business groups, colleges and universities, and research organizations. Sustainability is viewed as a key to success, and a deep commitment to assessment provides for ongoing evolution of every aspect of the program.

Programming provided to each community includes a week-long celebration of learning conducted by a National Team of researchers and engineers reflecting organizations from across the human space flight and space science communities. During the week, training is provided for up to 350 K-12 educators, 5000-8000 K-12 students are visited in the classroom, and 2-4 family science nights are held, each for 300-1000 parents and their children.

The Current Network

Journey through the Universe hopes to reach communities with limited human space flight and space science education resources, or those where resources are not utilized community-wide. This includes, but is not limited to, communities in rural settings, as well and low-income populations in urban settings. Currently eight communities are in the network: Nogales, AZ, Tuskegee, AL, Washington, DC, Altamont, KS, Marquette, MI, Broken Arrow, OK, Muncie, IN, and Moscow, ID.

StarDate/Universo: Astronomy for the Masses

Sandi Preston and Mary K. Hemenway

University of Texas at Austin, McDonald Observatory, 2609 University, Suite 3.118, Austin, TX 78712, sandi@astro.as.utexas.edu

Each week, *StarDate*, the daily 2-minute astronomy radio program produced by the University of Texas McDonald Observatory, reaches 3.7 million people and *Universo*, the Spanish-language version, reaches 1.5 million people.

Additionally, a German-language version is produced and airs throughout Germany. Plans for a Southern Hemisphere version are underway. *StarDate* and *Universo* also offer a classroom component that is used by 750 teachers nationally, reaching over 750,000 students. The *StarDate* magazine has a circulation of 11,000. National distribution of *StarDate* is made possible by American Electric and Power and for *Universo* by Southwestern Bell Foundation. Local distribution for *StarDate* is made possible by small fees charged to radio stations. Radio listeners are among the 130,000 visitors that come to McDonald Observatory's remote site in west Texas each year. A new visitor center will open at the Observatory in December 2001. The new center is designed to serve 250,000 visitors a year and to be a hub for K-12 astronomy programs for teachers and students in Texas. Educational programs will align with both state and national science standards; these programs will serve a national audience though our web sites (http://stardate.org/ and http://radiouniverso.org/) and publications.

Session on Scientists' Participation in Education and Public Outreach

Context for the Panel on Scientists' Participation in Education and Public Putreach

The conference ended with a plenary panel session on Scientists' Participation in Education and Public Outreach. While the conference organizers identified this as one of the major themes that should be integral to and observable throughout all sessions, they also felt that its importance to the OSS program warranted a session devoted to this theme and that it would be an appropriate closing session for the conference.

Carl Pilcher, Senior Scientist for Astrobiology, Astronomy and Physics Division at NASA Headquarters, moderated the session. The panelists were Lucy Fortson, Department of Astronomy, Adler Planetarium and Astronomy Museum, and Department of Astronomy and Astrophysics, University of Chicago; Andrew Fraknoi, Director, Project ASTRO, Astronomical Society of the Pacific; Ramon Lopez, C. Sharp Cook Distinguished Professor of Physics, University of Texas at El Paso; and Charles McGruder, William McCormick Professor of Physics, Western Kentucky University.

The panelists were asked to give a brief description of their own experiences and their perspectives on key issues raised during the conference. This was followed by a discussion among the panelists and some time for questions from the conference participants.

For this volume, the editors asked Ramon Lopez to provide an overview of the topic, and he has done so in a lucid paper that assesses the OSS effort in establishing a community of space science educators. Dr. Lopez makes two key recommendations:

- The community needs to make better connections with the larger national science education reform effort.

- The community needs to develop a research literature in teaching and learning space science.

A paper by Andrew Fraknoi complements that of Lopez. Fraknoi describes the emergence of a new profession of space science educators and public outreach specialists, and the need for professional development opportunities for this community. He also lays out some provocative thoughts concerning the future of the OSS effort. Charles McGruder addresses one of the main threads of the conference – the response by the science community to diversity, including both outreach to underserved groups and workforce needs in the areas of science, mathematics, and technology. Lucy Fortson offers some personal reflections on the importance of involving research scientists in education and outreach, based on her own experiences over the past six years in a joint position at the University of Chicago and the Adler Planetarium.

At the end of the final session, participants had the opportunity to submit in writing brief responses to the following questions:

1. What are the one or two key issues that you heard at the conference that you considered most important?

2. Are there issues that are important to you that you did not hear discussed?

The results of this survey, as well as of the conference evaluation, are summarized in the letter from the editors on page xviii.

Space Science Education: The Emergence of a Professional Community

Ramon E. Lopez

Department of Physics, University of Texas at El Paso, El Paso, TX 79968, relopez@utep.edu

Abstract. The NASA Office of Space Science has set out on a bold venture in education by creating a network of space science educators, many of them with backgrounds in science. This professional space science education community has had considerable success in establishing itself, and it is beginning to make significant contributions to science education. However, in order to have a lasting impact, the community needs to connect better to the existing national effort in science education reform. The space science education community also needs to move toward a more scholarly approach to science education by developing a discipline-based research literature in teaching and learning.

Establishing Space Science Education as a Community

Since the late 90's, NASA's Office of Space Science has been promoting an ambitious program to involve the space science community in science education. To this end, OSS funded the creation of a set of "Forums" and "Broker/Facilitators" (Rosendhal et al. 2003). Together, the Forums and Brokers constitute the "Support Network." NASA/OSS also initiated a Minority University Initiative (MUI) to increase diversity in space science. Some of the MUI projects are focused on education (e.g., Lopez 2002). In addition, NASA missions devote considerable funds to education and public outreach efforts, known as E/PO. Taken together, members of the Support Network, the mission E/PO programs, and some of the MUI projects, along with a handful of other astronomers and space scientists working in education, have grown into a community of space science education professionals, many of whom have formal training in space science, and some of whom continue to be active scientific researchers. Overall, this community has proved to be very active, and it is having a growing impact.

In formal education, the community has focused on curriculum development and professional development of teachers. Magnetism, electricity, and the nature of the solar system are all topics that every student encounters in middle and/or high school. Collaboration between the NASA-funded Sun-Earth Connection Education Forum and the Lawrence Hall of Science (LHS) in Berkeley produced *The Real Reasons for the Seasons*, a middle-school curriculum package that teaches about seasons and the phases of the Moon. This excellent product is now in use by many school systems around the country since most middle schools include the solar system in the curriculum. Another education product, *Touch the Universe – A NASA Braille Book of Astronomy* (see Grice et al. on page 301), allows visually-impaired students to share in Hubble discoveries. Other education products are showing an increasing level of sophistication among the developers and represent a considerable improvement on previous NASA education materials.

Professional development activities have taken place all over the country, providing teachers with tools and training to make science exciting in the classroom. One particularly innovative professional development project has been the Space Science Institute workshops for scientists and E/PO professionals. This workshop has played a significant role in helping to develop a cadre of space science educators. Other successes have included a variety informal science projects, such as *Marsquest* and *Windows to the Universe*. An independent evaluation of the NASA/OSS effort, based on extensive interviews with a wide range of OSS E/PO providers, customers and others, has documented areas where real progress has been made (Cohen, Gutbezahl, & Griffith 2001), some of which have been referred to here. In general, NASA Education efforts of all kinds are improving because of the increasing knowledge-base of people in the space science education community.

This emerging space science education community has also begun to think of itself as a community, as evidenced by a presence at major scientific meetings. For several years now, the number of education sessions at American Geophysical Union (AGU) meetings has grown. During the Fall 1998 meeting there were no education sessions, but the Fall 2002 meeting boasted 12 education sessions, of which 8 were organized by members of the space science community. Similarly, at the Committee on Space Research (COSPAR) meeting in Houston in 2002, there were two space science education sessions. This was the first time that such special sessions were organized at a COSPAR conference, and an education session is scheduled for the 2004 COSPAR meeting in Paris. Most importantly, the NASA E/PO conference in Chicago, which is the motivation for this paper, drew close to 300 participants. And at that conference a new, all-electronic, peer-reviewed journal, the Astronomy Education Review (http://aer.noao.edu/) was announced. The ingredients for a truly professional Space Science Education community – people, funding, venues and publications for the dissemination of knowledge – are all in place.

Connecting to the Reform Movement

The science education efforts by the space science community are part of a larger, national science education reform effort. Among the key documents for the reform movement are the *Benchmarks for Science Literacy* (AAAS 1993) and the *National Science Education Standards* (NRC 1996). Together, these documents are referred to as "the standards" and science education that tries to implement the guidelines and philosophy in the documents is referred to as "standards-based." States have written their own standards, based (more or less) on the national standards, and these state standards are the basis for assessments of student progress as mandated by the "No Child Left Behind" legislation.

While space science educators have been cognizant of standards, and education products such as those described in the section above are standards-based, there is still a need to better connect the space science E/PO effort to other large-scale education efforts. As of yet there has been little connection between space science education and the NSF systemic initiatives or the new Math/Science Partnerships. Models for systemic change exist, and hundreds of schools systems are implementing district-wide change as described in *Science for all Children* (NRC 1997). Many of these school systems have received Local Systemic Change grants from the NSF, or are involved with the various NSF State, Urban, or Rural systemic initiatives. The space science community has in general not yet established effective partnerships with the school district reform movement, and organizations at the heart of this movement, like the National Science Resources Center, are not on the space science education radar screen.

One of the challenges of working with systemic change is that space science education must find effective ways in which to partner with school systems that have a much broader agenda than just space science. The recently released *The Sun to the Earth – and Beyond: A Decadal Research Strategy in Solar and Space Physics* (NRC 2002) calls for space science to make a contribution at the middle and high school levels, where appropriate space science (and physics) topics are found in the standards. Unfortunately, many well-meaning scientists would like to put space science in grades where it is not developmentally appropriate, particularly in elementary grades. For example, despite the cognitive and conceptual difficulties that young students have in creating a coherent conceptual framework for a spherical Earth (e.g., Vosniadou & Brewer 1992), one often finds space scientists pushing for the inclusion of the heliocentric model of the solar system in elementary grades. The space science education community needs to insure that the valuable time contributed by space scientists to education matches the needs of the education system.

One area in particular where the space science education community, as well as individual research scientists, can make a huge contribution in is the area of professional development for teachers. The major work in systemic change is to provide adequate professional development to teachers who are trying to move towards a more standards-based instruction (NSF 1997). This process often involves learning to use new instructional materials (containing new scientific content), along with new models for creating inquiry-centered learning environments in classrooms. Scientists can provide a crucial support element in that process (NRC 1997). The professional development provided to teachers to date in space science has had a good start, but there is a need to base more of what is done on the research literature on professional development (e.g., Loucks-Horsley et al. 1998; Mestre 1994; Shymansky 1992). By combining a solid knowledge of best practices in professional development and access to a scientific community, it is possible to make substantial, lasting contributions to systemic change in science education (e.g., Lopez & Schultz 2001).

Research in Physics Education: A Model for Space Science

The issue of becoming better grounded in research in professional development raises another, more fundamental issue – what about a research base in the teaching and learning of space science? While some research in this area does exist (e.g., Bailey et al. 2004), there is great need for a much more extensive research base in space science education. This research should be solidly grounded in a current understanding of cognitive science (e.g., Bransford, Brown, & Cocking 1999; Donovan, Bransford, & Pellegrino 1999). The growing number of space science education professionals should take on the mission of developing a discipline-based cognitive science that can serve as a research base, so moving space science education into the domain of a scholarly field of inquiry that can, and should, be housed in space science research institutions.

A model for such a discipline-based community exists in physics (McDermott 2001; Redish 1994). In fact, physicists contributed very early to cognitive science (e.g., Karplus 1997), and in some important areas, such as Expert-Novice differences, the study of problem-solving in physics has been a crucial part of the research base (see Chapter 2 in *How People Learn*; Bransford, Brown, & Cocking 1999). Currently, there are many physics education research groups at universities around the country, and a considerable body of literature has been developed (McDermott & Redish 1999). It is important to note that these groups are groups of scientists trained as physicists, but whose research area is physics education. They are housed in physics departments and they are, and should be, evaluated by the same standards as other fields of physics: publications, grants, invited and contributed papers at national and international meetings, and students graduated. In fact, in June 1999, the Council of the American Physical Society approved a resolution stating that physics education research is a legitimate subfield of physics, and that it should be conducted by physicists in physics departments. Similarly, space science education should be seen as an integral part of the overall space science enterprise.

As of yet the space science education community as a whole has not made the transition to a scholarly community, even though (as pointed out above) all of the prerequisites are in place. Many proposals and publications in space science education have few, if any, references to the research base, and, with a few exceptions (e.g., Zelik & Bisard 2000; Zelik et al. 1998), there are relatively few papers that could be called research papers. However, this is beginning to change. As space science education develops standard assessment instruments (e.g., Hufnagel 2002), opportunities will develop to do research on student thinking, understand what conceptual difficulties students have, and develop instructional experiences that allow students to more easily grasp the concepts. This initial work has been primarily in the content areas served by the introductory astronomy course (which has a huge enrollment – see Fraknoi (2002)), just as the initial forays in physics education research were in elementary mechanics (e.g., McDermott 2001) with its huge enrollment of engineering students.

Over time a more extensive research base in space science education (including astronomy education) will develop. It is likely that some of this research will center on the interpretation of imagery and visualization (e.g., Lopez & Hamed 2004), given the importance of these things to space science. There are also very likely areas of convergence with physics education research, such as student understanding of electromagnetic waves (e.g., Ambrose et al. 1999), or magnetism and magnetic fields. In fact, one of the recommendations of *The Sun to the Earth – and Beyond: A Decadal Research Strategy in Solar and Space Physics* (NRC 2002) is that several groups be established that would (in part) conduct active programs in space science education research.

In the short run, however, it is crucial that space science educators set a high standard for the work that they do. Proposals and papers that do not appropriately reference the literature should be rejected outright. We should strive to create original research and support vehicles like *Astronomy Education Review*. More conferences dedicated to space science education should be organized, along with sessions focusing on space science education research at the AGU and American Association of Physics Teachers (AAPT) meetings. And more papers that could be classified as research should be presented at such venues. In the end, as scientists, we must demand that research in space science education meet the same standards to which other research in space science is held.

Conclusions

The past few years have seen a huge growth in space science education, both in the scope of activity and in the overall effectiveness of those activities. However, these activities are still disconnected to some degree from systemic reform in school systems. The groundwork has been laid for a much more significant involvement in creating lasting change in science education, as called for by the recent NRC Decadal Survey. Models for a deeper involvement in reform exist and as the space science education community becomes more aware of these models it will contribute more to reform. The space science education community also needs to develop a more scholarly approach to its activities, and in particular, it needs to develop a literature on teaching and learning that meets rigorous standards for research. All of the ingredients exist for the creation of a new field of research that can be judged by the same standards that other fields are judged, and NASA/OSS can truly be proud of this development.

Acknowledgments. The author wishes to acknowledge conversations with C. Morrow, C. Narasimham, I. Doxas, E. Prather, and T. Slater. This work was supported by NASA grant NAG5-10407 and was also supported by the Center for Integrated Space Weather Modeling (CISM), funded by the STC Program of the NSF under Agreement Number ATM-0120950.

References

Ambrose, B. S., Heron, P., Vokos, S., & McDermott, L. C. 1999, Am. J. Phys., 67, 891

American Association for the Advancement of Science Project 2061 1993, Benchmarks for Science Literacy (New York: Oxford University Press)

Bailey, J. M., Prather, E. E., & Slater, T. F. 2004, Adv. Space. Res., in press

Bransford, J. D., Brown, A. L., & Cocking, R. R. (editors) 1999, How People Learn: Brain, Mind, Experience, and School (Washington: National Academies Press)

Cohen, S. B., Gutbezahl, J., & Griffith, J. 2001, Office of Space Science Education/Public Outreach January 2000-May 2001 Evaluation Report (http://spacescience.nasa.gov/education/resources/evaluation/NASA_Interim_2001_Report.pdf)

Donovan, M. S., Bransford, J. D., & Pellegrino, J. W. (editors) 1999, How People Learn: Bridging Research and Practice (Washington: National Academies Press)
Fraknoi, A. 2002, Astr. Ed. Rev., 1(1), 121
Hufnagel, B. 2002, Astr. Ed. Rev., 1(1), 47
Karplus, R. 1977, J. Res. in Sci. Ed., 14, 169
Lopez, R. E., & Schultz, T. 2001, Physics Today, 54(9), 44
Lopez, R. E. 2002, Voyages in Education and Public Outreach, 5, 8
Lopez, R. E., & Hamed, K. 2004, J. Atmos. Sol. Terr. Phys., in press
Loucks-Horsley, S., Hewson, P., Love, N.,& Stiles, K. 1998, Designing Professional Development for Teachers of Science and Mathematics (Thousand Oaks, CA: Corwin Press)
McDermott, L. C., & Redish, E. F. 1999, Am. J. Phys., 67, 755
McDermott, L. C. 2001, Am. J. Phys., 69, 1127
Mestre, J. P. 1994, Cognitive Aspects of Learning Science, Chapter 3 in Teacher Enhancement for Elementary and Secondary Science and Mathematics: Status, Issues, and Problems, ed. S. J. Fitzsimmons & L. C. Kerpelman (Cambridge, MA: Abt Associates, Inc.)
National Research Council 1996, National Science Education Standards (Washington: National Academies Press)
National Research Council 1997, Science for All Children: A Guide to Improving Elementary Science Education in Your School District (Washington: National Academies Press)
National Research Council, Solar and Space Physics Survey Committee 2002, The Sun to the Earth – and Beyond: A Decadal Research Strategy in Solar and Space Physics (Washington: National Academies Press)
National Science Foundation, Division of Elementary, Secondary, and Informal Education 1997, Foundations: The Challenge and Promise of K-8 Science Education Reform, NSF 97-76
http://cse.edc.org/pdfs/products/monographs.pdf
Redish, E. 1994, Am. J. Phys., 62, 796
Rosendhal, J., Sakimoto, P., Pertzborn, R., & Cooper, L. 2003, Adv. Space. Res., in press
Shymansky, J. A. 1992, Journal of Science Teacher Education, 3, 53
Vosniadou, S., & Brewer, W. 1992, Cognitive Psychology, 24, 535
Zeilik, M., & Bisard, W. 2000, J. of College Science Teaching, 29, 229
Zeilik, M., Schau, C., & Mattern, N. 1998, The Physics Teacher, 36, 104

Biography

Ramon E. Lopez is Professor of Physics and Space Sciences at the Florida Institute of Technology, and is the Co-Director for Education for the Center for Integrated Space Weather Modeling (CSIM), a Science and Technology Center funded by the National Science Foundation. He is a Fellow of the American

Physical Society and was awarded the 2002 Nicholson Medal for Humanitarian Service to Science. Lopez's current research focuses on magnetospheric storms and substorms, and making detailed quantitative comparisons between the results of global 3-D MHD simulations and observations during actual events. His activities in science education has included service as a member of the National Research Council's Committee on Undergraduate Science Education, on the AGU's Committee on Education and Human Resources, and as a member of the Board of Directors of SACNAS. He is a past Chair of the AGU's Space Physics and Aeronomy's (SPA) Education Committee, and he was Chair of the Panel on Education and Society for the Decadal Survey of Solar and Space Physics. He is the current Vice Chair of the Forum on Education, and will serve as Chair in 2005. As Director of Education and Outreach at the American Physical Society from 1994 through 1999, he was responsible for the Society's education programs, including the Teacher-Scientist Alliance Institute to mobilize scientists in support of systemic reform of science education across the country. Lopez is co-author of a popular book on space weather entitled *Storms from the Sun*, published by Joseph Henry Press in 2002.

Scientists' Participation in Education and Outreach

Andrew Fraknoi

Astronomical Society of the Pacific & Foothill College, 390 Ashton Ave., San Francisco, CA 94112, fraknoi@fhda.edu

I presume one of the reasons I am serving on this panel is that I have the privilege to be the Director of Project ASTRO, a program which partners volunteer astronomers with 4th - 9th grade teachers in their local communities in 12 regional sites around the country. You can read more about this project in a separate paper on page 289 of this volume.

Lessons from Project ASTRO

Many of the key things we have learned from nine years of doing Project ASTRO are in resonance with what you heard from Pinky Nelson, Phil Sadler and others at this meeting:

- The most important thing we can do is to help teachers (and students) to be good learners.

- It is best to do fewer topics in an astronomy course or unit, but to do them more deeply.

- If you are not aware of what students (and their teacher) are currently thinking about a particular topic, it is much harder to move them toward new thinking.

- Vocabulary counts: don't use jargon that your spouse's family would not understand.

- Empower those you work with to continue their exploration after you are no longer there.

An Impressive Beginning

Looking at the achievements of the NASA OSS E/PO Effort, both here at this meeting, and in the new Annual Report, it is impressive to see what has been accomplished. Jeff Rosendhal and all of you have many, many reasons to be proud. I think it is fair to say that no other program in the history of space science education has had as many resources and as great an effect as the one we are celebrating here. It is a bold experiment, and, happily, it is an on-going experiment, still adapting to its audience and its growing capabilities. I am impressed by:

- The breadth and scope of what you are doing.

- The willingness to make changes as you learn more about what works and what doesn't.

- The sincere effort of many parts of the Support Network to rethink the way things are being done (to be as effective as possible).

- The creativity of so many individual scientists and educators, trying to see what might work in the crazy educational environment in which we find ourselves in 21st century America.

Creating a New Profession

Not the least of the effects of your work is that a new cadre of "E/PO Professionals" is being created by this infusion of NASA OSS money. These are people whose entire or main line of work is doing education and outreach. Some have backgrounds in science, some in education, some in both. But it appears from discussions at this meeting that this group is beginning to think of itself as belonging to this new profession – a profession whose numbers were much smaller before your effort began.

Like any new profession, some of us have more enthusiasm than training. Some of us have been involved for a decade or more, but others have just recently joined the ranks. We have no agreed-upon credentialing in our profession, and no agreed-upon curriculum to take or give. So, many of us often just make it up as we go along.

Joining our ranks is very different from joining an established area of science. In E/PO there is no program to help new people get to know past practices and materials, or to learn the literature (which is, in any case, spotty and hard to find.) We hope the new journal, Astronomy Education Review, (discussed in a separate paper in this book) will help with this, but work published – or not published – a while ago may be harder to recover than work being done today. And, perhaps most significant, there is now little pressure on E/PO professionals to learn about work going on in other institutions before beginning their own work. These things are not easy to change, but will be absolutely necessary to change if we want to win greater respect for this new profession.

Can you imagine someone saying: "Well, I've been reading a bit about supernovae and have been thinking about them for a while. The whole field sounds like it needs a little straight-forward thought. So, while I haven't taken any courses in astrophysics and I don't really know the supernova literature, I think I'll do supernova research."

I hope that one of the ways we can help this situation is to set up many further opportunities for professional development for new people entering our field, so that they do not have to re-invent the wheel as they begin. On the other hand, the good news is that if anything is going to cause the academic space science culture in our universities to value education more, it is the work you are doing. You've got the ingredients that are needed – the involvement of scientists, the encouragement of a wide range of activities, and putting serious money where there were only words before.

Some Provocative Thoughts for the Future

So what about the future of your experiment? Here are a few provocative thoughts that might serve to spark some useful discussion at this meeting and as we return to our programs:

P-R Versus Education

In much of what you do, there is a tension between public relations and education. In public relations, we ask what our paying client needs to get done or sold or publicized, and then devise ways to get the consumer to accept the message or product. In education, we ask what the consumer needs, and try to get the paying client to meet that need. These two perspectives can clearly come into conflict. Those of you working for missions – look deep in the mirror and ask yourself: Would you give up your mission or project identification to do more effective education? Until some people have the courage to say yes to that question, the NASA OSS E/PO effort will remain more parochial and beside the point than it needs to be.

New Projects Versus Existing Ones

There is still too much emphasis in your system on creating something NEW and very specific to make your mark. Wouldn't it be wonderful if educators in your system got as many kudos for bringing good existing projects and materials to new audiences as they get for creating new materials or programs for the same small group of interested teachers and members of the public?

NASA Chauvinism

Although this is slowly changing, there is in many parts of your system a NASA-centrism or NASA-chauvinism that really doesn't seem necessary for a maturing organization. What I mean by NASA-chauvinism is that if NASA didn't do it, it really doesn't count or matter! This is still too often the case for web sites, written materials, or databases you produce. Ironically, you even exclude materials from others – including nonprofit organizations – that directly use NASA images or data. NSF-sponsored materials have no problems in mentioning the work of other organizations. (One presumes there is proper language that makes such recommendations those of the authors and not official NASA policy.)

Last-Minute Planning

Pinky Nelson reminded us that a typical NASA mission is not cobbled together in a garage overnight, but is planned and carefully built over a long time. Yet all too often the E/PO plan for a mission or project is indeed just thrown together at the last minute from whatever materials come conveniently to hand. There is often very little attempt to determine what is most needed or what could be the most effective. Nor is there much of an effort to seek the best partners for the work to be done. Instead, the work is either automatically farmed out to the group that did work in that subject area the last time or thrown at some local institution that is willing to make plans in a hurry. This, I gather, is as frustrating to the Forums and the Brokers as it is to outside groups hoping to work with NASA or looking to NASA, with its abundant resources, to take a real leadership role in education. It may even be frustrating to those who eventually have to do this work that was planned too quickly and with too little input. Clearly, we need to find ways to help E/PO proposers do this better – to have a way to reach out to a larger group of E/PO material and service providers.

Coordination Among NASA and Other Programs and Materials

As I talk with people at this conference, and those doing astronomy education around the country, it is clear that almost no one has a good grasp of the variety of programs and materials that NASA offers. Even veteran NASA education professionals can't keep track of who is doing what in E/PO in OSS and elsewhere. As a result, there is inevitable duplication and re-inventing the wheel. We must think of other ways that the entire system can encourage communication and coordination – internally and with the outside world.

Your resource database goes some way toward meeting this goal (and the new journal may help), but I believe that more resources should be brought to bear on organizing what has been done and is being done, rather than on creating even more separate programs and activities. Before a mission or a project prepares an E/PO proposal, wouldn't it be great if they could ask an E/PO guru or consult a really good information base of past practices: What's already been done (at NASA and elsewhere) in my arena and topic, and what is still really needed?

And, once materials have been produced (and tested), it would be good to find ways to get them more effectively in the hands of end users. Right now, the system for getting materials to those who need or want them can only be described as quantum mechanical. You never know where materials will pop up or who will get one and who won't.

More Evaluation

The NSF requires both formative evaluation (checking things as you develop them) and summative evaluation (checking things after you've developed them). Admittedly, good evaluation is time consuming and expensive. Yet testing one's materials and approaches under the supervision of people who really know education can improve them tremendously. It would be great if more evaluation help could be available throughout your system and project funds included both a requirement for and assistance with evaluation.

More Training

As I discussed earlier, both scientists and educators working for or with NASA could often benefit from further training in E/PO best practices, existing projects, lessons already learned, etc. While Cheri Morrow's 4-day seminars are excellent, they reach much too small a fraction of those in your network. Other, shorter, more local ways could be found to make sure that everyone doing E/PO is putting his or her talents to best use. There are people and organizations around the country that could help those within your network do such training.

Conclusion

In planning for the future, we must also understand that we may be overtaken by events beyond our control. We have heard, for example, how national and state tests may really reduce the amount of astronomy and space science being taught in this country. With an emphasis on basics, with a focus on biology, chemistry, and physics, it may be that "no student left behind" turns into "no astronomy left" in our nation's schools. The ongoing shortage of good science and math teachers is expected to get worse in the coming decade, and it is not clear whether our universities and colleges are really prepared to meet the growing need. Who then will be there to teach (or be allowed to teach) that module on the anisotropies in the cosmic microwave background that we have so lovingly developed?

It would be so good to put our "mission ego's" aside and work together (inside and outside NASA) on the much greater dangers and needs in education that face us all.

Selected Books

Percy, John, ed. *Astronomy Education: Current Developments, Future Coordination.* 1996, Astronomical Society of the Pacific Conference Series, Vol. 89. A useful and comprehensive introduction to current issues and projects in astronomy education at many levels; with good reviews, papers on individual program, and resource guides as of 1996. Order through the ASP Catalog at 1-800-335-2624.

Fraknoi, Andrew, et al, eds. *The Universe at Your Fingertips.* 1995, Astronomical Society of the Pacific. Features 87 exemplary astronomy and space science activities, selected from many sources, as well as resource guides and articles on effective teaching techniques.

Fraknoi, Andrew & Schatz, Dennis, eds. *More Universe at Your Fingertips.* 2000, Astronomical Society of the Pacific. Another collection of selected activities from groups around the U.S. and Canada, with additional resource guides and articles on effective K-12 space science education.

Fraknoi, Andrew & Schatz, Dennis. *El Universo a sus Pies.* 2002, Astronomical Society of the Pacific. Spanish translation of the favorite activities from the above two books.

Selected Articles

Fraknoi, A. "The State of Astronomy Education in the U.S." in Percy, John, ed. Astronomy Education: Current Developments, Future Coordination. 1996, Astronomical Society of the Pacific Conference Series, Vol. 89. Also available on the Web at:
http://www.astrosociety.org/education/resources/useduc.html

Adams, J. and Slater, T. "Astronomy in the National Science Education Standards," Journal of Geoscience Education, vol. 48 (no. 1), pp. 39-45 (2000).

Selected Web Sites

Astronomical Society of the Pacific Education Web Site

www.astrosociety.org/education.html This site has dozens of resource guides, complete back issues of the "Universe in the Classroom" newsletter on teaching astronomy in grades 3-12, information on Project ASTRO and Family ASTRO, sample hands-on activities, a solar system treasure hunt game, and much more.

Astronomy Education Review aer.noao.edu

A new electronic journal/magazine on astronomy and space science education and outreach, with research papers, articles, news, opinion, announcements and the opportunity to network with others doing E/PO.

Biography

ANDREW FRAKNOI is Chair of the Astronomy Program at Foothill College, and the Director of Project ASTRO and Family ASTRO at the Astronomical Society of the Pacific. He served as the Society's Executive Director for 14 years, and was the founding editor of The Universe in the Classroom, the ASP's newsletter on teaching astronomy in grades 3-12. For the last 21 years he has organized an annual, national workshop on astronomy and space science for teachers with the same title. He is the lead author on *Voyages through the Universe* (1997, 2000, Harcourt), one of the leading introductory astronomy textbooks and has written or edited over a dozen other books on astronomy and astronomy education. He has received the Annenberg Foundation Prize of the American Astronomical Society and the Klumpke-Roberts Prize of the ASP for his contributions to astronomy education. Asteroid 4859 was named Asteroid Fraknoi by the IAU to recognize his work in education and popularizing science.

Impressions from the OSS Conference on Education and Outreach

Charles H. McGruder

Western Kentucky University, 1 Big Red Way, Bowling Green, KY 42101, mcgruder@wku.edu

There are two major points I obtained from the OSS conference. First, the impression that OSS is spearheading a revolution that will transform America. What is the background of this impression? The economic well being of America depends primarily upon its innovation in science and technology. Advances in these interdependent areas however, depend upon the production of scientists and engineers. But, the USA finds itself in the situation of importing technical manpower to meet it needs. For instance 55% of the graduate students in physics are foreigners and in engineering it is 60%. If the vast majority of these new scientists and engineers stay in the USA for their productive years, then we do not have a problem; but if they do not, and presumably they will not as the economies of their native countries improve, we will not be able to maintain our economic well being. What is required to secure the future of this country is a turnaround in the attitude of young Americans toward science and engineering as a career goal. This is a challenge, which must be met. The reluctance of American youth is so strong that I suspect only the active involvement of scientists themselves with the youth can rectify the situation.

Secondly, the conference made clear the almost total lack of participation of minorities in general and African Americans in particular in the leadership (PIs and Co-PIs) of OSS space missions. But, in a few decades the so-called minorities will be the majority in America. Thus, it appears that the future technological workforce needs of America, a workforce which will be responsible for its future economic well being, can only be met by including minorities. In order to achieve this it requires both a revolution of thought and revolution in action. I believe that OSS E/PO can and should spearhead this transformation.

Biography

Charles H. McGruder is the William McCormick Professor of Physics at Western Kentucky University in Bowling Green, Kentucky. He joined Western Kentucky in 1993 after serving as Professor of Physics at Fisk University for three years. Dr. McGruder received his B.S. in Astronomy from the California Institute of Technology and his Ph.D. in Astronomy from the University of Heidelberg. Following postdoctoral studies at the University of Heidelberg, he spent eleven years in the Department of Physics and Astronomy at the University of Nigeria. Dr. McGruder's research interests are in the areas of extrasolar planets and the optical afterglow of gamma-ray bursts. He pursues these interests through the establishment of a network of robotic imaging telescopes. Dr. McGruder is past president of the National Society of Black Physicists.

The Importance of Involving Research Scientists in Education and Outreach

Lucy F. Fortson

Adler Planetarium and Astronomy Museum and Department of Astronomy and Astrophysics, University of Chicago, 1300 S. Lake Shore Drive, Chicago, Illinois 60605-2403, lfortson@adlernet.org

In this paper, I will discuss issues regarding the involvement of research scientists in Education and Public Outreach (E/PO) efforts. I will start with a discussion of definitions followed by reasons why it is important to involve research scientists in education and outreach. I will then describe characteristics of successful E/PO scientists followed by some of the many challenges faced by scientists who wish to participate in E/PO activities. I will conclude with some potential solutions and a number of questions that underlie the emergence of the new profession of E/PO scientists. This work is allegorical in nature and came about through personal experience and many discussions with colleagues in both the scientific and educational communities. Thus there are few references and many generalizations, and I apologize for this in advance. Nonetheless, given my particular experience over the past six years as a research scientist and professional E/PO scientist working in an astronomy museum, and given the nature of the conference in which a number of these ideas were presented, I hope it is of some value to the fledgling community of E/PO specialists.

Definitions

What does the phrase "Education and Public Outreach" mean? In my experience, "Education" typically means creation of a learning experience in either a formal or informal capacity through engaging in the formal educational system at a local, state or national level. This includes grades K-12, undergraduate and graduate students. For example, museum professionals engage in education in an informal way when school field trips that visit their museum use museum-based resources to enhance the student's learning experience but no sustained interaction between the museum professional and the visiting class takes place. Museum professionals engage in education in a formal way through sustained interaction with students and/or through teacher professional development. Museum resources are used to support this development and may include an experience for students/teachers in the museum or use of materials in the museum and/or classroom that are prepared by museum professionals.

"Public Outreach" means inspiring the public to learn about science through individual or group experiences that are primarily based on informal modes of communication. By definition, a museum engages in public outreach because its displays and programs "reach out" to the public as visitors. Some more subtle examples include work with community centers and after-school programs, media relationships, public lectures and courses and a temporary presence at state fairs, festivals and other community gatherings. Often, there is confusion between the terms "Education and Public Outreach" and "Public Information." Public Information is the factual information that is needed to describe to the public a mission or other scientific project. There can be a significant amount of overlap between Public Information and E/PO but I believe the community would be well served in making clear the distinction, as there are often different goals.

The National Science Foundation (NSF) has recently made an E/PO component imperative for any proposal submitted for scientific research. However, their definition of E/PO is based on the concept of "broader impacts" (NSF 2002a; NSF 2002b), which encourages "outreach" into underrepresented groups and other more classic E/PO activities but can also be satisfied simply, for example, through an innovative graduate training program. NASA takes the broader approach wherein E/PO is an activity that takes the scientist outside his or her normal sphere of influence (NASA 2002). The NSF is also beginning to draw a distinction between efforts that involve the Public Understanding of Science and those that are targeted towards the Public Understanding of Research (Field & Powell 2001). This distinction recognizes the important difference between educating the public about research science and educating them about basic science.

Why Involve Scientists in E/PO

Most research scientists are supported by public dollars. This is one of the most cynical reasons for convincing someone that research scientists are an important component in carrying out E/PO efforts. Nonetheless, this reason is compelling because it explicitly reconnects the scientist with the public in a contract that states, "the more the public understands science, the more money scientists will be given." Essentially, this reason can be thought of as "good PR" for science and scientists. Fortunately there are many more reasons with less hubris involved.

The most obvious reason to involve scientists in E/PO is because the scientist "knows" the science. This expertise is tremendously useful to an E/PO team even if the E/PO effort is not directly related to the scientist's research. In particular, it is the engrained facility with which the scientist is able to play around with different basic scientific concepts that allows an E/PO team to be innovative while retaining hard scientific accuracy. In addition, the scientist is experienced in critical thinking, which is directly related to the "inquiry-based" method of teaching that is advocated in the National Science Education Standards (NRC 1996).

One of the hardest aspects of fulfilling this "expert" role is when a scientist is asked to decide what the public really "should" know about a particular topic given all the compelling scientific information available on that topic. This requires that the scientist be more honest and less zealous about what is the most important science to distribute to the public (their particular experiment may not be). These decisions come about more in the informal education and public information sides of E/PO whereas in the more formal education projects, the National Science Education Standards should be used. This is not to imply that the NSES should not be used for designing a museum exhibit or a mission website. However, it needs to be recognized that often, projects of the latter type serve a wider range of audiences and also require pieces that contain more specific content than that found in the NSES.

As the expert, it is also appropriate for the scientist to take on teaching and mentoring roles in E/PO projects. The scientist can be involved as a teacher in a variety of ways: from minimal efforts where the scientist presents his/her research in a class to more intensive efforts that involve mentoring a teacher in a research project. Activities such as these provide opportunities for the scientist to act as a positive role model, so it is especially important that women and minority scientists are cultivated as participants. These are also some of the most powerful ways in which scientists can have a lasting impact on both implicit and explicit science education reform in our country. Another important aspect of scientists working directly with teachers and students is the inspiration that comes from the enthusiasm and passion that most scientists have for their research. However, one has to be careful with programs designed to use scientists as teachers of science outside the university environment – the scientist should never be seen as supplanting teachers but instead complementing and supporting education professionals.

Who is an E/PO Scientist

The role of the scientist in any Education and Outreach endeavor has multiple facets typically drawing on talents beyond those that are usually associated with established scientific capabilities. What this means is that not all scientists will automatically be successful E/PO practitioners. However, the possible ways in which scientists can become involved in E/PO are so numerous that most scientists can contribute to the effort in some meaningful way. While there are many papers that give a more detailed treatment of this topic (e.g., Bybee & Morrow 1998; Morrow 2000), I will give my own perspective on the characteristics of an E/PO scientist.

Typically, a successful E/PO scientist has at least one of the following qualities:

- Good communication skills
- Sensitivity to the needs of different audiences
- Enthusiasm/passion about science
- Willingness to take risks
- Respect for educators (and their research)
- Ability to generalize.

We also need to recognize that E/PO scientists fall on a spectrum that varies from spending the majority of their time on research/teaching, to spending the majority of their time on E/PO efforts. It is the breadth of this spectrum that gives such strength to the idea of involving scientists in E/PO work; professional E/PO scientists "talk the same language" as their research scientist colleagues and the education professionals. This provides a leveraging mechanism within the community of scientists participating in E/PO. However, it is the breadth of the spectrum that can also prove confusing to the current set of E/PO efforts. Perhaps a standard set of roles should be developed to provide flexibility of participation.

To define a standard set of roles, we need to acknowledge the simple fact that, with E/PO becoming an increasingly important element in the world of science, there is more demand on scientists' time to carry out this new task. In response to this new demand, scientists have had to consider many trade-offs. Some scientists have responded by creatively using their teaching role in a college or university setting to bring in E/PO projects. This clearly means that some trade-off in time has been made, and in some instances this can be detrimental to their career if the institution is not supportive of E/PO activities. Other scientists have literally created a new profession by becoming majority-time E/PO scientists (working in conjunction with NASA Headquarters, NASA missions, the Forums or Broker-Facilitators, the NSF Divisions, not-for-profit establishments and private enterprises). They have chosen to give up most or all of their scientific research and teaching time. Still other scientists have chosen another innovative career path by retaining their research component but trading their college or university teaching time for E/PO time. Each of these "species" (and their variants) has proven vital to the survival of the E/PO ecosystem – they all depend on one another, in ways which are sometimes subtle, to allow the whole to survive.

Challenges to the Successful Involvement of Scientists in E/PO

There are at least three cultures that give rise to challenges for the involvement of scientists in E/PO – those of academic science, K-12 education, and research funding agencies. The requirements of E/PO components for research grants are relatively inconsistent across the different funding agencies. This creates a large discrepancy within the academic system that can have many negative consequences. For example, review criteria for advancement along a career-path do not typically include involvement in E/PO activities and in highly competitive arenas there can be an implication that a scientist is less serious about his/her career if time is taken out to participate in E/PO. A scientist who is required by his/her particular funding agency to participate in E/PO activities can perceive him/herself to be – or actually may be – at a disadvantage with respect to colleagues. One obvious challenge inherent in the culture of academia is the pressure to "publish or perish." A related issue stemming from the academic culture and its misapprehensions of E/PO is a tendency for funding reviewers of researchers in innovative E/PO positions to indicate a lack of confidence that the research will be carried out.

The lack of consistency between the funding agencies is also borne out in what the definition of E/PO is, what is required of a scientist with regard to E/PO and when these requirements are enforced. As an example, NASA OSS responded when scientists and E/PO partners complained that it was a severe drain on resources for both the scientist and E/PO partner if they attached full-fledged E/PO proposals along with their science proposals in the beginning stages of the proposal process – and then were not awarded the grant. Now, detailed E/PO proposals come at later stages during the grant award process. The communication styles are also different between the cultures. Scientists tend to directly challenge methods and ideas while forcefully injecting their own thoughts into a discussion. While mostly meant in a spirit of debate, this can be very intimidating and abrasive, resulting in hurt feelings, perceptions of arrogance and generally a bad reputation for scientists' social skills. In addition, many scientists perceive education research to contain an uncomfortable lack of rigorous method, leading them to disregard important education research as a whole. While accusations of arrogance are completely justified in these cases, educational research is an area where a partnership between scientists and educators, joined with respect for each other's strengths, can benefit the whole community.

Another major issue that needs to be addressed by the E/PO community is the apparent conflict between the expectation of a scientist to develop programs centering on the details of their own (sometimes quite esoteric) research and the need of the education community for programs that deal with basic science concepts. There is a tremendous and ever-growing knowledge gap between the concepts involved in fundamental research and the science concepts that even average high school students are fluent in. How do you begin to describe the polarization of the cosmic microwave background radiation in a meaningful way to students who have little understanding of electromagnetism or the relationship between the sun and the "cosmos" – never mind a solid grasp of how the universe began. Research specific E/PO has its place especially in public information campaigns, but it is often difficult for scientists to accept that a major goal of an E/PO proposal should concentrate on delivering basic science concepts using their research goals and results as "hooks" where possible. However, it is important for the education community to recognize in turn that unless the new discoveries are taught as well, then we will always be playing a game of catch up. Once again, it takes a large amount of wisdom on the part of the scientist to know what aspects of their research are important to stress in an E/PO project and it takes a lot of trust and creative thinking on both sides of the scientist/educator partnership to get that information into a useable format.

Some Suggested Solutions

The hardest challenge, yet arguably the most critical, is to create an environment of explicit, and eventually, implicit support of the E/PO professional scientist within the academic and funding communities. To start this process, funding agencies and higher educational institutions should cultivate and support E/PO scientist professionals at all stages of their careers by creating a significant number of graduate, post-doctoral and in particular faculty fellowships and awards that involve a significant level of E/PO work. In addition, science museums and other appropriate non-profit institutions that are involved in E/PO should be encouraged to hire active research scientists. Faculty review criteria should include service to the community and post-doctoral candidates should be evaluated on their ability to contribute to E/PO efforts. A required class should be given for graduate students that will teach them best practices for E/PO work and can be combined with TA training. Additional professional development should be effectively provided for E/PO scientists at all points on the spectrum (whether interested non-professionals or persons interested in becoming professional E/PO scientists or those that already are). At least one professional degree-granting program should be established in the field of Education and Public Outreach for Science.

NASA and NSF should be commended for their recent efforts in bringing their science to the public. However, as noted above, unintended difficulties arise from a lack of consistency between these agencies vis-a-vis E/PO. While it is understandable that each agency would want to retain its own mission and identity, it would be beneficial to the overall effort if some cross-agency discussion could take place on E/PO and ideally create some coordination between the agencies.

Conclusions

The decline in science literacy that our country currently faces is an issue that can be helped by involving research scientists in E/PO efforts. However, the fundamental aspects of the discoveries made in science today are so far removed from the every-day experience of humans that it takes more and more effort and complete understanding of the fundamentals to convey these discoveries. This is not going to change and the divergence will most likely continue to grow. Thus, the partnerships created between research scientists and the education community during this critical time should become long-term partnerships that are second-nature to the process of science education and not just for the purpose of bringing our literacy levels back up again. It is with this philosophy that we can hope to change the public's knowledge and appreciation of science and inspire greater numbers of school children to become explorers of the universe.

I have also touched on a few of the cultural issues that need to be dealt with if this vision is to be attained. Many of these issues spring from very deep-seated traditions and assumptions about how science is done and who is qualified to be a scientist as well as what value science has to our culture as a whole. In pursuit of answers to these questions, many more begin to appear – only a few of which I repeat below:

- What is the overt relevance of science to the average American in today's world? We have moved from a society where the importance of knowing your science was rooted in knowing when to plant your crops to a society in which informed voting can alter the future health of the planet or the pursuit of the search for life elsewhere in the universe.

- What do we mean when we say the population in the United States is not scientifically literate? Should the general public know a determined body of scientific facts to be deemed scientifically literate or should they know the basics of scientific method and critical thinking?

- Who should determine which scientific topics the average American should know and to what level of detail should they know them?

- What is the role of informal institutions (such as museums and science centers) in science education? Where is the balance between disciplined learning and "hands-on" learning? Is it appropriate to ask that museums fill in for formal science education when museums rely on financial success to survive? Is the primary role of the modern science museum to educate or entertain?

During the June 2002 conference (sponsored by NASA's Office of Space Science) that generated these proceedings, these and many other questions percolated through the presentations and discussions. No firm answers were arrived at, but by defining the problems and continuing a dialogue, answers will eventually follow. Conferences such as these are an essential part of bringing together all members of this growing community to help devise the most efficient and effective way of bringing scientists into the E/PO efforts – and to keep the ecosystem healthy, happy and growing.

References

Bybee, R. W., & Morrow, C. A. 1998, Newsletter of the Forum on Education of the American Physical Society
http://www.spacescience.org/Education/ResourcesForScientists/Workshops/Four-Day/Resources/Articles/Roles_BM.pdf

Field, H., & Powell P. 2001, Public Understanding of Science 10(4), 421

Morrow, C. A. 2000, white paper
http://www.spacescience.org/Education/ResourcesForScientists/Workshops/Four-Day/Resources/Articles/Roles_M.pdf

National Aeronautics and Space Administration 2002, Explanatory Guide to the NASA Office of Space Science Education & Public Outreach Evaluation Criteria (Washington: NASA)
http://ssibroker.colorado.edu/Broker/Eval_criteria/Guide/

National Research Council 1996, National Science Education Standards (Washington: National Academies Press)

National Science Foundation 2002a
http://www.nsf.gov/pubs/2002/nsf022/start.html

National Science Foundation 2002b
http://www.nsf.gov/pubs/2002/nsf022/bicexamples.pdf

Biography

LUCY FREAR FORTSON has a Ph.D. in Physics, University of California, Los Angeles and a B.A. Astronomy and Physics, Smith College. She currently holds a joint appointment as a Senior Research Associate in the Department of Astronomy and Astrophysics at the University of Chicago and as Director of Astronomy at the Adler Planetarium and Astronomy Museum. Her research fields are Experimental Cosmic Ray Astrophysics and Gamma-Ray Astronomy with a focus on multi-wavelength analysis of Active Galactic Nuclei. Dr. Fortson's research work currently is conducted as a member of the VERITAS Collaboration – a ground based gamma ray telescope array that is being constructed on Kitt Peak in Arizona. She is carrying out a vigorous multi-wavelength campaign observing Active Galactic Nuclei (AGN) in optical and gamma ray wavelengths to understand the correlations in power output from these objects. Dr. Fortson's work at the Adler Planetarium has involved providing content for a wide variety of public education and outreach programs including exhibit design and show production, teacher workshops, demonstrations and adult education classes. Most recently she was the content director during the development of the new fully electronic Cyberspace Gallery. Her current responsibility as head of the Astronomy Department is to manage the nine staff astronomers and work with the other Adler department heads to bring accurate, appropriate, up-to-date content into all of Adler's programs and projects.

Poster Session

Summary of the Poster Session

Bernhard Beck-Winchatz

DePaul University, 990 W Fullerton, Suite 4400, Chicago, IL 60614, bbeckwin@depaul.edu

Cassandra Runyon

College of Charleston, Lowcountry Hall of Science and Math, 66 George Street, Charleston, SC 29424

At the poster session held on June 12, 2002, the richness and breadth of space science education and public outreach was evident in both the quantity and quality of the presentations. 112 poster papers were presented during the session, of which 67 were submitted for publication in this volume. Many of the themes that emerged during the oral sessions were also present in the posters. These include: effective partnerships between educators and scientists and the use of real space science data in E/PO; alignment of programs/projects/products with standards and benchmarks; professional development for educators in space science; engagement of diverse audiences; the development of a robust educational research base; and the viability of partnerships with organizations that have established national or regional infrastructures.

The National Science Education Standards and the AAAS Benchmarks emphasize the importance of authentic scientific investigations for both students and teachers. Space science is an exciting context for such investigations, and current data is readily available. Thus, many E/PO programs utilize some of the same data used by space scientists. For example, Sparks, Stoughton, & Raddick (p. 394) discuss how to engage students in astronomical research projects based on data from the Sloan Digital Sky Survey. Thomas (p. 402) describes a set of lab activities for community college students based on Mars Global Surveyor Data. The Mars Students Imaging Project (Klug et al., p. 323) is designed to guide students through authentic research experiences with Mars Odyssey data. Kuhlman et al. (p. 333) discuss an experiment that directly involves students and teachers in a Mars mission. Liggett et al. (p. 337) present strategies for using data from the Planetary Data System and Regional Planetary Image Facilities in education. Gelderman (p. 298) describes a network that partners research astronomers with high school students and teachers to engage them in authentic research experiences. Croft (p. 264) addresses the issue of integrating space science data in a classroom environment dominated by high-stakes testing.

One of the key implementation principles of OSS's E/PO program is the involvement of the scientists in ways that enhance core OSS research goals. Space scientists can make important and unique contributions, but time and resources available for E/PO are limited, and it is therefore critical to maximize the effectiveness and impact. Several presenters discussed programs that focus on the role of research scientists in E/PO. For example, Miner (p. 355) discusses the working relationship between the Solar System Exploration Forum and scientists in the Division for Planetary Sciences (DPS) of the American Astronomical Society. He also presents some of the result of an E/PO survey designed to measure the current involvement of DPS scientists in E/PO. Porro (p. 365) discusses the involvement of research scientists at MIT's Center for Space Research (CSR) in diverse E/PO projects, and stresses the role of the CSR's E/PO Office as an interface between the scientific community and formal and informal education communities. The importance of partnerships between professional educators and research scientists in professional and product development is highlighted in the contribution by Smith et al. (p. 388).

The need for good curriculum materials that are aligned with standards and benchmarks and support important learning goals was a common thread among the all conference sessions. Many of the posters dealing with formal education programs address this issue. Knudsen & Hammon (p. 326) discuss the National Education Standards "Quilts" developed at JPL, which are designed to help teachers link NASA E/PO materials to the standards. Several posters describe the development of standards-aligned classroom materials from mission specific information, e.g., Craig, Miller-Bagwell, & Larson (HESSI, STEREO, p. 262); Klug et al. (Mars Odyssey, p. 323); Kuhlman et al. (Mars Surveyor, p. 333); McCallister, Eisenhamer, & Eisenhamer (Hubble Space Telescope, p. 352); Ristvey & Behne (Genesis, p. 377); Sparks, Stoughton, & Raddick (Sloan Digital Sky Survey, p. 394). Because many museums, planetariums, and science centers conduct programs for teachers and students, alignment of educational content with standards and benchmarks is also a concern for E/PO programs that deal with informal education. Paul Knappenberger, President of the Adler Planetarium & Astronomy Museum, emphasized in his keynote address that any successful long-term strategy for improving math, science and technology literacies must include both formal and informal education and their interrelationships (Knappenberger, p. 139). Dusenbery & Morrow (p. 275) discuss this interrelationship in their poster on the *MarsQuest* exhibit. The *MarsQuest* education program includes workshops for master educators near host museums and science centers. To meet the needs of these educators, the exhibition's interactive experiences are linked with lesson plans that are aligned with the National Science Education Standards.

Several poster presenters discussed the importance of professional development opportunities for educators. Research shows that professional development for teachers is most effective if it is done in the context of materials teachers use in the classroom and combines current and relevant science content with reflections on student thinking and pre-instructional beliefs. For example, in a series of poster papers, Slater, Prather, and collaborators discuss the development of space science courses for teachers that use student-centered and inquiry-based curriculum materials, are sensitive to identified student difficulties, and develop teacher knowledge and skills in space science (pages 315, 359, 368, and 386; see also the panel presentation by Prather & Slater, p. 125). Several of the professional development programs presented during the poster session are modeled after the Solar System Ambassadors and Educator Fellows programs at the Jet Propulsion Laboratory. The GLAST Educator Embassador program (Cominsky & Plait, p. 260) trains master educators who help the GLAST E/PO team develop, test and disseminate educational materials. The SOFIA EXES Teacher Associate Program (Hemenway et al., p. 309) is working with a cadre of teachers who will promote astronomy within their communities and who will be prepared eventually for a flight experience on SOFIA (see also panel presentation by DeVore, p. 77). The Astronomical Society of the Pacific has expanded the ambassadors concept to the professional development of amateur astronomers (Fraknoi, Chippindale, & Bennett, p. 286; see also panel presentation by Bennett, p. 246). The Solar System Exploration Forum is collaborating with the Girl Scouts of the USA on professional development programs for Girl Scout leaders (Betrue, p. 249).

It is evident that as NASA's Space Science E/PO Program matures and expands, there is a growing need for a robust educational research base. This topic was extensively discussed in the session on science education research, and it was also addressed in several poster papers. For example, Offerdahl, Prather, & Slater (p. 359) discuss the results of a survey of pre-instructional student beliefs in astrobiology. Knudsen (p. 329) presents research-based recommendations for CD-ROM developers. Shipman (p. 384) provides insights into misconceptions introductory college students have about gravity.

Sakimoto (p. 380) points out that the space science missions currently envisioned by OSS span a time period during which a substantial turnover in the scientific workforce and changes in the Nation's demographics will occur. To prepare for this change, OSS is conducting targeted efforts to engage minorities in space science. Sakimoto discusses a grants program for minority universities and a partnership with professional societies of minority scientists. Several of the universities and professional societies OSS is partnering with were represented in the poster session. Johnson, Austin, & Zirbel (p. 312) discuss efforts at the City University of New York to engage minority undergraduate students in space science. Morris et al. (p. 175) have developed a partnership between NASA, minority serving universities, a science museum, and a predominantly Hispanic charter school in the Houston area with the goal of reaching out to minority and other underrepresented groups, and to encourage an ongoing interest in space science. Davis (p. 270) and Haro & Hund (p. 307) provide overviews of the missions and programs of the National Organization for the Professional Advancement of Black Chemists and Chemical Engineers and the Society for the Advancement of Chicanos and Native Americans in Science, respectively. Carruthers & Thomas (p. 257) describe space science education activities that engage minority students and their teachers and parents in the Washington, DC metropolitan area.

The development of space science education activities that are accessible to people with disabilities has become an important priority for OSS and others. The lack of individualized accommodations and adaptations often make it difficult for teachers and parents to engage students with disabilities in science and math. Consequently, disabled students often lose interest or are discouraged from pursuing these fields, even if they do show interest and talent. Grice et al. (p. 301) discuss the development of a Braille astronomy book that makes Hubble Space Telescope Images accessible to blind readers. The Genesis mission E/PO program (Ristvey & Behne, p. 377) is developing space science activities for blind students. The IDEAS grant program (Eisenhamer & Braedbury, p. 281) has funded several space science E/PO programs that target special needs communities. Finally, an OSS-wide "Exceptional Needs" working group has been formed to advise OSS and to broker partnerships between OSS flight missions and special needs communities (see Runyon, p. 457 in Appendix B).

In summary, the richness and scope of the astronomy and space science E/PO effort across the country was well represented through the poster session. The session was an important informal mechanism for communicating the engaging collection of lessons learned, project descriptions, and initiatives for the future throughout our very talented network of educators and scientists in attendance.

NASA/MSFC/NSSTC Science Communication Roundtable

M. Adams and D.L. Gallagher

NASA/MSFC/National Space Science and Technology Center, 320 Sparkman Drive, Huntsville, AL 35805, mitzi.adams@msfc.nasa.gov

R. Koczor

NASA/MSFC, Mailcode SD01, Huntsville, AL 35812

Abstract. The Science Directorate at Marshall Space Flight Center (MSFC) conducts a diverse program of Internet-based science communication through a Science Roundtable process. The Roundtable includes active researchers, writers, NASA public relations staff, educators, and administrators. The *Science@NASA* award-winning family of Web sites features science, mathematics, and space news to inform, involve, and inspire students and the public about science. We describe here the process of producing stories, results from research to understand the science communication process, and we highlight each member of our Web family.

The Science Directorate of Marshall Space Flight Center (MSFC) sponsored the first of our family of websites, *Science@NASA (SNG)* in May, 1996. At that time, we instituted an experimental Science Communications (SciComm) Process involving scientists, managers, writers, editors, and web technical experts. The close connection between the scientists and the writers and editors has assured a high level of scientific accuracy in the finished product. In addition, members of the traditional outreach activities of NASA, including Media Relations and the Education Office, were involved in the process and have benefited by a close working relationship with the scientists. Currently, the SciComm Roundtable meets each week to review *SNG* activities. Before stories are published, the Roundtable, as well as the principals quoted in the story, perform a final review. The efficiency of the roundtable is such that the review process normally takes 2 to 3 days. Additional reviews by appropriate project or program personnel have been incorporated into the Roundtable process, with some increases in elapsed time to publication.

Since the debut of the initial *SNG* website, we have added several more, for a total of six sites in the family. The websites each have a unique character and are aimed at different audience segments (or report different aspects of NASA science). Three sites are aimed at an adult, science literate audience: *SNG*, *Spaceweather.com*, and *Ciencia@NASA*. *Spaceweather.com* covers information about solar activity and its interaction with Earth's ionosphere and magnetosphere. *Ciencia@NASA* initiated in November, 2000, contains the Spanish version of SNG stories. Our most popular website, *Liftoff* to Space Exploration, is aimed at a high school audience. For the middle- or grade-school audience, we

have *NASAKids*, which includes puzzles, news stories, topics covering NASA's enterprises, and the NASAKids Club. For educators and students interested in classroom exercises, *Thursday's Classroom*, features lesson plans and classroom activities centered around *SNG* stories. URLs for each of the websites are as follows:
http://science.nasa.gov, http://spaceweather.com,
http://ciencia.nasa.gov, http://liftoff.msfc.nasa.gov,
http://kids.msfc.nasa.gov, http://thursdaysclassroom.com.

For the past five years, MSFC, the University of Florida School of Journalism and Communication, and Bishop Web Works have collaborated to perform academic research to better understand how individuals accept and process scientific information, acquired through the internet. Results of a reader survey (delivered via e-mail to *SNG* subscribers) show that 96% of respondents performed some activity as a result of reading science stories on our website (e.g., observing a meteor shower, looking for aurorae, discussing science and NASA with their children). Other results are as follows: 85% of respondents rated the quality of articles as good to excellent, 28% of respondents were students, 19% of respondents were teachers, 80% of the teachers said they used our materials in their classrooms, 68% of respondents said they read our stories at home, 63% of respondents said they passed on information from our stories to family and friends, 31% of respondents were from outside the United States, 25% of respondents were female.

Based on the following data, we believe that we have been very successful in transmitting NASA science to the public. In 2001, we registered hits[1], visits[2], and subscribers as shown in Table 1.

Table 2.   Metrics of the Science@NASA Family of Websites

	Hits	Visits	Subscribers
Science (adult content)	111,000,000	11,900,000	173,000
Ciencia (Spanish, adult)	4,061,000	403,000	13,000
Liftoff (high school and adult)	340,000,000	30,240,000	58,500
NASAKids (K-middle school)	70,150,000	3,240,000	33,400
Thursday's Classroom (informal ed.)	5,500,000	770,000	175,000
Space Weather (focused adult)	106,700,000	7,314,000	182,300

---

[1] A hit is any file that is transferred. So, for a typical web page, you download the page (1 hit) and any images on the page (1 hit for each image). A page with frames takes a hit for the frame, and one hit for each frame window, plus all the images.

[2] A visit is defined as contiguous hits from the same IP address. A new visit would require a delay of 15 minutes between hits.

Issues in Informal Education: Event-Based Science Communication Involving Planetaria and the Internet

M. Adams and D. L. Gallagher

NASA/MSFC/National Space Science and Technology Center, 320 Sparkman Drive, Huntsville, AL 35805, mitzi.adams@msfc.nasa.gov

A. Whitt

Fernbank Science Center, 156 Heaton Park Dr., Atlanta, GA 30307

Abstract. For the past four years the Science Directorate at Marshall Space Flight Center has carried out a diverse program of science communication through the web resources on the Internet. The program includes extended stories about NASA science, a curriculum resource for teachers tied to national education standards, on-line activities for students, and webcasts of real-time events. Events have involved meteor showers, solar eclipses, natural very low frequency radio emissions, and amateur balloon flights. In some cases broadcasts accommodate active feedback and questions from Internet participants. We give here examples of events, problems, and lessons learned from these activities.

Communicating science content through special events is especially popular with readers of *Science@NASA* (as evidenced by surges in web hits). In 1998, the first of many event-based activities was launched.

Tied to the peak of the Leonids meteor shower, a weather balloon carrying an image intensified video camera and a meteor sample-return package went aloft. The sample-return package contained aerogel[1], offering an opportunity to enlighten the general public about this interesting substance. In addition, the public were invited to submit meteor counts to *Science@NASA*, thereby involving the public in making scientific observations.

On two other occasions, in 1999 and 2000, we sponsored balloon launches associated with the peak of the Leonids meteor shower. Although it was cloudy locally, the balloon launch allowed viewers to see at least a few meteors. In addition, using an INSPIRE[2] receiver, we listened to the meteors, since meteors, especially bursters, produce very low frequency (VLF, below 10kHz) radio noise.

---

[1] Aerogel is a silicon-based solid with a porous, sponge-like structure in which 99% of the volume is empty space. When a particle, like a meteor, hits the aerogel, it buries itself, creating a carrot-shaped track up to 200 times its own length. For more information, see: http://science.nasa.gov/aerogel/

[2] INSPIRE – Interactive NASA Space Physics Ionosphere Radio Experiments is a non-profit, educational corporation whose objective is to involve high school students in observing all types of natural and man-made VLF noise with inexpensive receivers. For more information, see: http://image.gsfc.nasa.gov/poetry/inspire/.

From 1998-2000, more than 2.4 million people watched the live webcast or the replay.

In 2001, using NASA TV, the internet, and live audio commentary, we reported on Leonid meteor shower conditions across the country. An image intensified camera was positioned on the ground in Huntsville, AL with the constellation of Orion (and later, Gemini) in its field-of-view. M. Adams reported on Huntsville viewing conditions and identified constellations. D. Gallagher answered questions about meteor sounds and related topics. Electronic mail questions, submitted to Spaceweather.com, were answered live by Dr. Tony Phillips (from Bishop, California), via telephone link.

A total solar eclipse is another event which captures the imagination of the public and excites young people. In 1999 and 2001, in conjunction with planetaria in Atlanta, GA (Fernbank Science Center), Huntsville, AL (Von Braun Astronomical Society), and Western Kentucky University, we shared the experience of a total solar eclipse with readers of *Science@NASA*, Boy Scouts, and Girl Scouts. In preparation for the eclipse, astronomers at each venue supplied background information and led educational activities, e.g., mapping sunspot data to produce a graph of the sunspot cycle. During the eclipse itself, M. Adams, on location in Romania in 1999 and Zambia for the 2001 eclipse, was in telephone and internet communication with each of the planetaria, describing events as they occurred. Images of the partial phases of the eclipse and a graph of the air temperature, taken at ten minute intervals, were supplied to the internet.

Effective use of time for these types of events requires a detailed plan and extensive rehearsal with equipment. Problems almost always occur at a remote site, so it is important to arrive early, at least three days before the event. Especially in the case of a solar eclipse, when there is no time for mistakes, it is important to assign only one major task per person. Make a check list and use it. Finally, be certain to evenly share involvement between venues.

In addition, using planetaria and the internet to communicate science requires addressing and understanding their issues:

- Promoting and advertising the events
- Raising funds for the events
- Finding and serving under-served audiences
- Follow through, and
- Identifying unique or unusual science to tie with events

Overall, students and science attentive adults express a special interest in reading about real-time events; but they respond with an even greater interest when they can be actively involved (e.g., via email questions or through featured venues or by taking their own data) and when they can observe the process of science at work.

Student Participation in Prototype Mars Rover Field Tests

R. E. Arvidson

Washington University, Earth and Planetary Sciences, Brookings Drive, Campus Box 1169, St. Louis, MO 63130, arvidson@wunder.wustl.edu

C. D. Bowman

Raytheon ITSS, NASA Ames Research Center, M/S 269-3, Moffett Field, CA 94035

D. M. Sherman and S. W. Squyres

Cornell University, 426 Space Sciences Building, Ithaca, NY 14853

Introduction

In May of 2003, NASA will launch the next mission in the Mars program, the Mars Exploration Rovers. These twin rovers will explore the geology of the martian surface for approximately 90 sols. The Athena Science Team has been preparing for rover surface operations for the past 3 years using JPL's Field Integrated Design and Operations Rover (FIDO). The LAPIS program was developed as part of Athena education and public outreach, funded by the JPL Mars Program Office, to involve students actively in the testing of the FIDO rover.

Description

LAPIS originated in 1999 as a prototype active-participation educational program designed to involve small groups of high school students and their teachers in testing the prototype Mars rover, FIDO (Arvidson et al. 2000a). FIDO is a prototype rover for the 2003 Mars Exploration Rover Mission and is used to test operation concepts and to train the Athena Science Team to operate rovers and payloads in complex terrains (Arvidson et al. 2000b). The LAPIS students and teachers worked with Athena Science Team mentors to support these objectives developing a complementary mission plan and implementing an actual portion of the tests using FIDO and its instruments. The pilot program provided the basis for LAPIS II, implemented in 2000, and LAPIS III, implemented in 2001.

Over three years, the LAPIS program has provided a non-traditional setting for students to learn to work in teams and to use a variety of approaches to solve problems. LAPIS is designed to mirror an end-to-end mission: Geographically distributed groups of students form an integrated mission team and work together with Athena Science Team Members and FIDO engineers to plan, implement, and archive a two-day test mission, controlling FIDO remotely over the Internet and communicating with each other by email, the web, and teleconferences. Their teachers help coordinate the team and support the students' interactions with their mentor. To further broaden the impact of the program, the student team develops and maintains their own web site, communicating their activities and lessons-learned to other students and to the public

Objectives

The overarching goal of LAPIS is to get students excited about science and related fields. The program helps students explore future study and career goals in science, mathematics, and engineering by providing them with the opportunity to apply knowledge learned in school, such as geometry and geology, to a "real world" situation. Limiting the number of students in each group insures that the participants have continuous one-on-one interactions with teachers, Athena Science Team mentors, and FIDO engineers. Additionally, through the LAPIS program it is hoped that students will experience improved communication skills and appreciation of teamwork, enhanced problem-solving skills, and increased self-confidence. The reward for the teachers focuses on seeing highly motivated students approach and solve difficult problems. Additionally, teachers gain a greater understanding of space science and technology to bring into their science or mathematics classes. For the Athena Science Team mentors, it is an opportunity to inspire and encourage the next generation of scientists and engineers and to share the excitement of their work.

Future Implementation

LAPIS will act as a model for outreach associated with future FIDO field trials and the Athena Science Payload mission operations during the 2003 Mars mission. We intend to broaden the base of participation beyond the original four sites by taking advantage of the wide geographic distribution of Athena team member locations.

References

Arvidson, R. E., Bowman, J.D, Dunham, C. D., Anderson, R. C., Backes, P., Baumgartner, E., Bell, J., Dworetzky, S. C., Klug, S., Peck, N., Sherman, D., Squyres, S., Tuttle, D., & Waldron, A. M. 2000a, Eos, Trans. Am. Geophys. Union, 81, 113

Arvidson, R. E., Squyres, S., Baumgartner, E., Dorsky, L., & Schenker, P. 2000b, Eos, Trans. Am. Geophys. Union, 81, 65

Discovery Missions: Unique Approaches to Education and Public Outreach

Shari E. Asplund

Jet Propulsion Laboratory, 4800 Oak Grove Drive, Mail Stop 156-230, Pasadena, CA 91109, shari.e.asplund@jpl.nasa.gov

Introduction

NASA's Discovery Program is comprised of a series of low-cost, highly focused, competitively selected planetary science investigations. Discovery missions aim to enhance our understanding of the solar system by exploring the planets, their moons, and other small bodies using innovative approaches to assure the highest science value for the cost. Discovery, which began in 1992, represents a breakthrough in the way space exploration is conducted.

Description

Discovery was among the first NASA programs to require that an education and public outreach plan be part of every investigation. One of Discovery's supporting objectives is to increase public awareness of, and appreciation for, solar system exploration through exciting education and public outreach (E/PO) activities. Each mission is expected to spend one to two per cent of its total budget on E/PO. The Office of Space Science (OSS), which funds Discovery missions, strongly encourages space science researchers to engage actively in education and public outreach as an important component of their NASA-supported professional activities.

The Missions

Each of the eight Discovery missions selected to date has developed a unique approach to formal and informal education and outreach to the public. Our posters highlight some of the notable E/PO activities being conducted and planned by the six current Discovery missions:

Near Earth Asteroid Rendezvous, or NEAR, the first spacecraft to orbit and land on an asteroid, spent one year gathering scientific data from Eros. An extensive education and outreach program centered around orbit insertion included the following:

Student Press Conference: Twenty-five young journalists asked questions of mission scientists on topics ranging from the science and objectives of the NEAR mission to the prospects of finding organic life on asteroids.

Public Lecture: Co-sponsored by the Planetary Society, "NEAR's Tryst with Eros" provided an overview of the mission objectives as well as why we study near-earth asteroids.

Orbit Insertion Events: Live Coverage on NASA TV as scientists and engineers monitored the telemetry from the spacecraft and celebrated the successful insertion into orbit.

Space Day 2000: Comcast, Discovery Networks, and the Maryland Dept. of Education presented *Comcast – Discovery Mission 2000: Operation NEAR* with student press conferences, exploration station mini-missions, lesson plans for teachers and space facility tours.

Maryland Summer Center for Space Science: Thirty-two gifted and talented middle school students planned and launched a simulated Discovery Program mission, including the design and fabrication of instrumentation and construction of a scale model of a spacecraft.

Stardust will capture interstellar dust particles and comet dust from Comet Wild 2 and return samples to Earth. Stardust has established partnerships with many educational organizations, museums and science centers to provide opportunities for students, educators and the general public to learn about small bodies and the mission.

"Think SMALL In A BIG Way": A comprehensive Educator's Activity Guide for grades 5-8 focusing on asteroids, comets and meteorites, tied to mission events and correlated to national science education standards.

"Be a Spacecraft Engineer": This activity introduces students to elements of spacecraft design using the Stardust spacecraft and the International Space Station as examples.

Microchip Name Collection: More than 1,300,000 people have sent their names along for the ride to Comet Wild 2.

Genesis is a mission to collect solar wind samples and return them to Earth to answer questions about the birth and evolution of the solar system. E/PO efforts include:

Education Modules: Modules keyed to national science standards have been developed on Cosmic Chemistry, Dynamic Design, Exploring Origins and more for grades 4-14.

Genesis Education Development Centers: A nationwide network of teachers who pilot test activities and modules in their classrooms and conduct workshops for other teachers.

Public Outreach Modules: Modules on Sunlight and Solar Heat, Origin of the Solar System, and Atoms, Elements and Isotopes are designed to demystify science, math and technology for the general adult population.

Clean Room Technology: To highlight the unique environment of the ultra-clean room where the spacecraft was assembled, a video and accompanying Teacher Guide were produced. An Interactive Field Trip to the clean room is available at the Genesis web site.

Community Youth Organization: Genesis supports Boy Scouts and Girl Scouts with activities designed for scouts to earn badges and patches in space exploration and astronomy.

Comet Nucleus Tour, or CONTOUR, will encounter at least two diverse comets, taking high resolution images and analyzing dust and gas to answer fundamental questions about their composition.

CONTOUR Comet Watch: Coordinated global ground-based observations of CONTOUR comets for educational benefit and scientific purposes, modeled after the highly successful International Halley Watch.

CONTOUR Amazing Space: A five week workshop for K-12 teachers to create educational modules and develop classroom resources based on the mission.

CONTOUR Flight Deck: A competitive opportunity for students to participate in space science investigations by targeting spacecraft cameras during an Earth-Moon encounter.

MESSENGER, the Mercury Surface, Space Environment, Geochemistry, and Ranging mission, is a focused scientific investigation of Mercury and the forces that have shaped it. The E/PO strategy is to meet or exceed national education standards and to establish high level collaborations with individuals and institutional partners throughout the country.

The MESSENGER Classroom: High quality educational activities and programs are a major part of MESSENGER's E/PO strategy. Projects include development of K-13 educational modules, a national competition for students investigations, and interactive web-based explorations. Educator workshops will be conducted and engineers and scientists will work with educators to identify underlying key science issues and tie them into curriculum. Real time science data will be provided to both schools and museums.

Window on the Universe: MESSENGER scientists are Visiting Researchers in this Challenger Center community-based program designed to take students, educators and families on a journey through the universe. K-12 educators receive hands-on training on educational activities; students meet and talk with Visiting Researchers in their classrooms.

MESSENGER: The Movie: A documentary will take the public from the drawing board through the entire mission, showing the human story behind a decade of development on the quest to answer fundamental space science questions.

Deep Impact is the first experiment to crash a large object into the surface of a comet, creating a huge crater and revealing never before seen interior material for extensive study.

Small Telescope Science Program: A joint effort between amateur and professional astronomers and private observatories around the world to make ground-based, photometric CCD observations of Comet Tempel 1.

Amateur Astronomer Network: Amateur astronomers will partner with teachers to provide viewing of Comet Temple 1 for students and their parents and arrange public viewing opportunities before and during the impact.

Student Programs: Materials will be developed for K-12 students using space science themes and actual mission data to teach science, reading, math and language skills, in line with national education standards. Special programs will be designed for learning disabled and home school students.

Planetarium Show: The University of Colorado's Fiske Planetarium will develop a planetarium show featuring NASA's comet missions exploring the origins of the solar system. The program will be available to other planetaria and institutions.

Bringing the Mysteries of Space Down to Earth and Into Your Classroom

M. Baguio and M. W. Fischer

Texas Space Grant Consortium 3925 West Braker Lane, Suite 200, Austin, Texas 78759, baguio@tsgc.utexas.edu

W. T. Fowler

University of Texas at Austin Department of Aerospace Engineering and Engineering Mechanics

Introduction

Since 1989, the Texas Space Grant Consortium (TSGC) has conducted a multitude of K-12 teacher training workshops across the state of Texas. By integrating personal interaction with scientists, tours of state-of-the-art laboratories and hands-on activities, these workshops provide an avenue for teachers to further their professional development in science education. Though these workshops TSGC has brought the mysteries of space down to earth for tens of thousands of K-12 teachers, who in turn have impacted millions of students nationwide.

Astronomy in the Solar System

TSGC is currently conducting a series of workshops on Astronomy and Solar System education. Working with the University of Texas Astronomy department, Texas Extension Service 4-H and Youth Development program, Texas Education Agency Educational Service Centers, and Solar System Educators, TSGC is providing training and curriculum to 6th grade teachers. Teachers participate in astronomy experiments and demonstrations to increase their knowledge about inventory and exploration of the Solar System from historical perspectives, our modern picture of the solar system, and misconceptions about Astronomy and the Solar System. Teachers participate in activities such as components of the solar system, stars, galaxies, tools of astronomy, structure of the earth system, and physical characteristics of the planets to be incorporated into their classroom curriculum in engaging interactive methods. Teachers explore hands-on lessons that encourage cooperative learning and experiential education, all of which of which are aligned to both the Texas Essential Knowledge and Skill and the National Math and Science Standards.

LiftOff Summer Institute

The Liftoff Summer Institute, now in its twelfth year, is organized around space science or aerospace themes drawn from the many research and engineering programs of NASA. The goal of these workshops is to enrich teaching of math, science, and technology by providing educators with information, ideas, activities, and materials that can be used to augment their regular curriculum and shared with colleagues. Each year LiftOff focuses on a different theme, which have included Comets, Asteroid and Meteorites; Mars: Life in Extreme Environments; and Space Biomedicine.

The LiftOff workshops have shown that the excitement that teachers and students feel about space science and exploration can be tapped to enrich math and science classes. In addition, the workshops provide teachers the rare and for some, unique, opportunity to spend a week working with scientists and engineers involved in up-to-the minute missions and projects that are not yet well known to the public at large. In the words of one LiftOff participant:

> This was one of the best experiences in teacher education that I have ever had! I am leaving with a renewal of interest and enthusiasm. I will be singing your praises for a long time. Thank You!!!

Solar System Educators

Flying Swatters and Snowballs in Space? Moons of Jupiter? A partnership between TSGC and the Solar System Educator Program enables Texas teachers to expand their knowledge about astronomy. Workshops conducted at the Conference for the Advancement of Science Teachers, Rio Grande Valley Challenger Center, RGVA Science Conference, and LiftOff Summer Institute, provided teachers hands-on opportunities to make comets, track the moons of Jupiter, model aerogel, and increase their knowledge of space exploration missions such as Stardust, Chandra, Galileo, Cassini, and Deep Impact. The goals of the SSEP program is to inspire America's students, create learning opportunities, and enlighten inquisitive young minds by engaging them in the Solar System exploration efforts.

The Role of Amateur Astronomers in Informal Education and Outreach

Michael Bennett

Astronomical Society of the Pacific 390 Ashton Ave. San Francisco, CA 94112, mbennett@astrosociety.org

Abstract. Amateur astronomers represent a huge, largely untapped resource of energy and enthusiasm for doing informal education and public outreach in astronomy and space science. Research conducted by the ASP suggests that, of the roughly 50,000 astronomy club members and perhaps 100,000 "unaffiliated" amateur astronomers in the US, at least 5,000 – and possibly 10,000 – currently participate in some form of public outreach, reaching a wide variety of audiences through many different venues. No national programs in informal education are underway to support these efforts. The ASP proposes to develop a series of outreach materials and an improved infrastructure to support increased and improved public outreach by amateur astronomers.

Panel Remarks

The ASP believes that amateur astronomers represent a vast, largely untapped resource for expanded informal education and outreach in the United States. Let's look at the numbers.

How many US amateur astronomers are there? It depends to a great extent on how you define an amateur astronomer, and on how many people would characterize themselves as amateur astronomers, but there are some data.

The two major monthly magazines, *Astronomy* and *Sky & Telescope*, each have circulation of around 120,000 to 150,000. The publishers tell us there are relatively few overlapping subscriptions, perhaps 10-15%. So, approximately 200,000 to 250,000 people read one of the major monthlies.

Meade Instruments claims to sell roughly 1 million telescopes per year, but we suspect that most of those buyers are not, and probably won't become, active amateur astronomers.

The ASP's best guess, and it is little more than that, is that there are at least 100,000 – and maybe 200,000 – people in the US who are sufficiently interested and active to call themselves amateur astronomers.

Membership in astronomy clubs is much easier to quantify. The ASP's database of astronomy clubs – which we think is one of the most complete – currently numbers about 675 clubs. Average membership size is 75, yielding about 51,000 "affiliated" amateurs. The Astronomical League, a "club of clubs," has about 250 member clubs, with a total membership of about 20,000. So it's pretty clear there are between 40,000 and 50,000 astronomy club members in the US.

Traditionally, amateur astronomy has been about telescope making, telescope using, and sky-watching. Some amateurs, especially through the American Association of Variable Star Observers (AAVSO), have contributed observations used by researchers.

But a surprisingly large fraction of amateurs express their love of the hobby, and of astronomy, by sharing their enthusiasm with others, through a wide variety of outreach, educational, and public activities.

In 2000, under an NSF planning grant, the ASP began to study amateurs and public outreach. We conducted a web-based survey, well advertised in *Astronomy* and *Sky & Telescope*. Some 1100 people started the survey, and over 700 completed it. We followed up the initial survey results with six focus groups of around 10 participants each, held around the country.

Because of the obvious self-selection effects, the survey cannot be considered statistically valid. But the results are suggestive nevertheless.

Of the amateurs who completed the survey, nearly two-thirds are actively involved in some form of public outreach. Because this survey was aimed at amateurs who do outreach, that number is almost certainly higher than the overall average. In a smaller poll we conducted, an average of 20% of club members said they participated in some form of outreach, and that number is probably much more representative of clubs and club members overall. Even if only 15% of all club members participate in outreach, that's still almost 8,000 amateur astronomers conducting some kind of outreach educational activities.

Of the roughly one-third of survey respondents not doing outreach, nearly 50% indicated they would participate in outreach activities if the barriers to entry (primarily knowledge of what to do and self-confidence) were reduced.

Amateurs who do participate in outreach are surprisingly active, participating, on average, in some 18 outreach activities per year (35% of the respondents report doing outreach 3 or more times per month!). If these results are even close to accurate, amateur astronomers participate nationwide in nearly 100,000 outreach events per year!

What are these outreach activities? They cover the complete gamut of informal education in astronomy. When asked to check off all of activities they engage in, 82% reported they hold star parties, 81% reported they do class visits, 58% checked public presentations, 35% said club events, and 29% reported working with science centers, museums, or planetariums.

71% of the respondents say they serve the general public, 70% say they serve school children via classroom visits, 50% say they serve families, and 48% report working with community organizations.

Amateur astronomers who want to conduct outreach activities are almost all self-trained. Their efforts are almost all individual, although in some cases clubs have formed outreach committees to help coordinate their members' events.

While there have been a few informal outreach programs which include amateur astronomers among their participants, to our knowledge no national programs have been aimed specifically at amateurs.

The ASP's Project ASTRO has, over the past 10 years, trained a total of about 1200 astronomer/teacher partnerships. About half of the Project ASTRO volunteer astronomers have been amateurs.

NASA's Solar System Ambassador program supports about 300 volunteers, about 100 of whom are amateur astronomers. And NASA's Space Place emails items suitable for astronomy club newsletters to about 170 clubs.

The ASP is developing a program – as yet unfunded – in partnership with the Astronomical League to provide outreach kits specifically aimed at amateur astronomers, for use at star parties, class visits, astronomy club meetings, public talks, and with youth groups. These kits will each be built around an interesting topic, such as the search for life, searching for extrasolar planets, or the mysteries of black holes. Each kit will provide built-in training, and a variety of simple activities, manipulatives, images, sample scripts, resources, etc.

Distributed and supported at the club level, such kits may be used by several members of a club. They will be able to "mix and match" the materials to support different events and types of audiences.

The ASP believes that dissemination and use of such kits by the amateur astronomy community could substantively change the way amateurs do public outreach, the topics they tend to cover, and the number of amateurs who participate in outreach activities.

Working with Informal Education: A Partnership with the Girl Scouts of the USA (GSUSA) on a National Level

Rosalie Betrue

Solar System Exploration Education & Public Outreach Forum, Jet Propulsion Laboratory, 4800 Oak Grove Drive, Pasadena, California 91109-8099, California Institute of Technology, 1200 California Boulevard, Pasadena, California 91125, Rosalie.Betrue@jpl.nasa.gov

Introduction

Several years ago the NASA Solar System Exploration (SSE) Forum Forum met with the Girl Scouts of the USA (GSUSA) to see how best to work with Youth Groups. GSUSA was chosen for several reasons: 1. They are a diverse group – girls. 2. Their mission: to help girls be resourceful, independent women in society. 3. They are an international organization; providing us with the potential to sharing our knowledge internationally. 4. They are over 3.2 million American girls strong, as well as almost 1 million adult members – this presents us with opportunities for many, many high-leverage activities/events. 5. They base new programs on research and evaluation. 6. They are very pro-active when looking for funding.

GSUSA Criteria

The SSE Forum, on behalf of NASA's Office of Space Science, has since formed a partnership with the GSUSA. The partnership is on a national level. The GSUSA wants us to use two things as our driving force: 1) For the most part, the leaders do not approach the science badges, patches and try-its because they are intimidated with science – our task is to help them overcome their discomfort, and 2) help them in whatever way possible to fulfill their mission, which is to help girls be resourceful and strong.

To meet these needs the GSUSA was asked what we could do for them; these were their requests:

1. Show the girls the many careers possible for them in space science and in NASA and/or its centers.

2. Provide articles for the quarterly *Leader Magazine*.

3. Provide articles for the online newsletter – audience is leaders, parents and girls.

4. Provide space science content for their website.

5. Provide resources (speakers, scientists, handouts, etc.) for local events.

6. Find existing space science activities the girls can use to meet the requirement for earning badges (Jr. Girl Scouts), special interest patches (Cadettes and Sr. Girl Scouts), and try-its (Brownie level).

7. Provide content for special activities, events and products, such as their Space Science Kit.

8. Develop a training program for trainers of leaders and leaders. They in turn will teach the girls the activities at summer camps, camp outs, special events and troop meetings. Create "local" badges that can be offered as part of each training module for this training program.

Infrastructure

SSE Forum led the effort to develop an infrastructure that would work for NASA-OSS and the Girl Scouts.

- There are three teams: the training team, the activities team, and the written word team.

- There is one contact to interact with the one GSUSA contact.

- A database has been developed to track activities throughout NASA. The database will help the OSS Support Network join forces to best work with Youth Groups.

- To date, the teams are made up of members of these missions/programs: Deep Impact, JSC Astrobiology and Mars Outreach, Mars Outreach-ASU, Cassini, Galileo, Navigator, SSA Ambassadors, JPL Technology Program, Deep Space Network Program. The Association of Women Geoscientists, JPL Advisory Council for Women, the Microgravity team at Glenn Research Center and other members of the OSS Support Network have also expressed an interest in being part of this partnership.

Progress to Date

1. The training team has taken an existing educator workshop and modified it for Informal Education. This "Exploring the Solar System" workshop was tested in Dubuque, Iowa and in New Orleans in 2001. The workshops were enthusiastically received.

    We took "lessons learned" and modified the training for our first national GSUSA training workshop. This workshop took place April 19-23, 2002. Thirty trainers from all over the United States and Germany attended the training at the GSUSA Macy Training Center in New York. These trainers represented over 100,000 girls and 30,000 adult members. They will take these activities back to their councils; and those councils will share with neighboring councils. We will give six additional workshops in 2002.

2. We are now providing articles for the *Leader Magazine*. The first article, "Former Girl Scouts Skyrocket at NASA," appeared in the Winter 2001 issue. This article gives examples of careers in space science and also links the featured women to activities that will meet the requirements for badges, patches and try-its. We will continue to focus these articles, in some way, on careers in space science.

   The second article, "Space Exploration on a 'Small' Scale: NASA Discovery Missions," appeared in the Spring 2002 issue.

3. A Space Science Kit was developed with the Activities Team. The GSUSA asked for the space kit to focus on the Solar System. The Kit was expanded for special events and not just parent/child events. There were 3,000 kits assembled and sent to all councils. The featured career-woman is Dr. Claudia Alexander, a geoscientist and comet expert.

The Future

We will continue to use their requests/needs to build on this partnership. We will take the successes of this partnership and model programs for future Youth Group partnerships. The next targeted partnership will be the 4-H Clubs of America.

NASA Robotics Education Project

C. D. Bowman, F. Boyer, and J. Hering

Raytheon ITSS, NASA Ames Research Center, M/S 269-3, Moffett Field, CA 94035-1000, cbowman@mail.arc.nasa.gov

M. J. León

NASA Ames Research Center, M/S 258-2, Moffett Field, CA 94035-1000

Introduction

NASA's Robotics Education Project (REP) was created in 1998 to provide robotics-related educational courses and workshops, technical support and resources, and organization of activities and robotics competitions to enhance the (K-14) public's scientific and technical familiarity, competence, and literacy. NASA REP works to capture the educational potential of NASA's robotics missions by supporting educational robotics competitions and events, facilitating robotics curriculum enhancements at all educational levels, and maintaining a web site clearinghouse of robotics education information. REP seeks to "contribute to the future exploration of our Solar System through the development of an educated robotics technology workforce" (León 2000).

Competitions and Courses

REP supports high school student participation in FIRST (For Inspiration and Recognition of Science and Technology), a hands-on, tele-operation robotics competition. Through FIRST, high school students work with teachers and engineering mentors to learn and practice valuable skills in robotics, engineering, imagination, and teamwork. Many students who participate in FIRST are turned on to the world of science and technology and become motivated to pursue degrees in those fields in college. Through the ten NASA Centers, all of which participate in REP, NASA sponsored 134 teams in 2001 that affected more than 3570 students nationwide from Florida to Alaska and Hawaii. REP actively pursues outside collaboration, including finding partners for NASA-sponsored regional events and securing in-kind donations from national corporations.

REP also supports participation in the KISS Institute for Practical Robotics (KIPR) National Botball Competition, a hands-on, autonomous robotics competition open to middle school students nationwide. Through this program, students work with teachers and mentors to build a LEGO robot and program it to compete in an annual challenge. REP is also involved in the development of online robotics courses in association with major universities, and provides numerous public outreach presentations to corporations, schools, and political offices.

Web Site

To disseminate information and reach an even greater number of people, the Robotics Education Project maintains a web site (http://robotics.nasa.gov). This web site acts as a clearinghouse for information on robotics, competitions, lessons and activities, as well as broadcasting live and archived video feeds of the FIRST Robotics and Botball competitions and robotics workshops to viewers around the world. Through the web site, students and the public can submit questions about robotics or ask for help on a project or activity. The site is also updated weekly with robotics-related news stories and relevant links to other NASA sites and government, university, and private robotics research.

Future Implementation

REP provides the public with access to cutting-edge robotics information and resources, supports student participation in competitions and challenges, and works to leverage NASA's robotics missions in order to capture the imaginations of students and the general public. Over the next few years, REP will further support NASA's goal to "inspire America's students, create learning opportunities, and enlighten inquisitive minds" (NASA 2000) by expanding opportunities for involvement in robotics competitions, improving teacher access to standards-based, robotics-related curricular enhancements, and increasing the awareness of robotics as a portal to greater understanding of science, math, and engineering.

References

León, M. 2000, Robotics Education Project Management Plan FY 2001 (Moffett Field, CA: Robotics Education Project)

National Aeronautics and Space Administration 2000, Strategic Plan (Washington: NASA)

The International Space Station Amateur Telescope

Orville H. Brettman and Barry Beaman

Astronomical League, 13915 Hemmingsten Rd., Huntley, IL 60142, rivendell@worldnet.att.net

Purpose

The International Space Station Amateur Telescope (ISS-AT) is a project conducted primarily by amateur astronomers with multiple purposes, many with strong education and public outreach (E/PO) implications. While ISS-AT will tremendously advance amateur astronomy technology, it may prove to be of great value to the growth of space science and space facilities engineering. There are a variety of estimates as to the number of active or armchair amateur astronomers in the United States. Generally, 250,000 is the number most accepted. In addition, there is reason to believe that as many as 10 million individuals might have sufficient interest to occasionally access the ISS-AT. More importantly, ISS-AT can be a major incentive for elementary and secondary students to explore the possibilities of careers in space science and space facilities engineering. In fact, the ISS-AT Alpha Telescope, located at Sonoita, Arizona, has already produced more than two thousand photographs of astronomical objects, many of these made by students. At least one of these students used the obtained results to win a second place in a science fair. Would you expect that student to have developed a heightened interest in space science as a result of such success? Think of the impact that student's project would have had if it had been done using data collected from a space borne telescope.

Description

As currently envisioned, ISS-AT will consist of a network of ground-based telescopes and the Amateur Space Telescope (AST) located on the International Space Station (ISS). The ground telescopes will operate primarily in support of the AST, but can also operate independently on projects that do not require the sophisticated capabilities of the AST. The ground network will have six or more telescopes, including the Alpha Telescope, located in both hemispheres and widely separated to allow daytime classroom access to night-time observing across the globe. The network's nerve center will be the ISS-AT Operations Center at Dyer Observatory, Vanderbilt University in Nashville, Tennessee. The ground-based telescopes will be in real time contact with the Operations Center. Guest amateur astronomers or students (customers) will be able to operate the telescopes from their home or school. Satellite uplinks and downlinks will provide the communication links. The AST, however, must operate in a batch mode with the customer providing instructions that will be used by the Operations Center staff to create operating scripts. These scripts will be batch processed to the AST once daily. The resulting data will be batch downlinked, also once daily. After the Operations Center receives the data, it will be sent to the customers and placed on the web site.

Organizationally, ISS-AT is the brainchild of Mac Gardner, a retired Boeing engineer. ISS-AT was given life by interest shown in it by the Boeing Company, who, through Mr. Gardner, contacted the Astronomical League and asked the League to be the operational director of the project. Charles Allen, who was the League president at the time of the request from Boeing, asked Mr. Orville Brettman to chair the League committee tasked to develop ISS-AT. Since that time the ISS-AT Committee has grown to more than 80 members, all enthusiastic volunteers. Also, a number of organizations have shown interest in ISS-AT. In addition to Boeing and the Astronomical League, the Raytheon Corporation, Dell Computers, Dyer Observatory, Hughes Network systems, and several NASA Centers have made contributions to ISS-AT in the form of advice, small grants, or in-kind equipment donations. In addition, a number of telescope and accessory manufacturers have made in-kind donations to allow startup of the Alpha Telescope.

The AST

The AST is the heart of the ISS-AT project. It will be a 14 to 16 inch aperture Schmidt-Cassegrain telescope (SCT) with an f/50 focal ratio on a mounting design that will be determined by the installation location. Most design considerations are still under study. The ISS-AT Committee hopes it will be installed on an Express Pallet located on the zenith side of the ISS truss.

Though the high f-ratio will be a limiting factor for some experiments, the AST will still provide a wide range of observing opportunities to its users. A partial list of suggested activities follows:

- Synoptic observation of the planets
- Extended object imaging
- Survey of the Messier catalog
- Comprehensive high-resolution survey of galaxies
- Variable object program
- High resolution lunar reference atlas
- High resolution lunar libration images
- Photometric calibration sequences

Why Build It?

"Like buried treasure, the outposts of the universe have beckoned to the adventurous from immemorial times. Princes and potentates, political or industrial, equally with men of science, have felt the lure of the uncharted seas of space and through their provision of instrumental means the sphere of exploration has rapidly widened." Thus wrote the visionary astronomer George Ellery Hale in his 1931 essay "The Possibilities of Large Telescopes." The great telescopes that Hale brought into being were key to the discovery that the "spiral nebulae" were the vast assemblages that we know today as galaxies, and led Edwin Hubble, the Hubble Space Telescope's namesake, to discovery of the expanding universe. In the same spirit that Hale wrote on the possibilities of large telescopes some seventy years ago, this paper considers the possibilities of small telescopes in orbit about the Earth. "But the truly distinguishing feature of the ISS-AT lies not [in] its aperture, but in its unique openness and accessibility to every person on this planet." So said Richard Berry in *Perfect Optics in Earth Orbit*

The ISS-AT does offer tremendous opportunities for Education and Public Outreach in astronomy, space science, and space facilities engineering!

Space Science Education and Outreach Activities in the Washington, DC Metropolitan Area

George R. Carruthers
Naval Research Laboratory, Code 7609, Washington, DC 20375-5320, george.carruthers@nrl.navy.mil

Valerie L. Thomas
The LaVal Corporation, 2004 Clearwood Dr., Mitchellville, MD 20721, vthomas@erols.com

Introduction

We describe a program of space science education activities (including related areas of Earth science), for pre-college students, educators, and parents, that we have carried out during the past 10 years and are continuing at present. These activities have been performed and directed by Science, Mathematics, Aerospace, Research, and Technology (S.M.A.R.T.), Inc. and the Naval Research Laboratory (Space Science Division and Community Outreach Committee).

Other participating or collaborating organizations include the DC Space Grant Consortium, the National Technical Association, Inc., the National Society of Black Physicists, the National Air & Space Museum, NASA Goddard Space Flight Center, Minority University-Space Interdisciplinary Network (MU-SPIN), Exploratorium, Space Explorers, Inc., Sun Earth Connection Education Forum, and US Satellite.

Education and Outreach Programs

Current or recent past programs in which we are participating, or have participated, include the following:

Earth & Space Science Video Series

We directed and participated in producing a series of videos in Earth and Space Science, intended for pre-college students and their teachers. Students also served as "actors" in these videos, which were co-sponsored by NASA IDEAS grants. Master copies of videos produced to date have been provided to the Teachers' Resource Center at NASA Goddard Space Flight Center.

Courses in Earth & Space Science for DC Public Schools Science Teachers

Courses in the topic areas of Earth & Space Science for pre-college science teachers were created and offered in 1996 and 1997, in full-time, 2-week-duration summer sessions, and in the fall semester of 1999 in a 2 days per week, after-school (evening) session. This activity was co-sponsored by the DC Space Grant Consortium and IDEAS.

Monthly Saturday Workshops for Pre-College Students at the National Air & Space Museum

A series of monthly Saturday morning presentations in aerospace science and technology, intended for DC-area students, their teachers, and parents, was held at, and with the support of, the National Air & Space Museum during the 1992-1998 time period. The presentations were made, on a volunteer basis, by professional scientists, engineers, and technologists throughout the Washington, DC area (including some coming from as far as Hampton, VA).

Mentoring and Assisting Individual Students and Teachers

A major, informal part of our education and outreach activities has been in assisting students in the DC-area public schools in developing science fair projects and essay competition entries in areas of Space Science. This also included working with teachers to improve their classroom resources and ability to direct and guide the students.

Special NASA Education/Outreach Programs Associated with Space Missions

We worked in facilitating and assisting students and teachers in their participation in NASA-sponsored education activities, including those associated with specific space flight missions (such as MoonLink, associated with the Lunar Prospector mission, and NEARLink, associated with the Near Earth Asteroid Rendezvous [NEAR] mission), co-sponsored by the DC Space Grant Consortium and Space Explorers, Inc.

Summer Employment of Students

We provided summer employment at NRL for high school students interested in space science and technology, via the Department of Defense's Science and Engineering Apprentice Program (SEAP). Many of these students were also participants in science fair and essay contest projects during the school years, and were provided guidance and assistance at their schools and at NRL.

Visiting Scientists and Engineers Seminar Series

In 1999, a program for middle-school students was held at the S.M.A.R.T. Technology Learning Center in Washington, DC. NASA scientists and engineers gave engaging presentations on a wide variety of topics to students who were participating in the YMCA's YCARE summer program.

Signals of Spring Program

We partnered with U.S. Satellite on the Signals of Spring program for staff development of DCPS teachers in the 1999-2001 time period. Teachers were trained for an authentic science research program for middle and high school students, in which the students would use web-based satellite data and remote sensing technology to track the spring migration of large birds.

Solar Eclipse 1999 and 2001 Web-Cast Coverage

We assisted in, and facilitated, web-cast coverage and provided related educational activities and Family Night sessions, for students and the general public, in connection with the solar eclipses in September, 1999 and June, 2001 (neither of which was directly observable from the United States).

Current Status and Near-Future Plans

S.M.A.R.T., Inc. was recently awarded a grant, from NASA via the South-East Regional Clearing House (SERCH), to develop a series of four videos and associated "hands-on" laboratory activities, for middle school and high school students, on the topic areas included in the NASA Office of Space Science's four research theme areas: (1) The Sun-Earth Connection; (2) Solar System Exploration; (3) Astronomical Search for Origins and Planetary Systems; and (4) Structure and Evolution of the Universe. Each of these videos will involve the participation of teachers and students in at least one middle school and one high school in Washington, DC. We also participated in a more recent proposal submitted by the DC Space Grant Consortium, to NASA, in response to their announcement of an Aerospace Workforce Development Competition (in connection with NASA's National Space Grant College and Fellowship Program). The role of S.M.A.R.T. would be to develop, and teach, introductory and intermediate-level courses in Earth & Space Science, for students in the DCSGC-member universities in the District of Columbia.

The Educator Ambassador Program for NASA's GLAST Mission

Lynn R. Cominsky and Philip Plait

Sonoma State University, Department of Physics and Astronomy, 1801 East Cotati Avenue, Rohnert Park, CA 94928, lynnc@charmian.sonoma.edu

Abstract. NASA's Gamma-ray Large Area Space Telescope (GLAST) mission is being built for a planned launch in 2006. The primary goal of GLAST is to identify and study nature's highest energy particle accelerators through observations of active galactic nuclei, pulsars, black holes, supernova remnants and gamma-ray bursts. One of the components of the education and public outreach (E/PO) program of the mission is the GLAST Educator Ambassador (EA) program. This program, while new to high-energy astrophysics, is modeled after the Solar System Ambassadors and Educator Fellows programs at the Jet Propulsion Laboratory.

Overview of the Educator Ambassador Program

The GLAST Educator Ambassador Program consists of five master educators or curriculum designers who are helping the GLAST E/PO team develop, test and disseminate educational materials at regional and national venues. The EAs receive an annually renewable yearly stipend of $2500, travel expenses to the biannual summer training workshops and to the GLAST launch in 2006. The GLAST EAs were chosen from an outstanding pool of 40 applicants, who responded to advertisements in late 2001. The successful candidates were chosen based on their experience, ideas for developing and disseminating GLAST materials, enthusiasm and geographic diversity. Particular attention was paid to applicants with special abilities to reach underserved communities. The GLAST EAs selected in the first round were Tim Brennan (Vermont), Daryl Taylor (New Jersey), Michiel Ford (Kansas), Jason Smith (Maryland) and Teena Della (British Columbia, Canada).

Program Evolution

The applicant pool in the first selection round was so strong that we selected two educators to fill alternate positions. The two alternates were Robert Sparks (Wisconsin) and Rae McEntyre (Kansas). After the 7 EAs were chosen, and before the summer training session had occurred, many EAs immediately began to work on activities and dissemination for the program.

These initial interactions were so promising, that we convinced the SEU Education Forum to sponsor three additional EAs that could be trained along with those from GLAST. The role of these SEU EAs would be to support some of the smaller SEU missions with less extensive E/PO programs. In the winter of 2002, we issued a second announcement seeking applications: those chosen were Dr. Mary Garrett (Michigan), Christine Royce (Pennsylvania) and Tom Estill (California). A shift in the budgetary situation for the Swift gamma-ray burst mission (whose E/PO program is also led by Sonoma State) allowed the two GLAST alternates to be promoted to full-time EAs for the Swift mission. As a result, by the Summer of 2002, there were 10 EAs in the cohort, funded by these three different programs.

Summer Workshop 2002

The first EA training workshop was held at Sonoma State University (SSU) during the week of July 15-19, 2002. The week was devoted to educating the EAs about the SEU NASA missions and their science, testing and assessing activities, and a tour of facilities at the Stanford Linear Accelerator Center and campus, where GLAST and Gravity Probe-B are being built (respectively).

SSU personnel gave lectures on NASA, basic astronomical science, the electromagnetic spectrum, supernovae, black holes, and gamma ray bursts. Stanford GP-B personnel conducted a workshop demonstrating some of the strange effects of relativistic gravity. Other featured presenters included a teacher trainer from San Francisco's Exploratorium, representatives from the CHIPS mission, Lynda Williams (the Physics Chanteuse) and Gerson Goldhaber from Lawrence Berkeley National Laboratory, who explained recent observations of supernovae, dark energy and the cosmic microwave background. The EAs were also told about the GLAST Telescope Network (GTN), an SSU effort to establish a series of small ground-based robotic telescopes to observe active galaxies and gamma-ray bursts. Many workshop sessions focused on various GLAST, Swift and SEU mission activities designed for classroom use. A few of the EAs also shared activities that they have developed and successfully used in their classrooms.

Assessment of the week's events was performed by WestEd for GLAST and the Lesley University Program Evaluation Research Group (PERG) for SEU. Assessment included a pre-test on astronomy and space science, with the same questions repeated in a post-test, immediate assessment of each activity after it was performed, as well as extensive post-workshop surveys and interviews. Preliminary reports indicate the workshop was very successful. Most of the scheduled activities were well-rated by the EAs, including the SSU personnel lectures, the tour of SLAC and several of the classroom exercises. There was both positive and negative feedback on the activities, which will help in the development of future exercises as well as in planning the next workshop. We can honestly report that we learned as much from the EAs as they did from us.

Additional information

To learn more about the summer training workshop, the schedule of upcoming EA events, and the procedures for the next round of applications (in 2003), follow the links for the GLAST, Swift and SEUEF programs at the SSU team's E/PO website at http://epo.sonoma.edu.

Our Sun – the Star of Classroom Activities and Public Outreach Events

Nahide Craig, Anne Miller-Bagwell, and Michelle B. Larson

UC Berkeley, Space Sciences Laboratory, MC 7450, Berkeley, CA 94720-7450, ncraig@ssl.berkeley.edu

We will present innovative classroom activities for grades 8-12, as well as public outreach web-based resources featuring solar data, mathematics, and solar scientist interviews. These resources have been developed through partnerships between NASA scientists and educators. The classroom activities are well-aligned with National Science Education Standards. "Sunspots" – an inquiry-based resource – emphasizes mathematical connections through measurement, graphing, and analysis of satellite and student-acquired data. "Sunspots" has been successfully classroom tested in diverse school districts around the country. The resource incorporates background information, including the importance of the Sun in ancient cultures, a historical account of sunspots observations, and current NASA research. In addition, it includes guidance for safe sunspots viewing and a Java interactive research tool that allows students to analyze possible correlations between sunspots and x-ray active regions from Yohkoh images of the Sun. "X-ray Candles – Solar Flares on Your Birthday" allows students to discover the solar cycle by analyzing x-ray flare data and graphing the percentage of high energy flares over time. "X-ray Candles" was developed as a part of the HESSI E/PO effort and is featured on an upcoming episode of the emmy winning NASA Connect broadcast program. We will also present how scientists from NASA's STEREO mission are contributing to Education and Public Outreach through interviews incorporated in the high-visibility Eclipse 2001 webcast event, and through a STEREO website hosted by the Exploratorium.

These resources are produced by HESSI and STEREO E/PO and The Science Education Gateway (SEGway) Project supported by NASA's SR&T Program.

Seven Years of Growing Space Science Education

Nahide Craig, Anne Miller-Bagwell, and Isabel Hawkins

UC Berkeley, Space Sciences Laboratory, MC 7450, Berkeley, CA 94720-7450, ncraig@ssl.berkeley.edu

SEGway is a national consortium of science museums, research institutions, and educators who work together to present the latest space science research for students, educators, and the general public. Resources developed through this rich partnership can be used to enhance classroom science programs or as self-guided modules for public education and extended learning.

SEGway is an online educational resource center, adapting space science research and information for the benefit of broad audiences using Web-based learning technologies. SEGway is also a platform for showcasing cutting edge discoveries – providing proposal support, partner brokering, management and coordination for E/PO programs of NASA missions at UC Berkeley.

SEGway is currently is collaborating with the San Francisco Unified School District (SFUSD) through the University of California's Interactive University outreach program. to develop astronomy curricula in support of the District's Middle and High School science programs, which follow state and national science education standards. The SFUSD has functioned as a test bed for newly developed online lessons, establishing valuable local relationships with teachers, students and administrators who provide input, feedback, and evaluation.

SEGway is supported in part by the Office of Space Science's Information Systems research cluster – Supporting Research and Technology Program. Visit SEGway on the web at http://cse.ssl.berkeley.edu/segway.

Effective Use of Space Science Data and Research in the Classroom

Steven K. Croft[1]

Center for Educational Technologies, Wheeling Jesuit University, Wheeling, WV 26003, scroft@cet.edu

Introduction

Effective integration of new results in space science into formal (K-12 and university) and informal educational settings is a challenge for most individual research scientists as well as for research institutions such as the NASA centers. Part of the problem is the lack of familiarity on the part of most scientists trying to create educational materials with the challenges teachers face in a typical classroom setting. Some of the challenges that make it difficult for a teacher to use space science data and research results in the classroom include: the lack of special knowledge on the part of most teachers to understand the specifics of most cutting-edge research projects, the sheer volume of data and the huge sizes of some image files (especially Landsat and other Earth Enterprise image files), the standards-driven pressure to "teach to the test," and the already overflowing curriculum in most schools.

At the Center for Educational Technologies, we have been testing the effectiveness of various technology-based techniques for utilizing space science data and results in the classroom for over a decade. We have produced a series of National Science Standards-based products, including the web-based Exploring the Environment, the CD-based award-winning BioBLAST and Astronomy Village series, a series of Earth System Science online courses, and we have run teacher workshops in a wide variety of settings. We outline below a few of the important observations we have made while building and testing these products.

Teacher Training

Training in the basics of space science and the software used in image and data analysis involved in educational materials based on cutting-edge work is absolutely necessary. Most teachers do not have the time, resources, or specialized background to use the data or original reports themselves. The most effective means of training is face-to-face teacher workshops. However, the ability of teachers to effectively use materials in the classroom after a workshop depends largely on the time spent working with the materials. Half-day or one-day workshops are not really effective as the skills and information are only partially assimilated and quickly forgotten. Weeklong workshops are better, and month-long workshops are better still. The best approach is to bring the same teachers

---

[1]Current address: NOAO, 950 North Cherry Avenue, Tucson, AZ 85719, scroft@noao.edu

to a series of 3 to 4 annual workshops of at least a week's duration followed up by ongoing contact during the school year. In this way, the teachers gain the confidence and understanding necessary to use the materials in class. Unfortunately, this approach is very expensive and time consuming, and often not practical for teachers using educational software packages obtained through commercial or NASA distribution centers. In our software we provide a "Teacher's Tool Kit" which provides suggestions for classroom use, correlations with national standards, supporting content information and sample answers to all of the problems and questions posed in the software. Such a teacher's supplement is the minimum support necessary to provide the relevant knowledge necessary to successfully use the classroom materials.

Image Data

Images are among the most appealing products of space science research. However, simply providing online access to large libraries of scientifically significant and/or beautiful images does not automatically lead to effective class usage, because most teachers and students don't know what to look for in the images. Sometimes they don't even know what they are looking at, and the images become simply pretty pictures. Images are most effective when presented in small sets of related images supported by a lesson or set of objectives specifically built around the images.

Using Cutting-edge Research

Students usually cannot just jump into the data or images relating to cutting edge research problems – they don't have the background knowledge to understand the problem, much less the fine details. Providing teachers with specific content to pass on to the students is one approach, with the advantages and drawbacks outlined above. In multimedia products designed for self-guided or home use, the bridge between the students' knowledge and the research problem must be "hard-coded" into the software. The problem here is the trade-off between providing sufficient coded information and the cost of the project. In *Astronomy Village: Investigating the Solar System*, we used a two-step process that appears to work fairly well in the classroom. We introduce students to a specific topic (like life in the solar system) with an introductory module containing a series of learning activities in relevant background material. Then we provide a selection of modules built around a specific research questions and data sets that build on the information and skills given in the introductory module.

Open-ended Questions

Students are effectively drawn into research data when the questions posed to them are open-ended, that is, scientists are still actively researching the problem and there is no "right" answer at the back of the book.

Overflowing Curriculum

Getting new educational products into local curricula increasingly dominated by high-stakes testing is perhaps the greatest problem of all. One approach we have used is to correlate good space science research problems with specific national science and mathematics standards, and then show teachers how these research-related activities can replace existing curriculum activities by teaching the same standards. Properly designed open-ended research-based problems can also fulfill science inquiry standards in ways that traditional curricula often can't.

NOMISS: Integrating Space Science and Culture Through Summer Programs and Professional Development Experiences

Richard Crowe and Alice J. Kawakami

University of Hawai'i at Hilo, 200 W. Kawili St., Hilo, HI 96720, rcrowe@hubble.uhh.hawaii.edu

The University of Hawai'i at Hilo (UHH) offers the only baccalaureate astronomy degree program in the State of Hawai'i, and has on its campus the base facilities of several of the Mauna Kea observatories. As a result of funding through a NASA Minority University Education and Research Partnership Initiative grant, UHH is forging a unique partnership with Kamehameha Schools, the State Department of Education, the Institute for Astronomy, Gemini Observatory, NASA Infrared Telescope Facility, and Subaru Observatory. New Opportunities through Minority Initiatives in Space Science (NOMISS) is designed to engage a broad spectrum of participants, K-12 students and their teachers, undergraduate university students and their professors, and community and business partners by bringing together modern space science and concepts of Pacific sky lore and traditional Hawaiian knowledge. Through new instrumentation courses and new laboratory curriculum, as well as co-operative student internship and research projects with the observatories, the UHH undergraduate program will be ideally suited to provide the pre-professional training needed for students, including those of Hawaiian ancestry, to obtain careers in astronomy and employment in Mauna Kea observatories. A new summer astrophysics course includes observing and acquiring telescopic images from the summit of Mauna Kea. The NOMISS program is also focused on extending astronomy-related outreach to K-12 students and teachers, using curriculum that connects Hawaiian traditions, including celestial navigation, with the observational astronomy conducted by the Mauna Kea observatories. The ultimate aim of this is to encourage more students of Hawaiian or Pacific Island ancestry to enter careers in space science, as well as to increase awareness of astronomy within the Hawaiian community. To date, ten students in all (seven of them UHH astronomy or physics majors) have been employed as interns at the Mauna Kea Observatories. Four of the ten have interned at the Gemini Observatory, two at the NASA Infrared Telescope Facility, one at the Institute for Astronomy in Hilo, two at the Keck Observatory, and one at the Subaru Observatory.

In the summer of 2001, the NOMISS grant allowed the Department of Physics and Astronomy to offer, for the first time, a course in modern observational astrophysics, with emphasis on "hands-on" use of instruments to acquire data with research-grade telescopes atop Mauna Kea. Students gained on-site observing experience with CCD photometry and spectroscopy through direct acquisition and data analysis using modern laboratory data reduction software. Applications to stellar astrophysics were covered. A new observational astronomy text featuring the Mauna Kea Observatories was compiled and written by UHH professor William Heacox specifically for the summer course. The course was also offered during the 2002 Summer Session. Classes were held in the Astronomy department computer lab every weekday, in addition to 18 scheduled half-nights on the UH 0.6-meter (24-inch) telescope atop Mauna Kea. 14 students were enrolled; all completed the course. Two UHH instruments were mounted on the UH 24-inch during observing. The principal instrument used was an Apogee AP6ep 1024x1024 CCD; the AP6ep detector has 24-micron pixels, and when mounted on the 24-inch, images a field of view on the sky 9 arc-minutes across. Such a field of view is excellent for imaging prominent galaxies, star clusters and nebulae. Participating students moved the telescope to these objects, focused the object on the detector, collected the data, and later processed the images back in the lab or at Hale Pohaku using Maxim CCD or AIP. Processed images are displayed at http://hubble.uhh.hawaii.edu/images/index.html. During the last six nights of observing this past summer, the instrument used was a 1-Angstrom-resolution STIS-7/SBIG spectrograph. Spectra of many bright stars (from 1st to 5th magnitude), spanning a complete spectral class range of O-M (blue-red) and a broad luminosity class range of I-V (supergiant-dwarf), were obtained by the students.

NOMISS provides professional development sessions for its network of teachers from both public and private schools on Oʻahu and the island of Hawaiʻi. During the past 18 months, we have been working with a core group of 20 teachers at public and charter schools, as well as at the Kamehameha Schools and St. Louis High School. The NOMISS perspective on curriculum development is to provide teachers with rich experiences that relate to both the science of astronomy and the cultural heritage of the students of these islands. With that philosophy, we conducted five-day summer retreats (in both 2001 and 2002) on the Big Island of Hawaiʻi, which included excursions to cultural sites such as Cape Kumukahi and Mauna Kea. Participants were also able to obtain views of clusters and galaxies through the UH 24-inch telescope, primarily used for UHH programs. Also in the summer of 2001, Co-I Kawakami acted as a lead coordinator, while P-I Crowe served as an instructor, for "AstroVaganza - 2001," in which 60 teachers from Hawaiʻi and the mainland participated in a week-long program that blended curriculum and training in Polynesian navigation, Hawaiian culture and astronomy on Mauna Kea. During the 2001-2002 school year, three one-day workshops on cultural protocol were held. NOMISS teachers have thus gained a continuum of experiences that bring together knowledge of the indigenous culture and traditional science. Teachers now are designing and implementing curriculum activities to increase their students' learning about culture, math and science, particularly astronomy.

Through the efforts of NOMISS Coordinator Nathan Chang, a number of community organizations and individuals have supported NOMISS activities by providing various types of resources and donations. Through his participation in various community groups, Coordinator Chang has strengthened relationships and partnership with Kamehameha Schools. He has also set up speaking presentations with the three Rotary Clubs in Hilo, as well as at radio stations, and has helped prepare 8 newspaper articles. This has assisted the project with donations, and has resulted in much free publicity about the project and favorable public support. NOMISS was recognized as a significant initiative in developing a skilled workforce for the island with an award from the Hawaiʻi

Figure 1. Teachers preparing hoʻokupu for ceremony on Hilo Bay prior to Mauna Kea excursion.

Business Magazine's Targeted Industries Growth Report (TIGR), sponsored by City Bank. The TIGR award, presented to Crowe and Kawakami in 2001, is coveted by local start-up entrepreneurs. Partnerships with the observatory and business communities have aided NOMISS in making significant progress toward its project goals over the last 2 years.

The National Organization for the Professional Advancement of Black Chemists and Chemical Engineers

Darrell L. Davis

NOBCChE, 910 Bentle Branch Lane, Cedar Hill, TX 75104, darrelldavis2020@sbcglobal.net

An ad hoc Committee for the Professional Advancement of Black Chemists and Chemical Engineers was organized in April 1972. The establishment of the committee was assisted financially by a grant of $850 provided by the Haas Community Fund, and a $400 grant administered through Drexel University. The committee then surveyed Black professionals to ascertain their interest in establishing a formal organization dedicated to the professional advancement of Black chemists and chemical engineers. Enthusiastic questionnaire responses prompted the committee to expand and reconvene in September 1972 to set up a structured organization and to devise a means of securing funds to finance its development.

The NOBCChE Organization for the Professional Advancement of Black Chemists and Chemical Engineers was established to develop and carry out programs to assist African-Americans in realizing their full potential in the fields of chemistry and chemical engineering. African-Americans make up a very small percentage of the professionals in their respective disciplines. In spite of this acute under-representation of minorities in their ranks, most national professional organizations representing technical personnel have not adequately addressed the particular needs and interests of their African-American colleagues.

The organization's purpose is to maintain and support regional programs that assist African American scientists in fully realizing their academic and/or professional potential, introduce science and technology as viable professional goals to students on the elementary and high school levels, and encourage college students to pursue higher education in technical and scientific disciplines.

NOBCChE's objectives include recruitment and retention of African-Americans in science and engineering-related university programs, continued professional development of African-American scientists, and community involvement of professionals to provide essential role models within the African-American community. The organization depends upon existing talent among African-Americans in professional positions to aid in accomplishing program objectives. Membership in NOBCChE is offered to anyone who supports the mission and goals of the organization.

NOBCChE supports several programs within its organization, aimed at addressing the crisis of the under-education of African-American students in the country, and specifically the decline in the preparation of these students in mathematics and science education.

The local chapters of NOBCChE have provided the major mechanism for carrying out the educational objectives and goals of the organization, that is to provide the opportunity for introducing science and technology to students at all educational levels. A variety of programs are now in progress. Some of these programs are listed below:

1. Science Bowl Competitions
2. Science Fair Competitions
3. Career Day Participation at Secondary Schools
4. Adopt-a-School Program
5. Training secondary students for Proficiency Tests
6. Starlab Summer Science Program (six-week summer science program hosting 55 middle school and high school students with math and science aptitudes: Atlanta Chapter)
7. Super Science Saturday Program (day of science activities including interactive experiments, demonstrations and live animal exhibits for K-12; 300 hundred students participants: Atlanta Chapter)
8. Participate in National Chemistry Day with the American Chemical Society
9. Bowling for Kids in Science
10. Science Outing Reuben Fleet Science Center and Jet Propulsion Laboratory (San Diego Chapter)
11. Community Science Day

NOBCChE proposes to establish a Science Academy that will recruit, train, and retain minority students for the purpose of eventually becoming trained scientists. There are three primary goals for participants in this academy:

- to develop basic skills that will enable these students to succeed in the sciences
- to teach competitive scientific reasoning skills; and
- to improve the confidence and ability of these students.

Other objectives include reversing negative stereotypical images students may have of themselves, providing mentoring relationships with professional scientists, and counseling them so that they can improve themselves throughout the rest of the year. The goals are to be realized specifically through developing reasoning, math, science study, and computer skills and exposing these students to research experiences.

Developing a Simulation-Based Curriculum: Challenges and Opportunities

Isidoros Doxas

Center for Integrated Plasma Studies, CB 390, University of Colorado, Boulder, CO 80309-0390, isidoros.doxas@colorado.edu

One of the most troubling tendencies that science departments face today is an increase in the number of students that become disenfranchised of science (Seymour 1992, 1995). Increasing numbers of students become disenchanted with science after only their first or second science course, come to view science requirements as a burden, seek to take the minimum number of science hours required for graduation, and want nothing to do with science ever again.

One of the explanations that has been suggested for this problem is based on students' level of intellectual development (Piaget 1967). Students enter college more or less prepared for a higher education in the humanities. They have been largely taught the "grammar" of the humanities and usually require no more than a quick introduction before they can comfortably manipulate ideas (as opposed to fumbling with the structure of their arguments) in courses from History to Rhetoric. Unfortunately, the same is not true in the sciences. Any truly scientific discourse requires, of course, precise quantitative reasoning, and college freshmen just do not have the required math skills. We are therefore condemned to compete against courses in philosophy and political science that offer lively discussions on contemporary issues (courses that allow students to exercise their newfound ability to manipulate ideas) with science courses that are still largely concerned with the "drill" discipline required for teaching the "grammar" of science.

Simulation based curricula aim to increase student interest in introductory science courses for non-majors by putting the "grammar" of science in the background, so students can concentrate on its "prose." This is achieved by using the computational power placed at our disposal by today's PCs. The material is centered around a number of computer simulations that allow students to explore in relative depth problems that are complex enough to be of practical interest, while the simple physical principles behind those problems are presented in a just-in-time manner. This kind of computational capability not only helps us engage the students in true quantitative reasoning, it can also help us teach them the uncertainty inherent in current research topics, as opposed to the level of certainty that we accord more mature theories. We can thus help students graduate from the pervasive dualistic view of science (an answer is either right or wrong, and the job of the teacher is to teach us how to get to the right one, e.g., Nelson 1994) to more sophisticated intellectual models (Perry 1970, 1981) that allow room for critical thinking.

While promising to engage students at a higher cognitive level, simulation based curricula also present significant challenges. In order to study problems that are complex enough to be of practical interest, courses often need to deviate from the usual progression of concepts found in introductory science texts, and they occasionally even require concepts from another field. This presents a challenge to most faculty interested in using a simulation based curriculum, since departments are usually reluctant to rethink the sequencing and prerequisites of their courses. In addition, faculty interested in developing simulations for their courses quickly find that typical textbook explanations of even simple concepts like the seasons, or the phases of the moon, sweep a lot of critical details under the carpet, something that is not possible with a true simulation. Although the temptation to revert to a cartoon that exhibits the "correct" behavior is great, persistence with a real model can be very rewarding.

These challenges (and intellectual rewards) were faced during the development of the Solar System Collaboratory, a collaboration between four Colorado schools that aims to increase interest in science among non-science majors by engaging students in rigorous quantitative argumentation about topics of general interest, like the greenhouse effect. Over the past four years, Collaboratory participants have used simulation based modules to teach the basic science facts and concepts behind the greenhouse effect. Collaborative learning techniques were used throughout the module testing, and results were tracked using multiple triangulated methods. Student self perception of learning and attitudes towards science were tracked using questionnaires, actual use of the modules was tracked by direct observation and videotaping of students in class, and objective learning gains were tracked by matched before/after testing. Over 1500 students participated, although not all responded to all questions in the surveys, and the number of students interviewed, either individually or in focus groups, was approximately an order of magnitude lower.

Results show that students overwhelmingly (95%) perceive themselves as visual and/or hands on learners. Of those, 84% believe that computer-based simulations provide them with a visual and/or hands on learning experience. This data is based on students' perceptions of themselves as learners and their critical views of their learning environments, but they matter greatly; student self-perceptions and opinions are often a source of motivation for them. In addition, 98% of the students felt that they learned a lot from working with the modules. This self perception of learning is supported by normalized learning gains[1], which lie in the range 0.2-0.5. This is significantly higher that learning gains from standard lectures (e.g., Hake 1998), and are comparable to other pedagogically motivated curricula (e.g., Reddish & Steinberg 1999).

References

Hake, R. 1998, Am. J. Phys., 66, 64

Nelson, C. E. 1994, New Directions for Teaching and Learning 59, 45

Perry, W. G. Jr. 1970, Forms of Intellectual and Ethical Development in the College Years: A scheme (New York: Holt, Rinehart & Winston)

---

[1] Normalized learning gains are defined as [% score after]-[% score before]/(100-[% score before]), which is the gain in the actual score as a percentage of the maximum possible gain.

Perry, W. G. Jr. 1981, in The modern American college: Responding to the new realities of diverse students and a changing society, ed. A. W. Chickering (San Francisco: Jossey-Bass)

Piaget, J. 1967, Six psychological studies (New York: Vintage)

Redish, E. F., & Richard N. Steinberg 1999, Physics Today 52, 24

Seymour, E. 1992, Journal of College Science Teaching, 21, 230

Seymour, E. 1995, Science Education 79, 437

Making the Connection Between Formal and Informal Learning

Paul B. Dusenbery and Cherilynn A. Morrow

Space Science Institute, 3100 Marine Street, Suite A353, Boulder, CO 80303-1058, dusenbery@colorado.edu

The Space Science Institute (SSI) of Boulder, Colorado has recently developed two museum exhibits called the *Space Weather Center* and *MarsQuest*. It is currently planning to develop a third exhibit called *Cosmic Origins*. The *Space Weather Center* was developed in partnership with various research missions at NASA's Goddard Space Flight Center. These exhibitions provide research scientists the opportunity to engage in a number of activities that are vital to the success of these national outreach programs. The focus of this paper will be on how the *MarsQuest* project is making connections between informal and formal education and the roles that scientists can play in these endeavors.

Science centers and museums are playing an important role in the public's understanding of science through experiential learning. They serve millions of Americans. In 2000, over 120 million people visited the Association of Science-Technology Centers (ASTC) member institutions in the U.S. (ASTC 2001). An increasingly important part of the educational infrastructure is the number of programs that science centers offer for school children and teachers. It is estimated that in 2000 nearly 26 million schoolchildren were served by ASTC member institutions (ASTC 2001). Science Centers are playing an integral role in cultivating our country's science literacy.

In order to engage the museum-going public in the adventure of Mars exploration, the Space Science Institute of Boulder, Colorado developed *MarsQuest*. It includes a 5,000 square-foot traveling exhibition that is now touring the country, a 40 minute planetarium show, and a comprehensive education program. The exhibition will enable millions of Americans to share in the excitement of the scientific exploration of Mars and learn more about their own planet in the process. The associated education program will also be described, with particular emphasis on workshops to orient museum staff (e.g. museum educators and docents) and workshops for master educators near host museums and science centers. The workshops make innovative connections between the exhibition's interactive experiences and lesson plans aligned with the National Science Education Standards (NRC 1996). These exhibit programs are good models for actively involving scientists and their discoveries to help improve informal science education in the museum community and for forging a stronger connection between formal and informal education.

MarsQuest is currently on a three-year national tour, which is fully booked and scheduled to end in August 2003. It has been so successful that the tour manager, ASTC, has begun a waiting list of museums and science centers that are interested in booking the exhibition if a second tour is scheduled. The number of museums on the waiting list now exceeds the number of available slots. A second three-year tour would begin in early 2004, when Mars exploration will again be in the news. These ongoing and future missions create an unprecedented opportunity to use the public's excitement over Mars exploration to draw visitors to museums and science centers and generate interest in science in general.

MarsQuest has been a tremendously successful exhibit not just in terms of attendance, but in achieving its education goals. Randi Korn and Associates (RK&A) performed evaluations of *MarsQuest* at different stages in the project, including a final summative evaluation (Randi Korn et al. 2002). In the summative evaluation, RK&A found that all of the interviewees at two different host sites (Tucson, Arizona, and Hampton, Virginia) were able to articulate at least part of *MarsQuest*'s main message (Earth-Mars comparisons). With other science exhibits, RK&A has found that it is more common for visitors to be unaware that there is an overarching theme connecting exhibit components, or to be unable to describe the exhibition's main message. After visiting *MarsQuest*, all of the interviewees were also able to recall specific facts about Mars, especially information that compared Mars with Earth. For example, some visitors correctly indicated that some landscape features on Mars are much larger than similar features on Earth, while others noted differences between Earth and Mars in terms of size, temperature, and gravity.

SSI is committed to strengthening the infrastructure of informal science education through professional development activities. The *MarsQuest* exhibition provides a strong connection between formal and informal education. At each venue that hosts the exhibition, the Institute conducts separate workshops for museum educators, docents, and local teachers that are designed to inspire and empower participants to extend the excitement and science content of the exhibit and NASA's Mars Exploration Program into classrooms and museum-based education programs in an ongoing fashion (e.g. floor demonstrations and camp-ins). The workshops are well aligned with the best practices and standards for the professional development of educators (Loucks-Horsley 1998) as well as aligned with the National Science Education Standards (1997). These workshops were developed collaboratively by Dr. Cheri Morrow, Education and Public Outreach Manager of the Space Science Institute, and Sheri Klug, Director of the Mars K-12 Education Program at Arizona State University. The workshops are co-facilitated by Dr. Morrow and Sheri Klug. They desire that the workshops leave a legacy that remains after the exhibition leaves. This only happens if museum educators feel confident to include Mars classroom activities in their outreach program.

The Space Science Institute is also working to ensure mutually beneficial connections between *MarsQuest* host museums and the NASA Education infrastructure (e.g. JPL's Solar System Ambassadors and Solar System Educators, Educator Resource Centers, and Space Grant Colleges). The *MarsQuest* education program assembles a kit of exemplary educational materials (including both a Docent Guide and an Educator Guide), which are linked to *MarsQuest* and tailored for compatibility with host site educational programming. The workshops orient and equip museum staff adequately to enable ongoing implementation of educational programming enhancements linked to the exhibits. They provide opportunities for teachers in grades 4-9 to learn how to use exhibit components and related standards-based curricular materials with their students. SSI in partnership with TERC and JPL have support from NSF to create a Website, called *MarsQuest Online*, that will assist museum staff and teachers, in preparing for and following up with exhibit-related educational programming. Finally, the *MarsQuest* Education Program is forging new and lasting partnerships between exhibit host sites and Mars scientists.

References

Association of Science-Technology Centers 2001, ASTC Sourcebook of Science Centers Statistics (Washington: ASTC)

Loucks-Horsley, S., Hewson, P. W., Love, N., & Stiles, K. 1998, Designing Professional Development for Teachers of Science and Mathematics (Thousand Oaks, CA: Corwin Press)

National Research Council 1996, National Science Education Standards (Washington: National Academies Press)

Korn, R., et al. 2002, *MarsQuest* Summative Evaluation (Boulder, CO: Space Science Institute)

Developing Exhibitions through Public/Private Partnerships: A Case Study of the Space Weather Center Exhibit

Paul B. Dusenbery

Space Science Institute, 3100 Marine Street, Suite A353, Boulder, CO 80303-1058, dusenbery@colorado.com

Lou Mayo

Space Science Data Operations Office, Code 630, Goddard Space Flight Center, Greenbelt, MD 20771

In the Space Age, we are becoming more dependent on space-based operations for communications, navigation, weather reporting, treaty monitoring, scientific observation, and other critical activities. As a result, we are more susceptible than ever before to processes in our Sun-Earth environment. Severe space weather events, such as bursts of radiation and magnetic storms caused by the Sun's coronal mass ejections (CMEs), could impact satellite operations, harm astronauts, and result in power outages on Earth. In our technology-dependent society, the ability to predict which CMEs will reach us and when that will happen has become increasingly important.

Scientists now have access to the most powerful array of ground facilities and spacecraft ever assembled for studying the space environment. Sensitive telescopes focus on the Sun's many layers, spacecraft measure the plasma and magnetic fields of our geospace environment, a web of ground stations records the complex interaction between the Sun and our terrestrial environment, and computer models provide improved forecasting of space weather. The multi-agency National Space Weather Program (NSWP) encompasses the efforts of the U.S. researchers who are studying Sun-Earth connections and attempting to provide timely, accurate, and reliable space environment observations and forecasts.

The confluence of a number of successful space weather-related missions and research programs makes this an ideal time to inform the public about the drama of space weather and the value of space weather research. Science centers and museums are playing an important role in the public's understanding of science through experiential learning. They serve millions of Americans. In 2000, over 120 million people visited the Association of Science-Technology Centers (ASTC) member institutions in the U.S. (ASTC 2001). An increasingly important part of the educational infrastructure is the number of programs that science centers offer for school children and teachers. It is estimated that in 2000 nearly 26 million schoolchildren were served by ASTC member institutions (ASTC 2001). Science Centers are playing an integral role in cultivating our country's science literacy. Space weather is of particular interest for museums, science centers and planetaria because:

- Space weather impacts people's lives, both on the ground and in space;

- Space weather is a hot topic, with the Sun still near solar maximum;

- The general public is fascinated by space related subjects.

The Space Science Institute (SSI) in partnership with Sun-Earth Connection (SEC) Missions (e.g. ISTP, SOHO, ACE, HESSI, and IMAGE) and the GSFC/NASA Sun-Earth Connection Education Forum (SECEF) developed a traveling mini-exhibit about space weather that is now on its national tour to science centers, museums, and visitor centers. The Space Weather Center is a 1,000 square-foot, updateable, interactive exhibit. The exhibit was funded by grants from NASA and the National Science Foundation's Upper Atmospheric Research Section. GSFC is a leader in developing and managing space missions to study Sun-Earth connections. The Space Science Institute is a non-profit corporation whose mission is to integrate space science research with science education. SSI is an experienced developer of traveling exhibitions ranging from 5000 sq. ft. to less than 100 sq. ft. The Institute also has active programs in curriculum development, professional development for scientists and educators, and in space science research. SSI manages the national tour. The exhibit has traveled to eight science centers/visitor centers since it began its national tour in March 2000. It will be at the Kitt Peak Visitor Center during the fall 2002.

The size of the exhibit was carefully chosen to be consistent with the constraints of cost and marketing. Because large exhibits have fewer venues available to them, a small exhibit can be marketed to a much larger number of science centers. Besides the exhibit, museums and science centers receive the Solarscapes curriculum produced by the Space Science Institute, materials for teacher workshops, and education and public outreach products produced by individual Sun-Earth Connections missions. NASA and NSF provide SSI support to conduct classroom and museum educator workshops at each exhibit host site. The workshops are well aligned with the best practices and standards for the professional development of educators (Loucks-Horsley 1998) as well as aligned with the National Science Education Standards (1996). Dr. Cheri Morrow, SSI's Education and Public Outreach Manager, oversees the exhibit's education program. This program is another example of how the project is taking advantage of its partnership with GSFC.

The exhibition examines key concepts in space weather research illustrated by data from on-going missions in near real-time and shows visitors how space weather (disturbances in space driven by solar activity) plays a role in their everyday lives. Large photomurals of the Sun and Earth, graphics, interactive devices, models, and sound combine to create a total immersion experience, allowing visitors to explore the many realms of our Sun-Earth environment and to better understand what is meant by living in the Sun's atmosphere. The exhibition is organized into 3 main content areas: 1) Living in the Atmosphere of the Sun, 2) The Dynamic Sun, and 3) Earth: In the Path of the Storm.

At the start of the project, clear roles and responsibilities were defined for both SSI and GSFC partners. The development team for the exhibition was made up of SSI staff, NASA mission E/PO leads, NASA's Sun-Earth Connection Education Forum, & Condit Exhibits, Inc., an exhibit design and fabrication company located in Denver, Colorado. Mission E/PO leads at GSFC contributed text, graphics, video, as well as helped with selecting the design. SSI staff were responsible for creating the concept plan for the exhibition. They also selected the fabricator and the interactive devices. The Space Weather Center exhibition is an example of how multiple partners in the public and private sector can work together to create an effective outreach project that is national in scope.

References

Association of Science-Technology Centers 2001, ASTC Sourcebook of Science Centers Statistics (Washington: ASTC)

Loucks-Horsley, S., Hewson, P. W., Love, N., & Stiles, K. 1998, Designing Professional Development for Teachers of Science and Mathematics (Thousand Oaks, CA: Corwin Press)

National Research Council 1996, National Science Education Standards (Washington: National Academies Press)

IDEAS GRANT: An Opportunity for Creative Collaboration

Bonnie Eisenhamer and Heather Bradbury

Space Telescope Science Institute, 3700 San Martin Drive, Baltimore, MD 21218, bonnie@stsci.edu

Abstract. Since 1994, the Initiative to Develop Education through Astronomy and Space Science (IDEAS) grant program has provided funding for partnerships between astronomers/space scientists and the education and public outreach communities. The IDEAS grant program provides a unique opportunity for scientists, educators and public outreach professionals to work together and explore creative approaches to developing education and public outreach programs and/or products. Highlighting three exemplary IDEAS programs, the poster provides insight into how a small amount of funding and creative collaboration can have a significant impact on target communities.

The IDEAS Grant Program is one component of NASA's Office of Space Science (OSS) Education and Public Outreach (E/PO) Strategy. The program is administered by the Space Telescope Science Institute (STScI) on behalf of NASA OSS. As part of the overall OSS E/PO program, the IDEAS Grant Program provides start-up funding for innovative, creative education and public outreach projects that feature active collaboration between astronomers/space scientists and formal education/informal education professionals. The IDEAS objective is to enhance science, mathematics, and/or technology education in the United States for K-14 students, teachers, and the general public. The program promotes partnerships that explore new ways to translate astronomy and space science into contexts that will educate and stimulate the interest of students, teachers, and the general public.

Creative Collaboration

A key goal for the IDEAS Grant Program is to enhance science education through astronomy and space science via creative collaboration between professional astronomers/space scientists and professional educators/informal science professionals. Creative collaboration not only refers to the partnership formed by the team members of programs but also to the wider community of audiences and venues served. There are many possible untapped audiences and communities that could be targeted with an astronomy/space science program, for instance, museums, national parks, universities, and the Boy/Girl Scouts. Additionally, there are dissemination possibilities, including publication in professional education journals; workshops/trainings at professional organization events; and traveling/replicating exhibitions, that could amplify the effect a program has on a broader community.

Model Programs

There have been many successful IDEAS programs that have used creative collaboration to leverage their programs and impact science education. Examples of such programs can be found in classroom technology program, professional development workshop, and a large-scale general public exhibit.

In 1994, a $20,000 IDEAS grant was used to fund a proposal targeting classroom technology. The program provided an inner-city, minority public school the opportunity to develop a computer network infrastructure in partnership with a local university. The grant also provided funding for a T-1 line at the school. Using the T-1 connection, educators at the public school were able to

- use the Internet as a resource to develop classroom activities,
- allow students to explore the Internet in a structured setting, and
- partner with the local university to train teachers/students in technology management.

The program continues today with expanded goals and objectives, which include infrastructure, system management/maintenance, training, and classroom technology integration for 29 public schools. Over the years, the program has received additional funding from numerous sources including a $200,000 award from the State Board of Education.

Another 1994 proposal requested $20,000 in seed money for a feasibility study to develop a permanent education exhibit on the National Mall. The proposed exhibit, Voyage, would be a 1/10,000,000,000 scale model that included all solar system bodies and moons, where appropriate. The exhibit design would feature solar system objects displayed on pedestals with modified National Park Service wayside stations beside each object. The wayside stations would provide accompanying information about the solar system object in major languages and be wheelchair accessible. The content for the wayside stations would foster interdisciplinary, inquiry-based learning, stress conceptual development, and align with national education standards for K-12 space science education. After several years, the partnership of national organizations, including the Smithsonian Institution, NASA and the Challenger Center, received approval and additional funding to build the exhibit. The final exhibit premiered on the National Mall October 17, 2001. There is a visitor's activity guide that accompanies the exhibit as well as plans to develop additional supplemental materials for classrooms and families. Installation plans for the Voyage exhibit are available so that it can be replicated at other science museums, universities, planetaria, and other community organizations.

In 1995, a $10,000 grant for professional development provided a diverse team of university astronomers and education faculty, and secondary teachers the opportunity to design an intensive, short-term university level workshop. The workshop offered rural and inner city high/middle school earth science and physics teachers the chance to study both traditional and modern astronomy for either university or in-service (professional development) credit. Participating educators were immersed in a wide variety of activities, including:

IDEAS GRANT: An Opportunity for Creative Collaboration 283

- attending lectures from renowned astronomers and science education experts,
- making night time observations,
- learning how to use PCs and workstations to retrieve information via the Internet,
- observing demonstrations that used inexpensive supplies, and
- how to use a planetarium.

The program not only continues but has expanded from one workshop to three:

- a stars and planets workshop,
- a galaxies and cosmology workshop, and
- an astrobiology and the new world of advanced materials workshop.

The program's expansion not only provided more topics for educators but expanded the collaborative effort of the university by including the College of Science, the College of Earth and Mineral Sciences, Continuing Education, and the Space Grant Consortium. The program maintains itself through a nominal participant fee and by volunteers from the astronomy and science education communities.

The American Geological Institute's National Earth Science Week

Lynsey Ellis, Caitlin Callahan, Cynthia Martinez, and Michael Smith

American Geological Institute 4220 King Street Alexandria, VA 22302, lee@agiweb.org

As the caretakers of the Earth, we have a responsibility to learn about the processes that happen in, on, and around our planet. We study the Earth in order to know how best to use its resources and to gain an appreciation for its potential hazards. Together we share an obligation to educate ourselves in order to make well-informed decisions about the Earth and our environment. Earth Science Week is one way to meet this challenge and bring the study of the Earth into the mainstream.

On behalf of its member societies and the larger geoscience community, the American Geological Institute (AGI) established Earth Science Week in 1998 to raise public awareness and understanding of geoscience professions and the contributions Earth sciences add to society. Earth Science Week is a national Earth science outreach program with four main objectives: 1) to give students new opportunities to discover the Earth sciences, 2) publicize the message that Earth science is all around us, 3) encourage stewardship of the Earth through an understanding of Earth processes, and 4) motivate geoscientists to share their knowledge and enthusiasm about the Earth. The main motivation for Earth Science Week is to help people learn to appreciate the value of Earth science research and its applications and relevance to daily life.

Since its origin, Earth Science Week (ESW) has been held the second full week of October. Earth Science Week focuses on a different facet of Earth science each year to help all people gain a better understanding and appreciation of the natural world. "Water is All Around You" is the 2002 ESW theme, emphasizing the importance of the Earth's greatest natural resource. Water is important to all life on Earth and plays a vital role in the Earth System. The theme also coincides with the 30th anniversary of the Clean Water Act, marking 2002 as the Year of Clean Water, the United States Geological Survey (USGS) National Water Monitoring Day on October 18, 2002, and the publication of the AGI Environmental Awareness Series Booklet and Poster 'Water and the Environment'.

Earth Science Week targets members of all communities. Events in celebration of Earth Science Week are hosted on national, state, and local levels, from activities within the National Parks to contests, games, and field trips for schools and classrooms. AGI supports many of these activities by producing and distributing outreach materials, including information kits and posters relating to the year's theme. Each kit contains an ideas and activities booklet and a variety of useful posters, bookmarks, and information sheets. In 2001, AGI distributed 150,000 Earth Science Week posters and over 10,000 information kits nationwide. The National Park Service (NPS) is very involved in Earth Science Week programming. In 2001, the NPS provided outreach materials to all national parks and monuments. A NPS Earth Science Week website was also developed to highlight events and activities hosted within the parks (www.nature.nps.gov/grd/esw/). Activities in the national parks ranged from special programs in Acadia National Park in Maine to naturalist talks and hikes covering the geology or biology of Haleakala National Park in Hawaii.

Many groups and organizations also support Earth Science Week by hosting events to raise the local awareness of Earth science. The Austin Earth Science Week Consortium, a group of volunteers from city and government agencies, museums, nature centers, universities and different industries, is developing an annual Earth Science Week program in Austin. This program includes a career fair for middle-school students and a fundraising book drive for the Austin Public Library to purchase Earth science books. Also in Texas, the Houston Geological Society, working with other organizations, hosted a number of Earth Science Week events in 2001. Activities included the 2nd Earth Exploration Extravaganza, a day of family fun at the Houston Museum of Natural Science; an essay contest for middle-school students; classroom presentations in local schools; field trips; and, a special symposium entitled *Views of the Earth: An Earth Science Symposium*. In St. Petersburg, FL, the USGS Center for Coastal and Regional Marine Studies celebrated ESW by hosting its third annual open house, "2001: A Science Odyssey."

Plans for events are already well underway for 2002. One such event is the 'Taste of DC'. The 'Taste' is an outdoor food and music festival event held in Washington, DC in which local vendors can display their fare. An Earth science pavilion sponsored by AGI is also incorporated into the festival. At the pavilion, festival-goers can participate in water related activities, learn about Earth sciences, and pick up informative literature discussing Earth science and Earth Science Week.

Members of the Earth science community have helped to make Earth Science Week a recognized event by soliciting proclamations from city mayors and state governors. These proclamations have served to emphasize the importance of celebrating Earth Science Week. Seven states have issued perpetual proclamations: Alaska, Delaware, Illinois, Nevada, North Dakota, Oklahoma, and South Dakota. In addition, New Mexico has already issued a 2002 proclamation. If your state has not yet officially recognized Earth Science Week, you can work with your state geological survey to encourage your governor to proclaim ESW 2002 an event in your state.

To learn more about Earth Science Week and to order an information kit, please visit www.earthsciweek.org. This site contains helpful hints for planning activities, as well as activity ideas. Links to water-based resources and other information are also available.

Support for Earth Science Week is provided by the United States Geological Survey, the American Association of Petroleum Geologists Foundation, Member Societies of AGI, and individual contributors.

The Educational Activities of the Astronomical Society of the Pacific

Andrew Fraknoi, Suzanne Chippindale, and Michael Bennett

A.S.P., 390 Ashton Avenue, San Francisco, CA 94112, fraknoi@fhda.edu

We report on some of the educational programs and publications of the Astronomical Society of the Pacific (ASP) that may be of particular interest to scientists and educators who are involved with NASA E/PO activities. Founded in 1889, the ASP now has members in all 50 states and over 60 other countries, and is the largest general astronomy organization in the U.S. The ASP devotes a considerable part of its resources to education and outreach (in both the formal and informal arenas). As discussed below, the Society has several networks of educators and a range of publications and programs that could be of use to missions and projects that wish to test or disseminate their materials or activities more widely.

Project ASTRO Network

This national program, with 13 active regional sites around the country, links professional and amateur astronomers with local 4th - 9th grade teachers. Each site offers training workshops and a wide range of hands-on materials. A series of guides to doing inquiry-based activities, called *The Universe at Your Fingertips*, has been developed and is in wide use in classrooms and teaching training programs. (See separate poster.)
http://www.astrosociety.org/education/astro/project_astro.html

Family ASTRO

This new program is creating a novel series of kits with hands-on astronomy activities for families, and training Project ASTRO partners, planetarium and museum staff, and others to do local evening and weekend family events in a number of the ASTRO sites.
http://www.astrosociety.org/education/family.html

The ASP Catalog of Educational Products

Our non-profit catalog of educational materials for astronomy, earth science, and space science goes to 300,000 people around the world, particularly educators at all levels. Includes audio-visual materials, software, observing aids, resource guides, posters, children's books, and more. Materials from NASA projects have been part of the catalog for years. An improved version is now on the Web, at the new ASTRO-SHOP section of:
http://www.astrosociety.org

Mercury Magazine

The ASP's popular-level magazine covers astronomy, astronomy education, interdisciplinary issues, and Society news. The magazine began in the 1920s under the title Leaflets of the A.S.P. See:
http://www.astrosociety.org/pubs/mercury/mercury.html

Newsletter for Teachers

The Universe in the Classroom, our popular newsletter on teaching astronomy in grades 3-12 that began in 1984, is now hosted on the web. It's filled with astronomy articles, classroom-ready activities, and resources specifically aimed at teachers.
http://www.astrosociety.org/education/publications/tnl/tnl.html

Workshops for Teachers

Since 1980, the ASP has sponsored "Universe in the Classroom" workshops - 2 to 4 day sessions to help teachers in grades 3-12 learn more about astronomy and effective teaching techniques, with a special focus on hands-on activities. Between 100 to 200 teachers attend each summer; we have trained several thousand teachers from around the U.S. and Canada over the years.

SOFIA Education and Outreach

Together with the SETI Institute, the ASP is providing educational and public outreach programs for NASA's Stratospheric Observatory for Infrared Astronomy project. (See separate poster.)
http://www.astrosociety.org/education/sofia.html

Amateur Ambassadors Project

Amateur astronomers have been an active part of the Society's membership since 1889, and the Society works with amateurs and amateur clubs in Project ASTRO and other programs. We now have a program to determine what amateurs around the country need to be able to do more in education and outreach and to create materials to help them.

Symposia for College Astronomy 101 Instructors

Every three years or so, the ASP organizes a hands-on "Cosmos in the Classroom" symposium on teaching astronomy to undergraduate non-science majors. The next one is being planned for summer 2004 in cooperation with the New England Space Science Initiative in Education (NESSIE). The proceedings of the 2000 symposium are still available through the ASP Catalog.
http://www.astrosociety.org/education/cosmos.html

Community and Small College Survey and Network

We are identifying and surveying astronomy instructors at colleges that do not have research programs in astronomy, to encourage their fuller participation in the astronomy education community. Over 800 instructors have already joined. The survey can be found at:
http://www.astrosociety.org/education/survey.html

Resource Materials for Education

The ASP provides frequently-updated resource guides for educators at all levels, both in printed and web format. Examples include "Debunking Pseudo-science," "Web Sites for College Astronomy Instructors," "The Best K-12 Astronomy Activities on the Web," "Astronomy & Environmental Issues" and "Women in Astronomy."
http://www.astrosociety.org/education/resources/resources.html

International Awards

The Society offers 7 awards, include the Klumpke-Roberts Award for lifetime contributions to public understanding of astronomy, the Brennan Prize for outstanding high school teacher of astronomy, and the Las Cumbres Prize for outreach by an amateur astronomer.
http://www.astrosociety.org/membership/awards/awards.html

Project ASTRO: A National Network Helping Teachers & Families with Astronomy Education

Andrew Fraknoi, Erica Howson, and Suzanne Chippindale

Astronomical Society of the Pacific, 390 Ashton Ave., San Francisco, CA 94112, fraknoi@fhda.edu

Dennis Schatz

Pacific Science Center, 200 Second Ave., N., Seattle, WA 98109

About Project ASTRO

- links volunteer astronomers (professional & amateur) in one-on-one partnerships with 4th - 9th grade teachers in their communities

- partners receive training in 2-day workshops that provide them with a wide range of effective teaching techniques, hands-on activities, and useful resources.

- astronomers commit to making at least 4 visits to "their" classroom each year and work with teacher on a plan appropriate for their interests and the students' needs.

- a few astronomers assist youth groups and after-school groups, rather than formal classes.

- supported by the Informal Science Education Program of NSF, NASA's Offices of Space Science and Education, several foundations and corporations, and individual donations.

Expansion of Project ASTRO

- now operating in 13 regional sites around the country

- each site has a lead institution, which coordinates the local program (lead institutions vary from a university astronomy department to a space museum)

- each site is supported by a coalition of local scientific and educational organizations (astronomy departments, amateur clubs, museums, observatories, school districts, etc.)

- sites must find their own operating expenses and are all now self-supporting

- site leaders keep in touch through a national network (communicating via electronic messages, teleconferences, and an annual site-leaders' meeting)

Publications from Project ASTRO

- *So You Want to Start an ASTRO Site*: 4-page intro to what it takes to organize a site;

- *The Project ASTRO How-to-Manual*: 48-page booklet of practical partnership suggestions;

- *The Universe at Your Fingertips*: 813-page notebook: 87 hands-on activities, resource guides, teaching suggestions, etc. (most activities from other sources around the country)

- *More Universe at Your Fingertips*: 356-page supplement, with 27 other activities, guides.

- *El Universo a sus pies*: Spanish translation of the "greatest hits" from the above 2 volumes

The first of these is free; others available through the A.S.P. Catalog: 1-800-335-2624 or see the ASTRO SHOP at our web site: www.astrosociety.org

Family ASTRO

- new project to apply the ASTRO techniques and network to involve families in doing astronomy activities (began in April 2000, with support from NSF)

- four-year pilot program to test weekend/evening family events, materials, training

- developing take-home kits and games in both English and Spanish for doing astronomy

- we are working with both current ASTRO partners and with museum and planetarium staff

- the first two kits are being pilot tested in several regional sites in 2001-2002: "Night Sky Adventure" and "Race to the Planets"

- the next two kits will be "Moon Mission" and "Cosmic Decoder"

- A game company is helping us develop professionally designed, educational astronomy games that will encourage families to do astronomy in their discretionary time

Results

- Evaluation showed teachers involved in Project ASTRO taught significantly more space science (and more science in general) in subsequent classes.

- Partners report significant changes in their lives and work as a result of the ASTRO experience: astronomers changing their college teaching, several amateurs going into teaching as a career, and teachers who were afraid of science now taking the lead in science reform.
- Different sites are taking the program in interesting directions, including work with 4-H clubs, the Chicago Housing authority, Native American reservations in the southwest, etc.
- The ASTRO Network has now trained over 1100 teacher-astronomers pairs. We estimate that over 80,000 students have been affected since 1994.
- ASTRO materials are now in use in over 27,000 schools and science museums.
- The ASTRO network stands ready to work with NASA missions, projects, or centers who want a group of dedicated teachers and astronomers to try out something they are developing, or want to adopt the ASTRO techniques and activities in their own settings.

The Project ASTRO Sites and Their Lead Institutions

1. Boston (Center for Astrophysics & Boston Museum of Science)
2. Central New Jersey (Raritan Valley Community College)
3. Chicago (Adler Planetarium)
4. Connecticut (Wesleyan University)
5. Eastern Washington State (Washington State University, Pullman)
6. New Mexico (New Mexico Museum of Space History, Alamagordo)
7. Northwestern Michigan (Northwestern Michigan College)
8. Ohio (Ohio Aerospace Institute)
9. Salt Lake City (Hansen Planetarium)
10. San Diego (San Diego State University & University of California, San Diego)
11. San Francisco Bay Area (Astronomical Society of the Pacific)
12. Seattle (University of Washington Astronomy Dept.)
13. Tucson (National Optical Astronomy Observatories)

For more information and contact people for each site (or the national program) see:
www.astrosociety.org/education/astro/project_astro.html

The Silicon Valley Astronomy Lectures: An Example of Institutional Cooperation for Public Outreach

Andrew Fraknoi

Astronomical Society of the Pacific, 390 Ashton Ave., San Francisco, CA 94112, fraknoi@fhda.edu

Kathleen Burton and David Morse

NASA Ames Research Center

For the last three years, three astronomy institutions in the San Francisco Bay Area have been cooperating with the Public Affairs team at NASA's Ames Research Center to produce a major evening public-lecture series on astronomy and space science topics. Cosponsored by Foothill College's Astronomy Program, the Astronomical Society of the Pacific, the SETI Institute, and NASA Ames, the six annual Silicon Valley Astronomy Lectures have drawn audiences ranging from 450 to 1000 people, and represent a significant opportunity to get information about modern research out to the public.

Foothill College provides a large theater for the series and the audiovisual set up. NASA Ames designs, prints, and mails a postcard about each lecture to a joint mailing list of about 10,000 people. All four groups work together in deciding on speakers, pooling mailing lists, and doing publicity through the media (as well as their own web sites, membership publications, and e-mail exploders.) Free notices about the series appear in a number of local newspapers and sometimes on radio.

A special effort is made to invite teachers from around the Bay Area and some high school teachers bring entire classes of students. We also have an e-mail list of every astronomy instructor (from adult school to community college to university) in the area, who are given advanced notification of each lecture.

Below is a list of the speakers so far. One of the talks, by Vera Rubin, was a celebration of the centennial of the American Astronomical Society, and they paid the speaker's way to San Francisco. Geoff Marcy's talk, in which he made the first public announcement of two new planets, drew quite a bit of media coverage. There has been a particular focus on topics in astrobiology for the series, a field which is one of the areas of special expertise at Ames.

The series is perhaps most noteworthy for the smooth way in which the institutions involved pool their resources. NASA Ames does not have an auditorium large enough for the series. The other organizations do not have the public affairs capabilities and media access that NASA does. Ideas and contacts for speakers come from all the groups involved. The audience response has been very positive, with many people indicating how pleased they are to have a chance to hear the latest developments directly from the scientists involved. And the college administration is delighted (and occasionally surprised) to see more people coming to the college for astronomy lectures than for football games.

The Silicon Valley Astronomy Lecture Series: Lectures So Far

1999-2000 School Year

- October 13, 1999: Dr. Geoff Marcy (University of California, Berkeley): *Finding New Worlds Around Other Stars.*

- November 17, 1999: Dr. Jill Tarter (SETI Institute): *Making Contact: The Search for Extraterrestrial Intelligence.*

- January 26, 2000: Dr. Alexei Filippenko (U. of California, Berkeley): *Einstein's Biggest Blunder: New Discoveries about Cosmic 'Antigravity'.*

- March 1, 2000: Drs. Christopher McKay (NASA Ames) and Margaret Race (SETI Institute): *Missions to Mars: Exploring the Red Planet.*

- April 12, 2000: Dr. Sallie Baliunas (Harvard-Smithsonian Center for Astrophysics): *The Changing Sun and the Climate of the Earth: Why Louis XIV Had Cold Feet.*

- May 3, 2000: Drs. Jeff Cuzzi, Dale Cruikshank, and Jeff Moore (NASA Ames): *Cold Hard Worlds at the Edge of the Solar System.*

2000-2001 School Year:

- October 11, 2000: Dr. David Morrison (NASA Ames Research Center): *What Killed the Dinosaurs: The Asteroid Threat and What We Can Do About It.*

- November 15, 2000: Dr. Sandra Faber (University of California, Santa Cruz): *Images from the Hubble Space Telescope: How they are Changing our Perspective.*

- January 24, 2001: Drs. Peter Ward (University of Washington) & Frank Drake (SETI Inst.): *The Rare Earth Hypothesis: Are Good Planets and Life Hard to Find?*

- March 7, 2001: Dr. Pascal Lee (SETI Institute): *Finding Mars on Earth.*

- April 11, 2001: Dr. Greg Laughlin (NASA Ames Research Ctr.): *The Ultimate Fate of the Sun and the Solar System.*

- May 2, 2001: Dr. Vera Rubin (Carnegie Institution of Washington): *What's the Matter in the Universe? (A Talk on Dark Matter).*

2001-2002 School Year

- October 10, 2001: Dr. Chris Chyba (SETI Institute & Stanford University): *Life in the Universe: Is it Just Around the Corner?*

- November 14, 2001: Dr. Lynn Cominsky (Sonoma State University): *Exploding Stars, Blazing Galaxies, and Giant Black Holes: The Extreme Universe of Gamma-ray Astronomy.*

- January 23, 2002: Mr. Scott Hubbard (NASA Ames Research Center): *Following the Water: The New Program for Mars Exploration.*

- March 6, 2002: Dr. Debra Fischer (University of California, Berkeley): *Planets Beyond: The Search for Other Solar Systems.*

- April 10, 2002: Dr. Alexei Filippenko (University of California: Berkeley): *Why I Believe in the Big Bang: Evidence about the Origin of the Universe.*

- May 1, 2002: Dr. Gregory Benford (University of California, Irvine): *Navigating the Gulf: The Borderline of Science and Fiction.*

Designing Collaborative Learning into the Curriculum

Kathy Garvin-Doxas

University of Colorado, Alliance for Technology, Learning and Society (ATLAS) Institute, Evaluation and Research Group, CB-040UCB, Boulder, CO 80309-040, garvindo@colorado.edu

Barriers to Collaboration

The most common misconception held by students about collaboration is that it is "getting along" with others or essentially people skills. This focus on "getting along" as opposed to learning or exploring new ideas, leads to certain types of classroom interaction styles and behaviors that work against true collaborative learning. Our research shows that the most important of these interaction and behavioral outcomes in space science classrooms are: *1) Failure to use conflict and disagreement to enhance understanding.* Discussion that enhances learning requires the expression of multiple perspectives and trust in fellow class members. This is conflict in the more robust sense of the word. *2) Tendency to focus on task only.* There is a constant tension in any collaboration between task-orientation and relational-orientation (personal exchanges that build trust and confidence – knowledge of students as individuals rather than anonymity). Good collaborative learning requires a balance of both orientations. *3) Premature appeals to authority.* When students focus on getting along rather than on the expression of multiple view points, they consistently avoid conflict by appealing, at the first sign of disagreement, to an outside authority (e.g., a TA) to mediate their discussion by providing them with the "right" answer.

Another barrier to collaborative learning is students' belief that the ability to collaborate is innate, but can be enhanced through experience. Although students can recite the politically correct stance on collaboration (it is necessary, it works well, and I'm good at collaborating), we found that this deeply-held misconception about collaboration is actually very widespread among students. This leads to other behaviors that interfere in collaborative learning: *1) Not hearing instructions about collaboration.* Since they believe that you cannot learn to collaborate, they typically ignore any instruction given on how to function effectively in a group. *2) Failure to take responsibility for group outcomes.* Armed with the belief that collaboration cannot be taught, students fail to assume any responsibility for the way their group functions as well as the quality of the learning that takes place.

Collaboration is also stymied by the student belief that *science requires no discussion*. This is a direct result of the fact that they have been socialized to believe that science is either right or wrong. When seen this way, it reduces learning in the subject to memorization only. This view also reinforces a desire to avoid making public mistakes. *Students also tend to resist the strategies that are typically used to involve them* in class and in group discussions and collaboration. When teachers pose questions as a means of developing discussion, students feel that they are being manipulated and that the teacher is being dishonest because clearly, s/he knows the correct answer. This sense that teachers know the 'right' answer leads to an underlying assumption by students that s/he is purposefully withholding information. Recognizing these strategies as strategic, leads to a defensive classroom climate and a lack of open communication (Gibb 1961).

Creating a Collaborative Learning Environment

Students need to be *encouraged to build and maintain personal relationships* in their class and in their groups. Create a comfortable learning environment where students feel safe to misunderstand or not know something in front of their classmates (e.g., Barker, Garvin-Doxas, & Jackson 2002). This will enable students to have and resolve meaningful conflict. A high level of trust and safety are necessary for discussion. Students also need to have *explicit instruction on how to work together* on assignments in ways that enhance learning and understanding. Thus, you must make your goals for group work explicit. Are they working in a group to enhance their learning and understanding; are they working in a group because there is not enough lab equipment for everyone; etc.? Collaboration in the learning environment requires that you give students guidelines for working effectively together. You can assign and/or discuss the fluidity of roles they can adopt when working together. Give them opportunities to practice these various roles. Provide them with a sense of control by discussing what they can do if they feel a member of the group is not doing his/her share of the work. Students need you to *model good discussion strategies for them*. Share with your students the process you follow when you think through a question. How do you decide what the question is asking; determine the first step you will take; determine when you have answered the question? How do you know when you understand the material? As part of modeling, you should ask questions that help them work through problems on their own and that enhance discussion as part of interactive lectures. If we know ___, what do we need to find out next? How can we do that? In modeling correct discussion strategies, you are also sending them the message that the process of solving problems and answering questions is as important as producing the "right" answer. Finally, you must *debrief all group assignments with them before they leave class*. Students cannot be comfortable with the outcome of their collaboration unless they realize that they have learned and understood what you intended for them to. Help them summarize what they should have learned that day in class.

The teacher's role in a collaborative learning environment is that of a facilitator of knowledge through discovery rather than a transmitter of information. There are many things teachers can do both in terms of course material and their own behavior: 1) Teachers must explicitly state that learning scientific concepts comes from discussion, not simply memorization of facts. Students need to hear that to understand concepts, they must talk about them to one another and to the teacher. Students also need to be told that science means engaging in a process of observation, experimentation, and theory-building. 2) It is also important for the teacher to introduce the students to the type of science that they know, experience, and live because it enhances trust within the learning environment and provides students with a context for understanding the scientific process. Science is built on fundamental principles, but these principles may be challenged as new data becomes available (e.g., Kuhn 1970). 3) Teachers must model the scientific process for students. At the same time, it is important that the language used to model the process does not seem contrived or forced; something that contributes to a defensive communication climate and negates trust and openness (e.g., Barker et al. 2002). 4) Finally, assignments and activities must require collaboration. While simple assignments that work toward lower levels of understanding do not require collaboration, complex assignments that contribute to students' conceptual understanding and the ability to apply their knowledge to different situations are often enhanced by it. Collaboration is only successful when it is motivated by the nature of the assignment and/or activity.

References

Barker, L. J., Garvin-Doxas, K., & Jackson, M. J. 2002, in Proceedings of the 33rd SIGCSE Technical Symposium On Computer Science Education, p. 43

Gibb, J. R. 1961, Journal of Communication, 11-12, 141

Kuhn, T. S. 1970, The Structure of Scientific Revolutions (2nd ed.) (Chicago: The University of Chicago Press)

Building STARBASE: Robotic Telescopes for Hands-On Science Education

Richard Gelderman

Department of Physics and Astronomy, Western Kentucky University, Bowling Green, KY 42101, richard.gelderman@wku.edu

Abstract. Funded as a NASA Office of Space Science education program, STARBASE (Students Training for Achievement in Research Based on Analytical Space-science Experiences) consists of a network of networks, developed to connect high school students and teachers with cutting edge research. Standards-based assessment has become the major driver for secondary schools in most states, creating a need for hands-on, interactive learning experiences aligned with the National Science Standards and Benchmarks. STARBASE faculty participate in teacher training workshops and visit science classrooms in secondary schools. Preliminary evaluation provides evidence that our collaborative efforts are much more successful when a substantial effort is made to maintain communication with the high school science teacher, and when lesson plans and supporting materials are provided to the teachers.

STARBASE: Students Training for Achievement in Research Based on Analytical Space-science Experiences

STARBASE is comprised of a network of networks: telescopes, professional astronomers, high school science teachers and students. The early phases of the STARBASE project have concentrated on the establishment of a global network of research quality astronomical facilities. Each of the observatories is operable remotely via attended operation over the Internet. In the near future the observatory systems shall be upgraded to provide robotic control, executing scripted observations without real-time human oversight.

STARBASE is also a network of astronomy research institutions and university faculty participating in cutting edge astrophysical investigations. Current STARBASE partners include Western Kentucky University, South Carolina State University, Planetary Science Institute, and Francis Marion University. Current research projects revolve around photometric variability of quasars, individual stars, binary star systems, and variability due to transits of extrasolar planets. With our robotic telescopes and research projects coming into place, the focus of STARBASE is turning toward expansion of the network of high school students and teachers.

The Need for STARBASE

Across the nation, state departments of education are assessing the performance of the public schools through tests aligned with state and/or national standards. Unfortunately, very few teachers have meaningful experience with the hands-on, interactive, and student-centered pedagogical approaches suggested by the National Science Education Standards and the Project 2061 Benchmarks. One of the primary goals of STARBASE is that more high school students should be introduced to science through the same real-life experiences that encouraged our professional astronomers to work toward a career in science.

Examples of STARBASE at work

- Astronomy workshops, stressing topics covered by middle grades Standards
- Hands-On Universe (HOU; hou.lbl.gov) teacher courses, incorporating astronomical images into high school curricula
- Faculty visits to integrated science, chemistry, physics, and astronomy classes
- Faculty mentors assist students with analysis of astronomical images
- Sponsorship of and participation in science clubs and star parties

Preliminary Evaluation of STARBASE

Collaborations between teachers and researchers take serious effort. Contrary to expectations of most funding opportunities, there is no magic scenario that will be reproduced across the nation at some minimal cost to the adoptees. What can be provided by NASA (or some other leading national agency or organization) is the means to better communication. In the spirit of improving communication, we offer a list of some of the many lessons learned as STARBASE has grown to include more university researchers and more high schools.

Public school systems typically do not provide time/support for teachers to grow and develop. High school teachers have very little free time. They require ready-to-use materials and information that will directly improve their teaching. Materials should initially be useable as a cookbook, yet be sophisticated enough to include room for them to grow into the material and become experts.

Teachers are busy and drop out of sight easily; it takes a great deal of coordination to keep collaborations going. The most successful relationships include abundant organization and pre-scheduling to establish a workable strategy; along with enough mutual flexibility to pull it all off under real-life circumstances.

Mass mailings (either snail or electronic mail) are almost never successful. Even the most fantastic new curriculum resource won't be used if it does not receive a personal recommendation from a trusted colleague. Glitzy packaging is a waste if the teachers never look at the material being forwarded.

Classroom visits can make much more of an impact if the researcher reinforces the curriculum being covered. The classroom teacher and researcher need to establish appropriate topics to be covered, as well as the visit's context within the overall lesson plan. A visit to the classroom will be quickly lost in the chaos of the high school experience if the visit isn't promptly followed up.

There are no decent astronomy texts for high school students. Adopting college-level texts is almost never an adequate solution, because college texts do not attempt to address the National Science Education Standards. With nearly every state education system driven by the Standards, it would be a huge benefit to all if someone developed a general purpose space-science text aligned with the Standards. Teachers do not have room for a supplementary curriculum; however they typically need to be provided with enough information so that their students will be able to map out the goals of the lesson, locate the related information, and understand the connections between various topics.

Touch The Universe: A NASA Braille Book of Astronomy

Noreen Grice[1], Bernhard Beck-Winchatz[2], Ben Wentworth[3], and Michaela R. Winchatz[4]

[1] You Can Do Astronomy, 125 Jones Drive, New Britain, CT 06053, ngrice5456@aol.com

[2] DePaul University, Space Science Center for Education and Outreach, 990 W Fullerton, Suite 4400, Chicago, IL 60614

[3] Colorado School for the Deaf and the Blind, 33 N. Institute St. Colorado Springs, CO 80903

[4] DePaul University, Department of Communication, 2320 North Kenmore Ave., Chicago, IL 60614

Introduction

People who are blind or visually impaired are often at a disadvantage in astronomy and space science because of the ubiquity of important graphical information not accessible to them. Like their sighted peers, many blind students have a natural interest in space, which can motivate them to learn fundamental science, math and technology concepts and skills. However, the lack of appropriate K-12 resources makes it difficult for teachers and parents to engage students in science. Those who do show an interest are often discouraged from pursuing these fields (Scadden 1996). Thus, many blind students ultimately lose interest. While advancements in technology have made it much easier for people who are blind or visually impaired to pursue scientific careers, their numbers remain small (Jackson 2002, Sakaran 1995).

As a first step toward making current NASA resources more accessible to the blind and visually impaired, we have developed *Touch the Universe: A NASA Braille Book of Astronomy* (Grice 2002a), a book that is based on a series of Hubble Space Telescope (HST) images. To make the images accessible to blind readers, shapes and colors are represented by raised lines, symbols and other textures. Simple explanatory text is given in both Braille and large print. The images are organized in order of increasing distance from Earth and take the reader on a journey of discovery from our solar system "neighborhood" to some of the most distant objects ever observed.

Development of Tactile Images

One of the most challenging tasks in the development of the book was the selection and design of tactile versions of HST images. To avoid clutter, they needed to be simplified by eliminating less important elements without compromising those parts that are needed to convey crucial scientific content. This proved particularly difficult for images of diffuse gas such as nebulae and other objects without pronounced visual boundaries. We found the guidelines for the design of tactile graphics developed by several groups (e.g., TAEVIS 2002, Edman 1992, American Printinghouse for the Blind 1997, Levi & Rolli 1994, Eriksson 1999) to be very helpful. In addition, prototype images and explanatory text were carefully evaluated with students at the Colorado School for the Deaf and the Blind. Their detailed comments were used for refinements. This cycle of design, testing, and refinement was iterated until a satisfactory result was achieved, i.e., until the students found both images and text clear and understandable.

Production of the Book

Prototypes of *Touch the Universe* were first presented at a press conference and in a poster paper during the 2001 summer meeting of the American Astronomical Society (Grice & Beck-Winchatz 2001). Since the book was developed on a modest $10,000 Cycle 9 E/PO grant from Space Telescope Science Institute (STScI 2000), our original plan was to manufacture only a few hundred copies "by hand." (The production of thermoform images is a labor-intensive process, whereby each tactile illustration page is formed individually in a thermal vacuum oven.) However, the interest in the book, generated in part by the media coverage after the AAS press conference, far exceeded our expectations and production capabilities. With additional support from NASA, the book is now being produced and marketed by the Joseph Henry Press, a division of the National Academies Press.

Evaluation

With a few notable exceptions (e.g., Grice 2002b, Grice 1998, Edinboro University of Pennsylvania 2002), very few attempts have been made to produce tactile representations of astronomical images. Thus, a very important component of our project was the evaluation of *Touch the Universe* after its release. We recruited educators from state schools for the blind across the country, who evaluated *Touch the Universe* with their students, and gave us feedback in the form of pre- and post-tests and questionnaires. Not surprisingly, we found the student participants had little or no background knowledge in astronomy prior to reading *Touch the Universe*. The book clearly helped the students advance their knowledge about the Hubble Space Telescope and astronomical objects depicted in the book. There was a significant increase in specific and correct responses to the questions after the students read *Touch the Universe*. Students responded favorably to the organization of the book (i.e., description followed by pictures next to each other). Many students expressed their high interest and enjoyment in reading the book.

However, the book was primarily designed to make Hubble Space Telescope images accessible to the blind, not to teach fundamental astronomical concepts. Many misconceptions about the images in the book, which became evident from the evaluations, are rooted in the lack of prior exposure to astronomy and space science and not in the way information is presented in the book. Clearly, there is a need for more books and other adapted resources that focus on fundamental astronomy and space science concepts.

We also received important information on ways to make future NASA books on current astronomical topics even more effective. For example, it was noted by several students and educators that the lines and textures in the commercially available version of *Touch the Universe* need to be more pronounced to the touch, especially for students with limited sensibility in their fingertips (e.g., due to diabetes). Audio materials could be added for those students who have difficulty in reading Braille proficiently. Three dimensional models that go along with some of the tactile images could help students better understand the information contained in the images and develop their skills for interpreting two-dimensional projections of three-dimensional objects. There were many requests for other astronomical topics, as well as for books oriented toward specific grade levels.

Conclusion

One of the most important outcomes of the *Touch the Universe* project is the realization that there is a great need and interest in topics related to NASA and astronomy/space science, which is not yet fully being met by the OSS E/PO program. State schools for the blind, consumer groups like the National Federation of the Blind, and individual itinerant teachers are capable partners who can help flight missions and individual space scientists develop E/PO programs designed to make astronomy and space science more accessible to this community.

References

American Printinghouse for the Blind 1997
    http://www.aph.org/edresearch/guides.htm
Edinboro University of Pennsylvania 2002
    http://www.edinboro.edu/cwis/planetarium/web/home.htm
Edman, P. K. 1992, Tactile Graphics (New York: American Foundation for the Blind)
Erikson, Y. 1999, in Proceedings of the 65th International Federation of Library Associations Annual Conference
    http://www.ifla.org/IV/ifla65/65ye-e.htm
Grice, N. 1998, Touch the Stars (Boston: Museum of Science)
Grice, N., & Beck-Winchatz, B. 2001, Bulletin of the American Astronomical Society, 33, 809
Grice, N. 2002a, Touch the Universe – A NASA Braille Book of Astronomy (Washington: National Academies Press)
Grice, N. 2002b, Touch the Stars II (Boston: National Braille Press)
Jackson, A. 2002, Notices of the American Mathematical Society, 49(10),
    http://www.ams.org/notices/200210/comm-morin.pdf
Levi, F., & Rolli, R. 1994, Manual of Tactile Graphics (Torino: Zamorani)
    http://www.arpnet.it/tactile/Uk/manual.htm
Sakaran, N. 1995, The Scientist, 9(2):1
    http://www.the-scientist.com/yr1995/jan/sankaran_p1_950123.html

Scadden, L. A. 1996, in Proceedings of the RESNA '96 Annual Conference, ed. E. Langton (Arlington, VA: RESNA Press)
http://www.dinf.ne.jp/doc/english/Us_Eu/conf/resna96/page51.htm

Space Telescope Science Institute 2000, Cycle E/PO Grant Program
http://cycle-epo.stsci.edu/

Tactile Access to Education for Visually Impaired Students (TAEVIS) 2002, Tactile Diagram Manual,
http://www.taevisonline.purdue.edu/

Elements of Successful Websites for Educators

Art Hammon

Office of Education and Public Outreach, Jet Propulsion Laboratory, 4800 Oak Grove Drive, Pasadena, CA, 91101, California Institute of Technology, 1200 California Boulevard, Pasadena, California 91125, ahammon@jpl.nasa.gov

Abstract. NASA creates many websites intended for educators. These websites contain essential elements that may be valuable for educators. The success of a website depends on the building process that is used to create educator websites. By identifying frameworks for websites and the sequential steps that are part of the website design process, the use by educators of this electronic dissemination and information tool is increased.

Description

"As teachers move from the Industrial Age into the Information Age, the expectations and goals for students alter. Consequently, curricula are in a state of flux as teachers attempt to cope with the new demands of a world that seems to undergo a new technological revolution every decade." (Mandel, *Social Studies in the Cyberage*, Skylight Training and Pub., 1998, p. 4.)

There are many challenges to creating websites that will attract educators. The pressure of teaching schedules and the challenges of working with students limits time and energy needed to find and adapt websites for formal education use. Elements such as engagement and acceptance by the users are crucial when considering website design. Using the results of four workshops which elicited information about website design from educators and principles of professional website designers, we present a format and menu of website design.

Other aspects, such as integration in an educators personal toolbox, ownership and creation of Affinity Communities will be reviewed.

Applications

Websites include information, images and activities that engage educators and are useful to them in their classrooms. Websites also offer NASA an important dissemination venue for our products and public outreach. The benefits to teachers and their students are :

- increased contact with NASA products
- improved communication with educators

- sense of ownership of NASA materials if websites are interactive
- feedback that can be used for the modification of NASA education products
- qualitative and quantitative information for evaluation

Application Steps and Support Needed From Website

Engagement

Encouragement, Minimalist success, Multi-sensory delivery of information, Ease of navigation (3-D is emerging standard)

Acceptance by user

Technical assistance, Articulation with educational area of interest, Appeal to curiosity by dynamic changes over time, Search engine exists, Avoid plug-ins and down-loads if possible- these are blocked by firewalls in school districts

Integration into personal

Ease of implementation, Value added to resource toolbox, pre-existing resource (text, video, worksheet), Inquiry-based motif, Articulation with standards, Interactive activities, Simulations

Ownership by user

Instructional use of website with students, Professional development of educator, Collegial sharing

Affinity Community

Participants interact in realtime, Online sharing in a multiple-input, simultaneous mode. Networking and support

Synthesis by user

Value added to website through creation of complementary instructional materials by user. Sharing of ideas with website designers

Professional Development

Online instructional component results in enhanced level of competence for educator. Educator receives professional development certificate from NASA.

Society for the Advancement of Chicanos and Native Americans in Science (SACNAS)

Luis Haro

University of Texas at San Antonio, Division of Life Sciences, 6900 N. Loop 1604 W., San Antonio, TX 78249-0664, lharo@utsa.edu

Lin Hundt

SACNAS, P.O. Box 8526, Santa Cruz, CA 95061-8526

Mission

The mission of SACNAS is to encourage Chicano, Latino and Native American students to pursue graduate education and obtain the advanced degrees necessary for research careers and science teaching professions at all levels.

SACNAS K-12 Program

The goal of the K-12 Education Program at SACNAS is to ensure that elementary, middle and high school students from traditionally underrepresented minority backgrounds receive superior educational opportunities, role models and the encouragement needed to pursue careers in science, mathematics, engineering and technology.

The Need

Educators are a vital link to the early and lasting development of interest and success of underrepresented minority students in the sciences. Yet, teachers who serve these students are disproportionately under-prepared to instruct scientific and mathematical topics. The components of the SACNAS K-12 Program form a multi-faceted approach to addressing the needs of teachers serving our nation's minority communities.

SACNAS K-12 Teacher Workshops

The SACNAS K-12 Teacher Workshops represent a national effort to support superior pre-college education in the sciences for Native American/Alaskan Native, Chicano/Latino, African American and Pacific Island students. The Society recognizes the essential role that educators play in developing student interest and achievement in the sciences. Because of this, SACNAS focuses its K-12 Teacher Workshops on teacher professional development – especially in content area knowledge and inquiry-based methodology.

During the annual Workshops, pre-college educators participate in four days of dynamic hands-on, inquiry-based science, mathematics, engineering and technology sessions, discussion groups and collaborations. Teachers take part in workshops focusing on specific disciplines and on methods of meeting the needs of underrepresented minority students, while simultaneously assessing their own objectives and goals in relation to the methodologies and content presented. Nationally recognized organizations, government agencies and education leaders lead sessions addressing inquiry within scientific disciplines including health science, mathematics, space science, microbiology, chemistry and more.

The workshops take place as part of the annual SACNAS National Conference offering teachers the unique opportunity to meet nationally renowned scientists (and scientists-to-be) from across the country who share their commitment to the education of underrepresented minority students.

To ensure that educators are able to take part in the Workshops, regardless of their district's professional development budget, SACNAS makes financial assistance available to teachers working within high underrepresented minority communities. In 2001, 2000 and 1999, approximately 75% of all Teacher Workshops participants received financial assistance to attend.

SACNAS Biography Project

Highlighting the life stories and professional contributions of Native American, Chicano and Latino scientists, mathematicians and engineers, the SACNAS Biography Project was created for use in the K-12 classroom. "The purpose of the biographies is for the students to see themselves reflected in the lives that are presented here." – Dr. William Yslas Vélez, Biography Project Principal Investigator.

SACNAS Teacher-Scientist Partnerships Initiative

Launched at the 2001 SACNAS K-12 Teacher Workshops in Phoenix, the new E-mentoring Program connects the country's minority-serving K-12 educators with research scientists as partners in the development of superior pre-college science education. With the inception of this e-mentoring initiative, the Society deepens its commitment to K-12 education by pairing teachers and scientists in year-round collaboration using Internet and email technology. The project's purpose is to sustain the professional growth of a cadre of educators dedicated to outstanding, inquiry-centered science education for the nation's minority communities, extending participant's learning beyond attendance at the SACNAS K-12 Teacher Workshops. Major support for this initiative comes from the NCRR/NIH Science Education Partnership Award (SEPA) Program grant #R25 RR15649.

The SOFIA EXES Teacher Associate Program

M. K. Hemenway, J. H. Lacy, D. T. Jaffe, and M. J. Richter

Astronomy Department, University of Texas at Austin, Austin, TX 78712-1083, marykay@astro.as.utexas.edu

Introduction

The Echelon Cross Echelle Spectrograph (EXES) (Richter et al. 2000) is a principal investigator instrument being constructed at the University of Texas at Austin for SOFIA (Stratospheric Observatory for Infrared Astronomy). Realizing that the integration of science and technology is maximized in the development of a new scientific instrument like EXES, a teacher associate program was begun in January 1998 with a goal to prepare a cadre of teachers who will promote astronomy within their communities and who will be prepared eventually for a flight experience on SOFIA. By spreading the experience out over several years, the teachers observe the development and construction of a state-of-the-art astronomical instrument through many phases.

Teacher Associates

Thirty-two different teachers have each attended one or more of the 28 one-day sessions. Eight of the current associates have been with the group since the first meeting. At any one time, the group is limited to 20 teachers. Although the program was designed for teachers within a 100-mile radius of Austin, some others throughout the state have become regular participants (coming as far as Jefferson, Texas, over 280 miles away, and Brownsville, 320 miles). Most of the Associates are secondary school science teachers (grade 6-12), but a few are science specialists, for example, working with a Regional Service Center or the Texas Rural Systemic Initiative. These specialists multiply their experience by working with many other science teachers.

Meeting Structure

The regular meetings, six Saturdays per year, are structured to provide a blend of cutting-edge science and classroom materials. A typical program begins with an update from a team member on current developments on EXES and SOFIA. A tour and possible use of a research lab or an interactive presentation on a science or technology topic follows. Occasionally, the Associates share their special science/technology expertise with the group. Each meeting includes a mini-workshop where the Associates experience a hands-on activity suitable for use in their classrooms; they receive handouts and some materials. The Associates often discuss with each other how to modify these activities for their particular teaching situation, e.g. physics or chemistry classes. These activities follow National Science Education Standards (NRC 1996).

Meeting Topics

Some sessions have concentrated on instrument components such as the characteristics and testing of diffraction gratings, electronic interfaces, or monochromaters. Several sessions have dealt with optics at different levels from simple experiments with plastic lenses to using ZEMAX on a computer to do ray tracing to optical design. One session examined the art of writing proposals for observing time, while another examined FAA certification regulations and how EXES is responding to them. Astronomy topics have varied depending upon the topic's relation to SOFIA or Associates' questions (e.g., star birth, variable stars, expanding universe, string theory, and the colors of galaxies). Some sessions have examined nationally developed curricular projects such as *Hands-On Astrophysics* (AAVSO 1997) or *Active Physics* (Eisenkraft 1998). The Associates frequently pilot test materials being developed for the Texas Astronomy Education Center such as *Telescope Technology for Teachers* (Hemenway & Armosky 2000).

Field Trips

Included among the activities have been two field trips to visit the SOFIA aircraft in Waco where it is being modified, a presentation by the Associates at an American Astronomical Society meeting (Hemenway et al. 1998), visits by some Associates to McDonald Observatory or the IRTF in Hawaii where TEXES (the groundbased precursor to EXES) has had observing time. An Associates, Judy Ball, made a video of the McDonald experience that will shared with the other Associates. In July 2001, six Associates and two staff (MKH and JHL) went to McDonald Observatory for their own three-night telescope run on the 0.9 m telescope equipped with a CCD camera. Reduction of the data formed one of the Fall 2001 meeting activities for the entire group.

Other Outreach Activities

The science team has interacted with students from nine of the participating schools of the Teacher Associates. One "visit" was done through a telecommunications link between the University and a classroom in San Antonio from which it spread throughout the nation via other links as part of a high school distance education class in astronomy. Some Associates have made presentations at state science meetings or written newspaper articles about their experiences. Two of the Associates were hired as consultants for summer projects organized by the science team using other funding sources. In addition, the science team has given many public lectures and maintains a website at http://nene.as.utexas.edu/exes/.

Conclusion

Teachers have exhibited a great deal of loyalty to the program. Although they receive some mileage expenses and meals, there is no stipend in this program. They especially seem to value the informal time with the scientists and graduate students. The entire science team has been involved in regular interactions, from formal presentations or school visits to electronic mail communications or lunchtime conversations. The teachers have gained knowledge beyond the scientific concepts presented to them; they have learned about how the community of science operates on a day-to-day basis. It has been a rewarding experience for everyone.

References

American Association of Variable Star Observers 1997, Hands-on Astrophysics, (Cambridge, MA: AAVSO)

Eisenkraft, A. 1998, Active Physics (Armonk, NY: It's About Time)

Hemenway, M. K., Lacy, J. H., Jaffe, D. T., Richter, M. J., Green, K., Harkrider, J. L., Lutsinger, C. L., Noid, E., Penn, R., Shepherd, L., Suder, R., Tykoski, M. J., & Willis M. J. 1998, Bulletin of the American Astronomical Society 30, 1291

Hemenway, M. K., & Armosky, B. J. 2000, Bulletin of the American Astronomical Society, 32, 1559

National Research Council 1996, National Science Education Standards (Washington: National Academies Press)

Richter, M. J., Lacy, J. H., Jaffe, D. T., Greathouse, T. K., & Hemenway, M. K. 2000, in Proceedings of SPIE, 4014, 54

Engaging Minority Undergraduate Students in Space Science

Leon P. Johnson and Shermane Austin

Physical, Environmental and Computer Sciences, Medgar Evers College, CUNY, leon.johnson@verizon.net

Esther Zirbel

Engineering and Physical Sciences, College of Staten Island, CUNY

Overview

In New York City an ambitious effort is underway to increase the representation of minority students in Space Science. Faculty in the City University of New York (CUNY) created the New York City Space Science Research Alliance funded by the NASA Office of Space Science (OSS) Minority Initiative. The objective of the program is to create a CUNY-wide Space Science curricula that will build a Space Science major and encourage students in Math, Science, Engineering and Computer Science majors to pursue minors in Space Science. The potential for such an effort is realistic given the 200,000 enrollment in CUNY, an urban university with over fifty percent minority students. Lending considerable weight to the program are the Hayden Planetarium of the Museum of Natural History and Goddard Space Flight Center – National Space Science Data Center (NSSDC) and Minority University-Space Interdisciplinary Network (MU-SPIN).

Under the leadership of Dr. Neil Tyson, director, a flagship course, "Introduction to Space Science," has been developed and offered to high school and undergraduate students to stimulate interest in Space Science. Dr. James Thieman, NSSDC, and James Harrington, MU-SPIN, coordinated student activities leading to undergraduate research opportunities at GSFC.

Space Science Curriculum

The curriculum developed is intended to provide a solid foundation for Space Science majors with specially developed courses in Space Science covering Planetary Science, Earth-Sun Connection, Stellar Structure, Galactic Structure and Evolution of the Universe. The curriculum is augmented with newly-developed courses in Observational Astronomy, Planetology and Computer Applications in Earth and Space Science. These courses coupled with standard requirements in Physics, Mathematics, Chemistry and Computer Science lead to a BS degree in Space Science under the CUNY Baccalaureate Degree Program. A subset of Space Science courses will allow students in mathematics, science, engineering and technology (MSET) majors to earn minors in Space Science. Experimental courses based on NASA missions and utilizing resulting datasets have been developed to attract new students to the program. A major task in the first phase of this initiative is building the Space Science minor. A feature of the program is the exposure to minority faculty (role models and mentors). Workshops led by peer tutors in the gatekeeper courses (Calculus Sequence, Calculus-based Physics, General Chemistry), hands-on inquiry-based instruction and collaborative learning exist in many courses and help to facilitate student progress. Research and research-related activities are required for all students.

Research Opportunities for Undergraduate Students

An important component of the program is providing research and mentorship for MSET undergraduate students interested in Space Science. Research opportunities at NASA centers, and other universities as well as CUNY, have been initiated and enhanced through partnerships with GSFC, JPL, GISS as well as South Carolina State University (SCSU). Already over a dozen students have participated in research opportunities in areas ranging from Quantum Computing, Solar Physics, Active Galactic Nuclei, HII Regions and Supernovae Remnants. In order to insure successful student preparation and participation, various activities have been developed including recruitment and screening, trips to GSFC, assistance in application preparation, providing CUNY pre-Summer mentors and requiring a course on image processing and data analysis. Students chosen become a recruitment vehicle that can be used in the future; the participating students will form a CUNY cluster, even post-graduate, to conduct recruitment sessions for new students.

Strengthening Faculty Mentorship

Preparing students for 21st century Space Science requires insuring that faculty mentors have the requisite skills to engage in research of interest to NASA and be able to transfer these skills to their students. To achieve this, a faculty development MU-SPIN-sponsored workshop on Astronomical Image Processing was given at South Carolina State University during the Summer, 2002. The instructors from SCSU and Medgar Evers College developed a curriculum whose topics include: CCDs, HST, Image Processing, UNIX, IRAF, SDSDAS and IDL. Faculty participants included astrophysicists, astronomers and computer scientists from SCSU, CUNY, and Norfolk State University. CUNY participants were able to utilize the skills learned for more effective mentorship of students seeking research opportunities at NASA centers. The workshop is being replicated at CUNY for wider involvement of faculty and students, particularly at community colleges.

Leveraging Public Outreach and K-12 Pathways

Colleges in CUNY already host a number of K-12 programs funded by NASA which provide a natural pipeline for the Space Science program. The Aeronautic Education Laboratory (AEL), the Science Engineering and Mathematics Aeronautics Academy (SEMAA) and the NASA PACE award support space science-related activities for middle school and high school students. These students, already excited by their enrichment experience, represent future Space Science majors and minors and therefore, a program of Space Science undergraduate presentations to these students have been recently initiated.

Lessons Learned

- Direct contact with NASA scientists, astronauts and exposure to NASA missions via a CUNY student trip to GSFC and Hayden Planetarium provided a motivating factor for involvement in the Space Science program.

- The self-development of student clusters for education and research resulted in a peer support group.

- The Space Science minor is very popular since many students with an interest in space science are reluctant to change their major, especially if they are juniors or seniors.

- Stipends for selected students to pursue research activities during the academic year and during the summer allows students to concentrate on their academics instead of seeking part-time employment in unrelated menial jobs.

- Flexibility in course structure and content, as well as the integration of NASA research and data analysis techniques into curricula, appear to be both academically challenging to the student as well as giving them a direct idea of the work NASA scientist are engaged.

- Faculty mentorship for all students is extremely important.

- There is a need to increase computer activities for all students.

- Strengthening the pipeline requires increased involvement of community college faculty and students.

The Invisible Universe Online: A Distance Learning Course on Astronomical Origins for Teachers

John Keller, Edward E. Prather, and Tim F. Slater

University of Arizona Steward Observatory, Tucson, AZ 85721, jkeller@as.arizona.edu

Introduction

As our scientific knowledge base is rapidly changing, there is a clear need to provide teachers with up to date content information as well as sound and useful strategies to include this information in the classroom. One approach that is working well is to use NASA-sponsored science as a foundation on which to build graduate courses for teachers. Delivered to in-service teachers via the Internet, the 15-week long course, *The Invisible Universe Online: The Search for Astronomical Origins for Teachers*, covers the long chain of events from the birth of the universe in the Big Bang, through the formation of galaxies, stars, and planets. The course curriculum focuses on the scientific questions, technological challenges, and space missions pursuing the search for origins. Overall the course is designed to be aligned with the goals and emphasis of the NRC *National Science Education Standards* (Slater 2000; NRC 1996). The course goals are:

1. Develop scientific background knowledge of astronomical objects and phenomena with peak emissions outside of the visible region of the electromagnetic spectrum

2. Understand contemporary scientific research questions related to understanding: a. how galaxies formed in the early universe b. how stars and planetary systems form and evolve

3. Describe strategies and technologies for using non-visible wavelengths of EM radiation to study various phenomena

4. Integrate the related issues of astronomical science, technology, societal issues, and career guidance for classroom teaching

5. Develop specific strategies for implementing concepts in the National Science Education Standards related to "invisible" astronomy and the search for astronomical origins

Each week, teacher-participants in the course uses a series of classroom-tested, inquiry-based hands on lessons. They also complete readings and have discussion based on one of the following: *Universe: Stars and Galaxies* (Freedman & Kaufmann 2001), *National Science Education Standards* (NRC 1996), *NASA Office of Space Science Astronomical Search for Origins Online Library* (Danner 2003), and other readings as assigned.

The course topics and activities listed by week are anticipated to be:

1. Introduction to the Astronomical Search for Origins;
2. Beyond Visible Light;
3. Temperature, Blackbody Radiation, and Wien's Law;
4. Doppler, Gravitational, and Cosmological Red Shifts;
5. Atmospheric Extinction and CCD Detectors;
6. Molecular Clouds and the Interstellar Medium;
7. Protostars and Stellar Evolution;
8. Circumstellar Disks and Extrasolar Planetary Systems;
9. Mid-Term Examination and Term Paper;
10. Protogalaxies;
11. Galaxies, AGN, and Seyfert Galaxies;
12. Super Massive Black Holes;
13. Super Structure Scale of the Universe;
14. Astronomy, Technology, Society, and Careers;
15. Final Exam and Final Project Due.

When NASA E/PO funds are available to subsidize this course, this course allows in-service teachers the opportunity to earn three graduate credits at a total cost of $100. When the course is run as a self supporting course, for tuition alone, the course costs each teacher-participant approximately $450. This is one of five astronomy for teachers courses currently offered by the CERES Project at Montana State University (Slater et al. 2001).

References

Danner, R. 2003, NASA Office of Space Science Astronomical Search for Origins Online Library
http://origins.jpl.nasa.gov/library/

Freedman, R., & Kaufmann, W. J. 2001, Universe: Stars and Galaxies (New York: WH Freeman Publishing)

National Research Council 1996, National Science Education Standards (Washington: National Academies Press)
http://books.nap.edu/html/nses/html/

Slater, T. F., Beaudrie, B., Cadtiz, D. M., Governor, D., Roettger, E. R., Stevenson, S., & Tuthill, G. F. 2001, Journal of Computers in Mathematics and Science Teaching, 20(2), 163

Slater T. F. 2000, Physics Teacher, 38(9), 538
Slater, T. F. 2000, Physics Teacher, 38(9), 538

Additional Information: This course is being developed and initially offered with the support of the SOFIA and SIRTF E/PO Programs. Please direct correspondence to the first author. Teachers can enroll in these online courses through the distance learning program at Montana State University (Slater et al. 2001), which can be found online at http://btc.montana.edu/nten/.

Image Processing Experiments for the Classroom

Walter S. Kiefer and Kin Leung

Lunar and Planetary Institute, 3600 Bay Area Blvd., Houston TX 77058, kiefer@lpi.usra.edu

Exploring the Solar System is a science enrichment program taught at LPI for gifted fifth-grade students (Kiefer et al. 2004). As part of this semester-long course, we teach a series of three computer labs on digital image processing. Images of the planets obtained by various NASA spacecraft have had a key role in the overall exploration of the Solar System. The purpose of these labs is to teach the students the basic nature of digital images and how computer processing of images can aid in the interpretation of images. The labs described here represent our experience developed from teaching this course on 16 occasions since 1992. These labs are appropriate for upper elementary and middle school students. With suitable choices of imagery, the labs could focus on the Solar System, geography, earth science, environmental science, or astronomy. The activities described here can be performed on personal computers. We use Paint Shop Pro® for these labs, but the instructions given here can be modified for use with other image processing software.

Each of the computer labs is about 45 minutes in length and begins with a formal instruction segment and ends with a less structured exploration segment. For simplicity, we focus on black and white images during the first lab and for part of the second lab. The remainder of the second lab and part of the third lab is devoted to studying color images. We also devote a portion of the third lab session to 3-D stereo imagery.

Black and White Images

We begin with an image of a familiar object, usually an image of one of the instructors. Using the Magnifying Tool, have the students blow the image up to high magnification. They will see that images are composed of individual rectangular blocks, known as picture elements, or pixels for short. Return the image to its normal size. Activate the Color Palette Tool and the Eyedropper Tool. Have the students place the mouse on a dark region of the image and note the intensity level (for black and white images, the R, G, and B values shown in the Color Palette window are identical and indicate the brightness level). This should be a low number. Now move the mouse to a bright region of the image. What happens to the intensity value? What do the students think will happen in an intermediate gray region? This relationship between brightness and numerical values is known as the gray scale. In the common GIF image format, brightness levels range from 0 (black) to 255 (white).

Because computer images are essentially arrays of numbers, they can be modified using mathematical operations. We introduce the students to two basic types of image modifications. The first and easiest type is brightness and contrast enhancement. Activate the Histogram window. Upper elementary students are familiar with histograms in the form of bar charts. The histogram is a very useful tool for understanding the brightness and contrast of an image. In the Colors pull-down menu, there is an entry for Adjust Brightness and Contrast (there are also a lot of other Adjust options, which can be ignored for the purpose of an introductory lab). Have the students explore the effects of changing Brightness and Contrast separately. How does each affect the image? Do the changes in the image's intensity histogram reflect the changes in the image? As a challenge, provide the students with an image that has been deliberately darkened and see if they can recover an improved version of the image.

Digital filters are a second type of useful image processing tool. These filters can systematically alter an image. In the Effects pull-down menu, explore how the various types of edge enhancement and sharpening filters modify an image. In this context, "edge" means any boundary between distinct brightness structures in an image. Can the students think of any reason why such filters might be useful to geologists (e.g., enhancing the visibility of faults or channel systems in an image)? Allow the students some time to experiment on their own with the effects of these filters.

Color Images

For labs using color images, you will want to use 24 bit color images (JPG or TIF formats). Such images build color from three color channels: red, green, and blue (RGB). Place the program in Eyedropper mode. Scroll the mouse across the image and notice how the R, G, and B intensity values change in regions of different colors. In the Colors pull-down menu there are entries for Split Channels and Combine Channels. Splitting separates the color image into its three component channels. For an introductory lab, stick with the RGB option. If you are careful about adjusting the sizes of the images, you can display all four images (color plus the 3 RGB channels) on the monitor at once. Does the appearance of the image in the various channels make sense? For example, why does the polar cap of Mars appear bright in all color bands? In images of the surface of Mars, why do red dust and gray basalt have different appearances in the various color bands? What does Jupiter's Great Red Spot look like in the various color channels?

You can use the Combine Channel function to reassemble the various color bands. End the lab in a memorable way by combining the images in an unusual color sequence. For example, can your students make Jupiter have a "Great Green Spot"? Can they turn Mars into the "Purple Planet"? In some versions of Paint Shop, you will have to turn off the "Sync Blue and Green to Red" option in the Combine Channel function's dialog box.

Stereo Images

We devote a portion of the final lab session to "three dimensional" stereo images. Such images can be very useful in visualizing the nature of planetary surfaces. We often organize this in a compare and contrast format. For example, we contrast the appearances of the volcanos Mount Saint Helens on Earth and

Olympus Mons on Mars. The different appearances of these volcanos reflects differences in the types of magma that formed each structure. Similarly, we compare the appearances of the surface of the Moon as seen on Apollo with the surface of Mars as imaged by Mars Pathfinder. We use the Neotek viewing system for viewing these images, although a similar activity could be performed using the common red-blue stereo glasses.

Image Sources

Images for these labs can be downloaded from a variety of sources, depending on the focus of your course. A basic selection of images for these labs is available from LPI (LPI 2004). Additional planetary images are available from the Planetary Photojournal (JPL 2004). Geography and environmental science courses could use images of the Earth from space (NASA-JSC 2004). Astronomy courses can use Hubble Space Telescope images (STScI 2004). Three-dimensional stereo images of the planets are available from LPI (LPI 2000). In all cases, set the copy protection on the images to Read Only prior to the start of the lab session.

Acknowledgments. This work was supported by NASA Contract NASW 4574. Lunar and Planetary Institute Contribution 1099.

References

Jet Propulsion Laboratory 2004
    http://photojournal.jpl.nasa.gov/
Kiefer, W. S., Herrick, R. R., Treiman, A. H., Thompson, P. B. 2004, page 130 of this volume.
Lunar and Planetary Institute 2000
    http://www.lpi.usra.edu/research/stereo_atlas/SS3D.HTM
Lunar and Planetary Institute 2004
    http://www.lpi.usra.edu/education/imagelab/imagelab.html
NASA-Johnson Space Center 2004
    http://earth.jsc.nasa.gov/
Space Telescope Science Institute 2004
    http://hubblesite.org/newscenter

Pre-College Students Contribute to the Cassini-Jupiter Millennium Flyby

Michael J. Klein

Jet Propulsion Laboratory / Caltech, Pasadena, CA 91109, mike.klein@jpl.nasa.gov

James P. Roller

Lewis Center for Educational Research, Apple Valley, CA 92307

When the Cassini spacecraft flew past Jupiter in January 2001, not only were scientists able to collect high-resolution data on Jupiter's radiation belts, but students and their teachers across the U.S. also had an opportunity to do real science. Using the Goldstone-Apple Valley Radio Telescope (GAVRT) science education partnership, the research team collaborated with teachers and students to perform a series of ground-based observations of Jupiter that were coordinated with spacecraft observations.

The GAVRT antenna, formerly known as DSS-12, was decommissioned from the Deep Space Network (DSN) in 1996. Led by a team of visionary scientists, educators, engineers, and community volunteers, the antenna found new life as an educational tool that offers teachers and students across the country a unique opportunity to experience the scientific process, as well as contribute directly to important, current research. The GAVRT project involves three partners; the National Aeronautics and Space Administration (NASA), the Jet Propulsion Laboratory (JPL) and the Lewis Center for Educational Research (LCER) in Apple Valley, California.

One highly successful curriculum of the GAVRT partnership is called Jupiter Quest, a hypothetical mission to Jupiter or one of its moons. Students measure the radio emission from the Jovian atmosphere and its radiation belts using the radio telescope (controlled via the Internet and GAVRT Mission Control in Apple Valley) and use the information in their mission plan. To participate in the program, qualified teachers receive a week of training at the Lewis Center, or Auburn University, or Penn State University

Jupiter Quest was selected as the pilot project because it contributes to an existing scientific study of Jupiter, and the microwave observations from the GAVRT team have scientific value. Educators on the team also recognize that students would be familiar with the bright planet Jupiter. Curriculum materials are developed to match National and State Science Standards and all learning modalities, including tactile, can be addressed.

GAVRT-trained teachers become active partners on the GAVRT team. Lewis Center staff are available to assist as teachers prepare their classes to conduct their observations with the radio telescope. The students plan for their observing session and learn calibration techniques that are required to ensure the quality of their data will be high. Teachers and students are supported "on line" during the observing sessions and they participate in telephone conference calls with GAVRT scientists at other times. Newsletters are distributed monthly to keep the GAVRT community informed of team results and new classroom materials, and to announce opportunities to participate in new curriculum projects. LCER staff are always available to help meet teachers needs.

The flyby of the Cassini spacecraft past Jupiter in January 2001 provided a unique opportunity to study Jupiter's radiation belts with high spatial resolution using a passive microwave radiometer that was built into the Cassini Radar Instrument. In a coordinated series of space-based and ground-based observations, named the Cassini-Jupiter Microwave Observing Campaign (Cassini-JMOC), Jupiter was observed at radio wavelengths during the Cassini encounter from November 2000 through March 2001. Cassini-JMOC had two objectives: (1) use ground-based observations to achieve in-flight calibrations of the Cassini radar receiver and thereby enhance the Cassini science at Saturn and Titan; (2) use the Cassini radar receiver to map Jupiter's radiation belts at a frequency above 10 GHz and thereby derive the spatial distribution of very high energy electrons (>20 Mev) for the first time.

The spacecraft successfully mapped the radiation belts and the GAVRT team contributed to the calibration, which is expected to achieve an absolute accuracy of at least 3 percent. The GAVRT team also monitored the intensity of the radiation belts at a lower frequency, where they detected several intensity variations of approximately 10% occurring on timescales of a few days. Similar short-term variations have been previously reported, but none with the precision of the Cassini-JMOC results. Approximately 2300 students at 26 schools in 13 states participated in the Cassini-JMOC observations. The preliminary results have been reported in *Nature* (Feb 28, 2002) and at two international conferences.

GAVRT began as an outreach project primarily focused on providing a unique science educational experience for students. The project quickly evolved to become an interactive team of students, classroom teachers, educators at the LCER, and NASA/JPL scientists and engineers who are using multiple pathways of communication to achieve a common goal. That goal is to involve students in the process of doing science as well as learning about science and to help students realize their contributions are valued. The enthusiasm and thoughtful questions that students bring to the regularly scheduled teleconference calls with scientists are appreciated. Some questions are very profound! Teachers report that students appreciate that GAVRT data are not tossed in the trash at the end of the semester and that their results will most likely be published in professional science and education journals.

The evolution of GAVRT will continue into the future. Teachers will come forward with ideas and needs for new curriculum materials. Students will continue to ask new and challenging questions. Scientists will propose research projects that will offer new opportunities for teaching and learning. The educational research team at LCER will be studying the impact of the project as it strives to promote science literacy, support better understanding of the scientific community, and excite students about learning.

The JPL contribution to this paper was performed at the Jet Propulsion Laboratory, California Institute of Technology, under contract with the National Aeronautics and Space Administration.

Involving Students in Active Planetary Research Using 2001 Mars Odyssey THEMIS Camera Data: The Mars Student Imaging Project

S.L Klug, P. R. Christensen, K. Watt, and P. Valderrama

Arizona State University, Mars Space Flight Facility, Moeur Bldg. Rm. 131, P.O. Box 85287-6305, Tempe, AZ 85287-6305, sklug@asu.edu

Introduction

Arizona State University (ASU) and the Jet Propulsion Laboratory (JPL) have partnered to create the ASU Planetary Imaging and Analysis Facility and Advanced Training Institute (PIAFATI) and Mars Education Center which is located on the ASU campus. This facility is available to professional scientists, graduate, undergraduate, community college, and 5th-12th grade students nationwide to target, acquire, process and analyze images of Mars.

The ASU Mars Education Program is the lead facilitator in the development and implementation of the *Mars Student Imaging Project (MSIP)*. This project was created using the heritage of the ASU Mars Education Program, which:

- has 11+ years of Mars education and outreach experience (Mars Observer, Mars Pathfinder, Mars Global Surveyor, Mars Odyssey, and includes the upcoming Mars Explorer Rover missions);

- is partnered with the JPL Mars Exploration Program to conduct national/regional teacher training events;

- has developed and continues to develop educational curriculum and products based on the Mars exploration process and Mars mission data results; and

- is housed within the 3000 sq. ft. ASU Mars Education Department located within the Mars Space Flight Facility on the ASU campus in Tempe, AZ.

The *Mars Student Imaging Project* will move students from studying science in a historical manner (i.e., text books) or passively observing scientists in action, to active participation in one of the still-remaining frontiers of science: The exploration of another planet – Mars. Dr. Phil Christensen, Principal Investigator for the THEMIS (Thermal Emission Imaging System) camera recognized the opportunity to involve students as actual researchers and has allocated a percentage of the THEMIS visible camera images to be targeted, acquired, processed, and analyzed by 5th grade through community college students participating in *MSIP*. The *Mars Student Imaging Project* will:

- engage students in research and inquiry-based learning that is aligned with the National Science Education Standards;

- involve approximately 150 student teams (8-200+ students/team) per year via on-site visits to the Mars Space Flight Facility at Arizona State University in Tempe, AZ or through distance learning at their home institution, and hundreds of thousands of students through the *MSIP* Archive Mission available on the *MSIP* website and CD;
- actively recruit national participation by underrepresented student groups (i.e. minority, rural, inner city, and female);
- provide mentoring through interaction with THEMIS team members and ASU Mars Education staff; and
- model the use of the scientific method and the process of authentic scientific research, and the use of actual mission data in a near real-time environment.

Educational Foundation

The *Mars Student Imaging Project* is founded upon solid educational principles. *MSIP* is structured heavily around the educational and psychological theory of constructivism. The constructivist theory basically states that students learn by comparing new experiences to their existing schema. In practice, constructivism is hands-on activities carried one step further. Rather than simply presenting a series of individual hands-on activities, constructivism suggests that you supply the students with a single, overarching framework of investigation that models the schema that each student possesses internally. In this way, you have assisted the process of reconstructing their schemata by giving them a shell to hold their new experiences. Using these techniques, retention is greatly increased – the students don't have to study when they already understand and can use and reinforce the learned material in a real-world situation.

Implementation

The *Mars Student Imaging Project* utilizes constructivist learning by providing an overarching project (the exploration of Mars) that starts by incorporating many of the normal subjects that teachers are already tasked to teach and students are expected to learn at the *MSIP*-appropriate grade levels (e.g. scientific and earth processes). Adult mentors of *MSIP* teams and participating students are supported in *MSIP* through their use of the free, downloadable *MSIP* curricular guides that are Standards-aligned and have been reviewed by NASA. Additional Mars online resources, such as the *Mars Activities – Teacher Resources and Classroom Activities Guide* and the *Mapping the Surface of a Planet* curriculum are also free and available at the *MSIP* website – http://msip.asu.edu.

The academic support for the students and the logistical support for the teachers provided by *MSIP* will not only help to increase all participants' knowledge in planetary science, but will create a sense of ownership, especially in the case of the students' learning, and provide a pathway of active engagement for them that will extend their science literacy beyond the usual classroom experience. The *MSIP* experience will also facilitate in the development of core learning skills such as, synthesis and analysis, critical thinking and problem solving skills, and working within a team environment.

The *MSIP Teacher Guide* provides support for the mentor leading a *MSIP* team by helping to provide the new infrastructure of the new learning that will take place as the students advance through the project. There is no prior knowledge in planetary science necessary to be a *MSIP* facilitator. The curricular materials, including the *MSIP Resource Guide*, coupled with interaction with the *MSIP* staff, allow anyone desiring to be a facilitator to do so. The *Mars Student Imaging Project* is also very adaptable to interdisciplinary studies and can be readily conducted outside a science class environment. The *Mars Student Imaging Project* is student-centered and student-driven.

The students become the scientists and readily accept this responsibility. They are encouraged to seek their own solutions to problems and questions that arise within the project. The *MSIP* Student Handbook guides the students through the scientific process, allowing them build their knowledge base about Mars, planetary exploration and authentic research processes. The students then create their own scientific questions and formulate their own hypothesis as to what might be feasible answers. The student team concludes the project with the submission of their results for review and publication of their findings on the *MSIP* website.

The National Education Standards "Quilts": A Display Method Aiding Teachers to Link NASA Educational Materials to National Education Standards

Rebecca Knudsen and Art Hammon

Jet Propulsion Laboratory, 4800 Oak Grove Drive, Pasadena California 91109, California Institute of Technology, 1200 California Boulevard, Pasadena, California 91125, Rebecca.Knudsen@jpl.nasa.gov

Introduction

The National Science Education Standards (NRC 1996) and National Council of Teachers of Mathematics Principles and Practices for School Mathematics (NCTM 2000) were produced to meet a growing concern about consistency and comprehensiveness in American education. They represent a comprehensive and definitive framework for the creation and implementation of K-12 science and math education. The format of the publications is scholarly and well written. To many educators, the publications in their book form are difficult to use. A matrix format developed at Jet Propulsion Laboratory places these standards in a format that can be navigated intuitively by teachers seeking to integrate NASA materials into the structure of standards based education.

The development of rubrics for the placement of products based on readability, conceptual development, safety, room logistics and materials available have been developed to complement the quilts. The Overarching Science Topics matrix developed by Dr. Robert Gabrys, Education Director, Goddard Space Flight Center, and forum directors and scientists completes the suite of tools needed to fully articulate NASA products to the national science and math standards.

Description

The Standard "Quilt" is a project of the Educational Affairs Office of Jet Propulsion Laboratory and the Solar System Exploration E/PO Forum. The word "Quilt" implies a display of conceptual and process information which produces unique "squares" from the fabric of content and the thematic organizing of those materials. The document addresses three needs:

1. Users of curricular materials should be helped to seek them based on the standards being taught and age span of students.

2. Producers of curricular materials should display their materials in ways that show their placement in relation to the standards.

3. The display of content standards should be articulated to the thematic way in which they will be presented to students.

The standards "Quilt" displays the standards in three ways:

1. The content subject which is the focus of the curricular material

2. The thematic way in which the presentation of the content is displayed

3. The grade spans used by the NSES are blocked as rows.

A beta version of the science quilt was created in CD-ROM format and distributed at science education conferences and workshops nationwide, as was a beta version of the math quilt in hard copy format. Recipients of the quilt were surveyed for feedback and suggestions on these beta versions. The results of the follow-up surveys of the quilt users have indicated a high interest in this product.

Due to the high interest in the quilt, NASA has asked that it become more comprehensive in scope and be developed with the participation of the entire community. Several issues concerning the selection, placement, and display of products on the quilt are currently being addressed.

The development of rubrics for the placement of products based on readability, conceptual development, safety, room logistics and materials available have been developed to complement the quilts. Microsoft Word readability was chosen as a convenient method of readability assessment. Using the Flesch-Kincaid Grade Level Score method, the rating is based on sentence length and number of syllables. Conceptual levels were derived from chapters in *Benchmarks for Science Literacy* (AAAS 1993). The sections on Safety, Room Logistics and Equipment were extracted from Environmental Protection Agency publications and NSTA position papers (EPA 2001; NSTA 1985, 1990, 2000).

The Overarching Science Topics matrix developed by Dr. Robert Gabrys, forum directors and scientists completes the suit of tools needed to fully articulate NASA products to the national science and math standards. This document lists all the major science concepts needed to present the mission science concepts under the forums of the Office of Space Science. A committee, directed by Dr. Robert Gabrys, of GSFC scientists and forum directors, contributed to the creation of this document.

The most current version of the quilt is available online as a testing bed for further evaluation. The reasons for the selection of a web-based format are based on the following assumptions.

1. It is efficient and cost effective – no paper products to ship, distribute or be left behind by educators.

2. Since the activities can be quickly accessed by standards and can be printed as needed and possibly edited, teachers may be more likely to use them.

3. It allows maximum distribution and access to educators as Internet connection has become a top priority for schools nationwide.

An initiative for the verification of product placement and rubrics, as well as the continued evaluation of the quilt, will be undertaken Summer '02 by the JPL education office. Help will be elicited from educators with a wide array of experience in this initiative. The results will provide valuable information for the entire NASA community in their future efforts to align NASA resources to the national educational standards.

References

American Association for the Advancement of Science Project 2061 1993, Benchmarks for Science Literacy (New York: Oxford University Press)

Environmental Protection Agency 2001
http://www.epa.gov/grtlakes/seahome/housewaste/house/products.htm

National Council of Teachers of Mathematics, Inc. 2000, Principles and Standards for School Mathematics (Reston, VA: NCTM)

National Research Council 1996, National Science Education Standards (Washington: National Academies Press)

National Science Teachers Association 1985
http://www.nsta.org/positionstatement&psid=18

National Science Teachers Assoiation 1990
http://www.nsta.org/159&psid=16

National Science Teachers Assoiation 2000
http://www.nsta.org/positionstatement&psid=32

NASA Educational CD-ROMs – Research and Evaluation

Rebecca Knudsen

Jet Propulsion Laboratory, 4800 Oak Grove Drive, Pasadena California 91109, California Institute of Technology, 1200 California Boulevard, Pasadena, California 91125, Rebecca.Knudsen@jpl.nasa.gov

Introduction

Since the introduction of educational technology in the K-12 classroom, educators have been flooded with a seemingly endless flow of educational CD-ROMs. NASA has contributed to this new trend in educational technology by developing a wide range of some of the latest, most high-tech CD-ROMs in the industry. In this rapidly changing field, every advance seems to promise new hopes and success, and the analysis of the previous educational software models seems irrelevant. As a consequence, virtually no research has been done to determine the usefulness of educational software packages in the K-12 classroom, neither within NASA nor the larger educational community (Ehrmann 1994). Thus, developers who invest their time and money into these educational CD-ROMs are left to make educated guesses as to what features and qualities of software the educational community prefers and uses.

Literature Search

While existing research on the use of CD-ROMs in the K-12 classroom is sparse, the literature search indicated certain trends in CD-ROM use. Educators seem to be interested in CD-ROMs that can be used in a variety of contexts, that are interactive, that promote learning in their students, and that match their curriculum standards. The types of CD-ROMs most commonly found in the classroom are (1) electronic encyclopedias, (2) research/reference materials, and (3) artwork/image CD-ROMs (PrimeArray Systems, Inc.; Becker, Ravitz, & Wong 1999). Educators typically use educational CD-ROMs outside of the classroom to create instructional materials and gather information for lessons, with a small percentage using them to create multimedia presentations and access model lesson plans. The most common CD-ROM-based assignments educators were found to give to their students were practice drills, research assignments, problem solving, and data analysis (National Center for Education Statistics 2000).

The most astounding conclusion, and perhaps the reason that so little research has been done on this topic, is that a very small percentage of our nation's educators are consistently using educational CD-ROMs in their instruction. Teachers face an overwhelming number of barriers when attempting to implement this form of educational technology. Some of these barriers include (1) high costs of the CD-ROM packages, (2) preparation and utilization time, (3) curriculum standards and testing pressures, (4) perceived poor quality of the

CD-ROMs, and (5) technical difficulties (Hoff 1999). Due to this void in research on the utilization of educational CD-ROMs, it was understandably difficult to find anything conclusive about the use of NASA's educational CD-ROMs. It was necessary, therefore, to conduct our own study.

Study Methodology

The study consisted of both quantitative and qualitative research. After a series of collaborations with a number of NASA's current educational CD-ROM developers, a Likert scale questionnaire was carefully constructed to elicit quantitative data. The questionnaire was put on-line, and an invitation to respond was sent by email to approximately 4000 educators. These educators were chosen on the basis of having received one of the following CD-ROMs: *Visit to an Ocean Planet* (TOPEX/Poseidon Mission), *Winds of Change* (NSCAT Mission), and *Ways of Seeing* (Cassini Mission). To elicit a higher response rate, educational posters and CD-ROMs were offered to those who completed the questionnaire.

Qualitative data was gathered in a series of telephone interviews with the questionnaire respondents who had indicated they were willing to participate. All interviews were recorded to ensure quality and accuracy. Both forms of data collection were considered in the data analysis and conclusions of this study.

Demographics

The email invitation to respond to the questionnaire successfully reached 3216 educators, due to incorrect or changed addresses. 817 of these educators responded to the questionnaire, a response rate at 27%. The highest percentage of respondents were teachers of grades 9-12 (44%), followed by 5-8 (36%), then K-4 (20%). The majority of respondents indicated they taught science (58%), with specific disciplines in general science, earth/space science, biology, environmental science, physics, physical science, and chemistry. There was a fairly even distribution among respondents' years of classroom experience, indicating that educators from all levels of experience use educational CD-ROMs. The distribution of educators from rural, urban, and suburban communities was also fairly even, and the majority of respondents (66%) were from schools with middle socioeconomic levels.

Further investigation of the respondent demographics reveals discouraging findings about the state of technology in our nation's schools. First of all, the majority of respondents (58%) had only 1-3 computers in their classroom, and an additional 5% had zero computers. This finding has little variance across locations, with the exception of rural schools that have slightly lower percentages of classroom computers. As one would assume, the 10% of schools with high socioeconomic levels had higher percentages of classroom computers, but there was little difference in percentages between those whose levels were middle or low. Another problematic finding is that a significant percentage of respondents (13%) had never received technical training during their careers, and the large majority of respondents (73%) had never received NASA training. This indicates that at least 73% of educators that received an educational CD-ROM from NASA were given no training or instructions on how to use it.

General Findings

When asked where educational CD-ROMs were generally found, respondents rated the Internet as the most commonly used source, above retail stores, training/workshops, and technical resource advisors. Respondents indicated that their selection of CD-ROMs is based primarily on recommendations from others, but also indicated that the alignment of the CD-ROM content with their curriculum was an important factor. Sustained use of the CD-ROM, however, depends primarily on whether or not the CD-ROM promotes learning in their students.

While educational CD-ROMs can be tailored for students to use with little or no direction from the teacher, the majority of respondents (67%) indicated they prefer CD-ROMs that are made for teacher applications. They also indicated that, given a choice, they prefer to use CD-ROMs in their instruction to the Internet. As for tutorials/usage instructions, preferences were split between placing them in a supplementary booklet (49%) or on the CD-ROM itself (46%).

Highlighted CD-ROM

For the remainder of this analysis we will focus on the *Visit to an Ocean Planet* CD-ROM (from the TOPEX/Poseidon Mission at Jet Propulsion Laboratory), due to the fact that the large majority of respondents were recipients of this particular CD-ROM. Unfortunately, 30% of these recipients never used this CD-ROM after receiving it. When asked why not, responses fell into 3 general categories: Intent to use in the future (not enough time, not to appropriate unit yet), technical difficulties (hardware problems, defective CD-ROM, no display mechanism for entire-class viewing), and dissatisfaction with the CD-ROM in general (did not align with curriculum, not grade level appropriate, difficult and time consuming to use). One response typified the sentiments of many: "The format is too inflexible to allow incorporation into my lecture. The material is very informative ... but we prefer our students to work with data on an inquiry approach, not just scroll through a presentation."

Those who used the CD-ROM, however, seemed to hold it in higher regard. The attitude assessment showed that most were very satisfied with the quality of the CD-ROM content and felt it was a valuable educational tool. Among their favorite aspects of the CD-ROM were the interactive segments, the multimedia components, and the overall accurate, in-depth information. Many indicated that the CD-ROM "adds a dimension to learning that cannot be obtained from textbook only," and that it "immediately gets the students involved." However, this positive assessment does not translate directly into high levels of usage. While respondents indicated that they used the CD-ROM most often for classroom demonstrations, the mean score for frequency was only "sometimes." Under the range of "rarely" fell other categories of use, such as research for student projects, personal preparation/planning, individual student free time activities, and computer lab activities.

Discussion

The intent of this paper is to offer research-based recommendations for those educational CD-ROM developers who truly wish to improve the state of education. Based on this study, including the literature search, quantitative, and qualitative data collection, I would like to make the following recommendations.

Educators are faced with immense time constraints. Therefore, educational CD-ROMs should be designed in such a way that they require a minimal amount of preparation time to be used in the classroom. CD-ROMs should also be designed so that they can be used with a limited number of computers, as well as old, slow, unreliable hardware. To assist educators in preparing for standardized tests, the content of the CD-ROM should not only correlate with the curriculum standards teachers must adhere to, but should be appropriate to replace required curriculum. It is also vital that the CD-ROM contain appropriate, educationally valuable content. The CD-ROM should offer some level of interactivity that allows the student to participate, become engaged, be assessed, and receive feedback on their progress. These factors, combined with quality training for all CD-ROM recipients, will not ensure widespread use of a CD-ROM program, but dramatically increase its likeliness.

References

Becker, H. J., Ravitz J. L., & Wong, Y. 1999
    http://www.crito.uci.edu/tlc/findings/computeruse/html/startpage.htm
Ehrmann, S. C. 1994, in Valuable Viable Software in Education: Case Studies and Analysis, ed. P. Morris, S. C. Ehrmann, R. Goldsmith, K. Howat, & V. Kumar (New York: McGraw Hill)
Hoff, D. 1999
    http://www.edweek.org/sreports/tc99/articles/curr.htm
National Center for Education Statistics 2000,
    http://nces.ed.gov/surveys/frss/publications/2000102/
PrimeArray Systems, Inc.
    http://www.primearray.com/DigitalSchools/overview.html

SNOOPY: A Novel Payload Integrated Education and Public Outreach Project

Kimberly R. Kuhlman[1], Michael H. Hecht[1], David E. Brinza[1], Jason E. Feldman[1], Stephen D. Fuerstenau[1], Thomas P. Meloy[1], Lucas E. Möller[2], Kelly Trowbridge[3], Jessica Sherman[3], Louis Friedman[4], Linda Kelly[4], Jeffery Oslick[4], Kevin Polk[4], Collin Lewis[5], Csaba Gyulai[5], George Powell[5], Anna M. Waldron[6], Carl A. Batt[6], and Martin C. Towner[7]

[1] *California Institute of Technology, Jet Propulsion Laboratory, 4800 Oak Grove Dr., Pasadena, CA 91109, kkuhlman@jpl.nasa.gov*

[2] *Moscow Junior High School, Moscow, ID 83843*

[3] *Lansing High School, Lansing, NY 14882*

[4] *The Planetary Society, Pasadena, CA 91106*

[5] *Visionary Products, Inc., 11814 South Election Drive, Suite 200, Draper, UT 84020*

[6] *Nanobiotechnology Center, Cornell University, Ithaca, NY 14853*

[7] *Planetary and Space Sciences Research Inst., Open University, Walton Hall, Milton Keynes, U.K.*

Introduction

As scientists and engineers primarily employed by the public, we have a responsibility to "communicate the results of our research so that the average American could understand that NASA is an investment in our future..." (Goldin 1999). Not only are we employed by the public, but we are also the source of inspiration for future generations of scientists and engineers. Student Nanoexperiments for Outreach and Observational Planetary Inquiry (SNOOPY) is an example of directly involving students and teachers in planetary science missions.

The Mars Environmental Compatibility Assessment (MECA) Student Nanoexperiments

The MECA Student Nanoexperiment Challenge was a partnership between MECA, The Planetary Society (TPS) and Visionary Products, Inc. (VPI). The MECA instrument suite, developed at the Jet Propulsion Laboratory (JPL), was scheduled for launch aboard the canceled Mars Surveyor Lander 2001. The MECA Patch Plate was designed to expose various materials to the Martian environment and be observable by the Robotic Arm Camera (RAC). Students 18 years of age and younger from around the world were invited to propose experi-

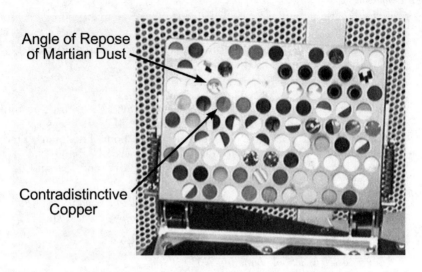

Figure 1. Nanoexperiments in the MECA Patch Plate (Kuhlman et al. 2002)

ments that were consistent with MECA's Mission: *to help us better understand how humans will be able to live on Mars.*

Each nanoexperiment was required to fit into single MECA Patch Plate (Figure 1) hole, 1 cm in diameter and 1 cm deep, have a mass of 3 g or less, require no power, and require only a single image by the RAC. The students were asked to submit both a short proposal and a prototype of their experiment for judging by The Planetary Society and the MECA Project team at JPL.

Sixteen entries were received from seven countries: Canada, Australia, Brazil, Israel, Japan, the United Kingdom and the United States. Two nanoexperiments were chosen for flight based on their alignment with MECA's goals and feasibility of fabrication: the Angle of Repose of Martian Dust and Contradistinctive Copper. These experiments addressed the behavior of windblown Martian dust on surfaces and the oxidation of different textures of copper. An alternate student nanoexperiment to investigate the behavior of spacesuit materials on Mars was selected during the MECA Student Nanoexperiment Challenge and was incorporated into a generic version of a future payload shown in Figure 2.

SNOOPY - Payload Integrated E/PO

An important goal of this project is to provide an opportunity for students to publish their work and results in the scientific literature. The students presented the results of their calibration experiments at the 33rd Lunar and Planetary Science Conference (Möller et al. 2002; Sherman et al. 2002). Should SNOOPY eventually fly, the data returned will be released to students and teachers as soon as it is released to the SNOOPY team. In the meantime, the students will publish their calibration results in the scientific literature and on the web.

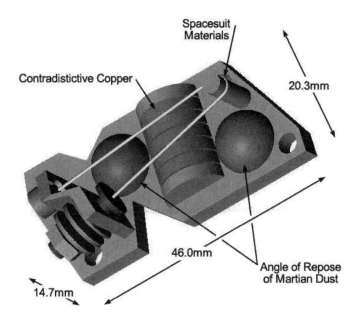

Figure 2. The SNOOPY payload (Kuhlman et al. 2002)

Another important goal of the SNOOPY project is the production of curricular materials aligned with the national and state learning standards through which students and teachers around the world can reproduce the nanoexperiments and perform their own calibration experiments. SNOOPY educational materials will be available through the The Planetary Society and JPL web sites (www.planetary.org and jpl.nasa.gov) in conjunction with the launch of the Mars Exploration Rovers in 2003.

References

Goldin, D. 1999, Testimony before the Committee on Science, U.S. House of Representatives
    http://legislative.nasa.gov/hearings/gold4-28.html
Kuhlman, K. R., Hecht, M. H., Brinza, D. E., Feldman, J. E., Fuerstenau, S. D., Friedmann, L., Kelly, L., Oslick, J., Polk, K., Möller, L. E., Trowbridge, K., Sherman, J., Marshall, A., Luis Diaz, A., Waldron, A. M., Lewis, C., Gyulai, C., Powell, G., Meloy, T., Smith, P. 2002, in 2002 IEEE Aerospace Conference Proceedings, 1, 317
Möller, L. E., Kuhlman, K. R., Marshall, J. R. & Towner, M. C. 2002, in Lunar and Planetary Science Conference XXXIII, #2015.
Sherman, J., Trowbridge, K., Waldron, A. M., Batt, C. A. & Kuhlman, K. R. 2002, in Lunar and Planetary Science Conference XXXIII, #1955.

The research described in this paper was carried out at the Jet Propulsion Laboratory, California Institute of Technology, under a contract with the National Aeronautics and Space Administration.

NASA Planetary Data: Applying Planetary Satellite Remote Sensing Data in the Classroom

Patricia Liggett, Elaine Dobinson, Douglas Hughes, Michael Martin, Debbie Martin, and Betty Sword

Jet Propulsion Laboratory, 4800 Oak Grove Dr., Pasadena, CA 91109, pat.liggett@jpl.nasa.gov

Introduction

NASA supports several data archiving and distribution mechanisms that provide a means whereby scientists can participate in education and outreach through the use of technology for data and information dissemination. The Planetary Data System (PDS) is sponsored by NASA's Office of Space Science (NASA 2004). Its purpose is to ensure the long-term usability of NASA data and to stimulate advanced research.

In addition, the NASA Regional Planetary Image Facility (RPIF), an international system of planetary image libraries, maintains photographic and digital data as well as mission documentation and cartographic data (USRA/LPI 2003). There are ten US and 8 international RPIFs, each facility is a reference center for browsing, studying and selecting photographic and cartographic materials from its general holding of images and maps of planets and their satellites taken by solar system exploration spacecraft. Each RPIF is staffed to assist scientists, educators, students, media, and the public in ordering materials for their own use

Both the PDS and the RPIF, while chartered primarily to support the scientific community, have been providing support to education and public outreach. Both provide access to and distribution of images and information about the planetary science resident in their sites. PDS provides web access to its archives of data and several tools that provide a non-science user to view images and learn more about them.

The Planetary Data System (PDS) consists of seven nodes each containing a specific portion of the planetary data sets. The Imaging Node at JPL contains images of the planets. The Atmospheric Node at New Mexico State University contains non-image atmospheric data. The Geosciences Node at Washington University in St. Louis Missouri contains data used in study of the surfaces and interiors of terrestrial planetary bodies. The Rings Node at the NASA Ames Research Center in San Jose contains data sets relevant to planetary ring systems. The Small Bodies Node (SBN) at the University of Maryland specializes in data concerning asteroids, comets and interplanetary dust. The Planetary Plasma Interactions Node at UCLA contains data related to the study of magneto-spheres and interactions with the interplanetary plasma. While each is accessible from the World Wide Web, each is geared primarily for supporting scientific research.

A working group has been formed to look at possibilities for leveraging off the products and capabilities, both at the PDS and the RPIFs. While many of these products and capabilities were developed initially to support the science community, leveraging NASA OSS can further utilize these existing programs, engage the science community in education, and provide the education community with additional support in teaching sciences in all classrooms.

Description

There are many challenges to getting the science data collected from the planetary missions into a form and format that can be used by educators in the K-12 classroom. The Planetary Data System (PDS) archives and distributes scientific data from NASA planetary missions, astronomical observations, and laboratory measurements and includes tools and technologies for accessing data by educators and the general public (NASA 2004). Likewise the RPIFs, whose general holding contains images and maps of planets and their satellites taken by solar system exploration spacecraft, provide mechanisms for accessing information (USRA/LPI 2003). Yet more is needed and more is being done to make science data not only available but also compatible with the lesson planning of the educator and usable within the learning environment of the classroom.

Applications

Both PDS and the RPIFs include images and related documentation available for educators in K-12. Packages such as *Welcome to the Planets* (NASA 2003a), which contains a collection of many of the best images from NASA's planetary exploration program, can be accessed via link from PDS. The collection has been extracted from the interactive program *Welcome to the Planets*, which was distributed on the Planetary Data System Educational CD-ROM Version 1.5 in December 1995. It has also been updated with the addition of more recent images and can be ordered online. *Welcome to the Planets* contains images of the planets and comets as well as of the spacecraft. In addition, annotation is provided with the images to better understand the image contents. Tools such as *DataSlate* (NASA 2003b) are available for viewing, measuring, and interpreting data from the PDS and other NASA sources. *DataSlate* is part of the CASDE educational tools that facilitate intuitive, graphical search of large image maps, and other visual data sets. It allows the quick comparison of data of different types covering the same region. *DataSlate's* power derives from a data structure that co-registers data of different types. As one moves about a particular dataset, all other datasets are kept in spatial synchronization. The user can click from a natural color image to an infrared image or to a radar image and examine the same region at these different points in the spectrum. This tool can be acquired via the web at http://casde.jpl.nasa.gov/dataslate/.

A tool developed at the USGS for PDS allows the user to access a PDS planetary image and select portions of the image for immediate display or download to a file that can be further manipulated in other image processing and analysis tools. It is located at a web site designed by the Planetary Data System's Imaging Node to provide a global point and click system for exploring various planets. The web page uses PDS images plus software known as *MapMaker* (USGS 2004) to make its maps.

Each of the seven nodes that make up the Planetary Data System, provides access to local tools and information related to the data in the archive. However, more is needed to help the teacher to use the tools and technologies in the classroom through training of the teachers as well as well developed, intuitively designed, and well documented products that can be readily used by the students. In addition, all tools developed for the classroom need to meet the education standards and curriculum needs of the educator.

Benefits to teachers and students

The ability to access "live" data from a repository utilized by the global science community lends meaning to the involvement by the student. In addition, use of "live" space science data helps to inspire and encourage the student to engage in scientific thought and potentially to consider further science study. Applications such as these have been useful not only for teaching and understanding space science but also in teaching fundamentals such as reading and basic research techniques. It also helps to provide the student with a sense of place and purpose within the larger context of space and science. Finally, technologies and capabilities such as these encourage the use of government-provided information and resources, which in turn helps to encourage the government agencies to provide more support to education applications.

Upcoming activities

A working group is being formed to address greater use of PDS, RPIFs and application technologies to support educators use of PDS and RPIFs. This group will look at the available tools and technologies for using planetary data in the classroom. A study will be done of the teachers needs and expectations for using planetary data and related technology in the classroom, as well as determine how best to apply planetary data in education and how technology can be used to help present this data. In addition, the distribution and training in the use of planetary data and the current set of tools and capabilities will be examined to develop a strategy that increases and enhances the use of the PDS and the RPIFs in the education process. The working group will look at the lessons learned by the Earth Science education community to gain from what they have learned and experienced.

New capabilities on the horizon

The *Dear Mars* project (Wilf & Hughes 2001), which is currently under earlyphase prototyping, will allow students to interact with a Mars's mission in real-time using a natural language interface to send emails to robotic instruments on the surface of Mars. Other new technologies are in development that will allow easier access to the PDS from other products. The *Planetary Product Server* currently under development will allow image data to be accessed for many available image-processing tools. Use of tools and technologies originally developed for Earth data processing and analysis will be examined for possible interfacing to the Planetary Data System.

References

National Aeronautics and Space Administration 2003a
http://pds.jpl.nasa.gov/planets/welcome.htm

National Aeronautics and Space Administration 2003b
http://catalog.core.nasa.gov/core.nsf/item/400.1-53

National Aeronautics and Space Administration 2004
http://pds.jpl.nasa.gov/

United States Geological Survey 2004
http://pdsmaps.wr.usgs.gov/maps.html

Universities Space Research Association/Lunar and Planetary Intitute 2003
http://www.lpi.usra.edu/library/RPIF/RPIF.html

Wilf, J., & Hughes, D. 2001, NASA Science Information Systems Newsletter, July 2001

The Wisconsin Idea: Bringing Knowledge to the Community Beyond the Campus

Sanjay S. Limaye and Rosalyn A. Pertzborn

Office of Space Science Education, University of Wisconsin-Madison, 1225 West Dayton Street, Madison, Wisconsin, 53706, SanjayL@ssec.wisc.edu

Abstract. The increasing expectation of leading research universities for effective outreach efforts aimed at reaching broader audiences has created the need for flexible programming that reaches geographically distributed and diverse communities. Recently, the University of Wisconsin-Madison has made major strides in addressing the educational needs of the broader community through a new funding initiative under the Wisconsin Idea Program. The Wisconsin Idea support has been crucial for the success of the K-12 education and public outreach programs in Earth and Space Science carried out by UW-Madison's Office of Space Science Education.

Introduction

The University of Wisconsin-Madison has a long history of outreach efforts that date back to the nineteenth century when the Washburn Telescope was built on the campus in 1882, followed by the call by UW President Charles Van Hise to bring the fruits of the knowledge to the community beyond the campus, the "Wisconsin Idea." The Wisconsin Idea still thrives and the campus vigorously supports outreach efforts undertaken by Office of Space Science Education (OSSE) through support from the Wisconsin Alumni Association and the University of Wisconsin Foundation, the Speakers Bureau and specific programs such as College for Kids, Pre-college Educational Opportunities Program for Learning Excellence (PEOPLE), High School Summer Science Institute, NASA/QEM Summer High school Apprentice Research Program – an eight week long residential program for minority high school students selected through a national competition, Summer Undergraduate Research Experience for undergraduates, Engineering Expo and Campus Open House days. Space Science outreach efforts aimed at the local and regional community have been conducted through OSSE and the UW Space Place, an off-campus facility that serves as a small science museum and a venue for public lectures and workshops for teachers and school students. A new initiative from the UW-Madison campus has enabled us to reach communities beyond Wisconsin through the national network of UW-Madison alumni clubs and the School of Education with the theme of "Our Home in the Universe" with the sub-theme, "Habitability of the Earth." Key programs include the following:

Exploring Mars and Other Worlds – A Summer Workshop for Minority Middle School Students

In this three week long program in summer mornings, minority middle schoolers from Madison area are exposed to collaborative activities and acquire, hone and exercise science, math, writing, communication and social and leadership skills, facilitated by ample supply of nutritious snacks. The space science component is introduced through finding a habitable place in the solar system or other planetary systems through (library or web research), discovering means of reaching that world through rocket building and launching activity, and exploring a scale model of the planet's terrain through robotic vehicles, culminating in a presentation to the parents in a graduating ceremony. Initially based on the Planetary Society's Red Rover, Red Rover Project, the workshop has evolved each year to expand the scope and breadth by making use of selected space science videos, field trips to campus labs and telescopes. The workshop has consistently received superlative evaluations for the last six years. OSSE has successfully conducted this workshop at schools in Milwaukee, with occasional use of the Distance Learning labs in Madison and Milwaukee.

A High School Course in Astronomy and Astrophysics

In most Wisconsin High Schools, Astronomy has not been offered as a course, and only a few have after-school Astronomy clubs. With the recent publication of the National and Wisconsin Science Education Standards, many school districts are adopting their own detailed standards and modifying curricula by introducing Space Science content. OSSE successfully approached a local high school about this possibility. A new course in Astronomy/Astrophysics for high school seniors featuring current space science themes is being offered in Spring 2002 at West High School. OSSE is coordinating support from the UW-Madison faculty and staff to pilot this course with the hopes of making it available to other schools.

However, schools and school districts face several challenges, the major hurdle being finding teachers who are able to teach the new content. At the high school level, there is thus a growing need in Wisconsin for developing a cadre of teachers who have at least some content background in basic Astronomy and Space Science concepts and themes. This need was further highlighted by the relative lack of use of a recently constructed remotely operable observatory by the Madison Metropolitan School District (MMSD), despite having held training workshops in the use of the facility. Simply creating the observatory was not enough, as the teachers struggled with its incorporation in the curriculum for lack of adequate background knowledge.

With a view towards providing professional development in space science, OSSE has begun a new effort that includes a summer workshop as well as hands-on experiencing astronomical observing for a select group of teachers. The workshop included presentations by scientists on topics relevant to a candidate high school space science curriculum and opportunities for teachers to share their ideas and proposed plans for implementing the space science content by developing curriculum units. The teachers toured several professional and amateur

Figure 1. Students in Mr. Hanna's 4th grade class at Reilly Middle School in Chicago using the Water Vapor Sun Photometer developed by OSSE to measure the atmospheric vapor in collaboration with the GLOBE Program

observatories in the vicinity of Madison, including MMSD's remote observatory, which they are likely to use in the coming academic year. The effort will continue for the next two years with periodic group meetings as well as another workshop next summer.

Astronomy in Middle Schools

Madison Schools now have access to a recently-developed, CCD-camera-equipped, remotely-accessible telescope. OSSE provides professional development and technical support to area teachers through amateur astronomers.

Distance Learning Programs

OSSE offers two Web Courses in Earth and Space Science directed at teachers through the University of Wisconsin-Madison.

Teacher Professional Development in Earth and Space Science

OSSE offers two Distance Learning Courses via the web in Earth and Space Science directed at teachers through the University of Wisconsin-Madison through asynchronous, supervised learning.

Public Programs

OSSE also offers Public Lectures around Wisconsin and participates in On-the-Road events through the On-the-Road Program and the Speakers Bureau in Wisconsin and neighboring states. Recently OSSE has begun developing posters for distribution to schools featuring Earth and Space Science themes.

The Imagine the Universe! E/PO Program

James C. Lochner

Laboratory for High Energy Astrophysics, NASA/Goddard Space Flight Center, Greenbelt, MD 20771, lochner@lheapop.gsfc.nasa.gov

Introduction

Since 1996, the *Imagine the Universe!* E/PO program has been bringing information and curriculum support materials to upper middle school, high school, and lower undergraduate students and their teachers on topics in the Structure and Evolution of the Universe (SEU). The Imagine E/PO program consists of a web site, a series of posters and information/activity booklets, and a repertoire of educator workshops. We involve both scientists and educators in the development and testing of the materials. We describe here the various aspects of this program.

Web Site

The *Imagine the Universe!* web site (http://imagine.gsfc.nasa.gov/) grew out of an early effort in 1996 by Dr. Laura Whitlock to develop a "High-Energy Astrophysics Learning Center" based on the science and activities taking place in the Lab for High Energy Astrophysics (LHEA) at NASA/Goddard. Much of the foundation of the present Imagine site was put in place at this time, with scientists in the Lab participating in writing articles. The site underwent a redesign and name change in 1997 to its present structure and major features. In 2000, the site was given a new look as a result of recommendations by an external review.

The *Imagine the Universe!* web site consists of a collection of articles about topics in SEU science, with an emphasis on high-energy astronomy, including the types of objects studied (ranging from the sun and black holes to active galaxies), techniques and tools used by astronomers studying these objects, and the outstanding questions and mysteries of the field. This material is presented at Introductory (grades 7-10) and Advanced (grades 11-14) levels. The site also gives descriptions of the satellites and instruments that have been used to study these objects. In addition, the site also contains news of discoveries, profiles of scientists who work in the LHEA, and Special Exhibits on topics ranging from x-ray transients to anticipating the science from the upcoming missions. The site also includes a "You be the Astrophysicist" feature in which users solve astrophysical problems using the same data as astronomers, such as determining the size of a star, the mass of a black hole, or the velocity of a galaxy. The Web site also includes the following features:

Lesson Plans and Educator Resources. Each of the articles includes linkages to math and science standards, a quiz, and a "Try This" activity which further extends and explores each topic. In addition, the site contains a collection of lesson plans developed with educators. These lessons teach math and science concepts using topics in astronomy, space science, and high-energy astronomy as engagements. Students explore the properties of supernovae, time orbital periods of x-ray binaries, determine the sizes of stars in eclipsing binaries, construct images from digital data, and more, using data collected from high-energy astrophysics satellites.

Ask a High-Energy Astronomer. This "Ask an Expert" service has been run since 1996, with scientists in the Lab fielding questions from users on a wide variety of topics in astronomy. The most interesting questions and answers are posted on the Imagine site by topic, with links from the relevant articles. The service currently receives 50 questions per week, and the archive of questions now numbers about 500.

The Imagine CD-ROM. We capture the web site annually onto a CD-ROM, along the StarChild web site (http://starchild.gsfc.nasa.gov/) and a year's collection of the Astronomy Picture of the Day (http://antwrp.gsfc.nasa.gov/apod/). Taken together, the sites on the CD offer material for every grade level and the general public. The CD is distributed at educator meetings, by email request, and by NASA CORE. We distribute 20,000 CDs annually, with a large fraction going to educator workshop providers.

Evaluation. We performed an independent review of the site in 2000, which resulted in a re-design in the look and layout of the site without changing its content. We also interviewed a group of educators familiar with the site to assess the use of the site in the classroom. This was followed up by an on-line survey. We found that half of repeated users of the site use the lesson plans and science articles, and most new users would likely use them. In addition, despite growing internet connectivity, most users want us to continue to provide the annual CD-ROM of the site, as this insures for them the most reliable access.

Posters and Information Booklets

We have also published a series of posters and information booklets. These posters and booklets extend the information on the web site for topics of particular importance in astronomy and the education curriculum. Topics have included the life cycles of stars, black holes, gamma-ray bursts, and the properties of galaxies and the dark matter problem. The booklets provide background for the teacher, and activities to use with the students. We distribute 10,000 posters and booklets annually.

Educator Workshops

We have also developed workshops which introduce and train educators on the use of the web site and the posters/information booklets. These workshops often debut at the National Science Teacher Association National Convention in the spring of each year. They are presented at other regional educator conferences and for educator groups coming to NASA/Goddard Space Flight Center throughout the year.

Involving Teachers and Scientists

We have successfully utilized teacher interns during the summer to develop lesson plans and activities for the web site. In addition, educators have written the activities and, more recently, the text for the information booklets. Educators have also developed and participated in premiering the workshops. We find that teachers presenting workshops to their colleagues aids in addressing many of the questions teachers have about applying the materials.

In addition, the DePaul University Space Science Teacher Consultants recently reviewed and field tested the lesson unit, "X-ray Spectroscopy and the Chemistry of Supernova Remnants." Much of the interaction with them was performed via email and teleconferences. Their recommended revisions greatly extend the utility of the lesson.

Scientists in the Lab not only participate in Imagine's "Ask an Expert" service, but also by writing and updating articles for the site, and contributing results of their research for news articles. Scientists also volunteer to be interviewed for our Scientist Profiles. A few scientists also work with teacher interns, and review text and activities developed by the teacher interns. The Lab recognizes the importance of scientist involvement in our education efforts and encourages participation by allowing all scientists to set aside a small portion of their time for E/PO.

Customer Interviews to Improve NASA Office of Space Science Education and Public Outreach Leveraging Success

Leslie L. Lowes

Jet Propulsion Laboratory, 4800 Oak Grove Drive, Mail Stop 180-109, Pasadena, CA 91109, Leslie.L.Lowes@jpl.nasa.gov

Gloria Jew

Geebs Technical Communications, 401 N. Lincoln St. Burbank, CA 91506

Introduction

Leveraging with organizations that serve our customers and focusing on the needs of those organizations are two prime elements of the NASA Office of Space Science (OSS) Education and Public Outreach (E/PO) Strategy. Understanding the needs of respected organizations who serve the formal education community, as well as areas of informal education such as museums, planetariums, and youth groups, serves as a basis for a successful relationship. On behalf of NASA OSS, the Solar System Exploration (SSE) Education and Public Outreach Forum has conducted a series of customer interviews with representatives from leading organizations who serve some of the audiences we wish to reach.

Rationale

"One of the biggest challenges faced by the OSS E/PO Program is the lack of direct knowledge of the needs of [formal and informal] education" (Cohen 2001). A basis for successful relationships and effective programs is that of a good understanding of the needs of the communities with which we wish to work, and insight into the realities of the environment in which those communities conduct business and design programs for their audiences. Our immediate purpose is three-fold: 1) to enlighten these external communities about NASA OSS E/PO infrastructure, capabilities, and assets, 2) to understand the communities' needs, discuss how OSS assets can be of benefit, as well as provide supporting information for determining measures of success in working with these communities, and 3) provide this knowledge for use as a gateway for strategic planning. We selected individuals for on-site interviews based on the breadth of their organization's understanding and relationship with their customers and their synergy with OSS E/PO. We asked the individuals to represent the needs of their general community, including but not limited to their organization. We selected organizations with which NASA OSS has newly developing or previously unexplored relationships at the thematic level.

Participants

To date, we have interviewed representatives who are leaders in five external organization types: (1) Formal Education (K-12), Alan Gould of Lawrence Hall of Science (LHS) and Tyson Brown of National Science Teachers Association (NSTA) SciLinks; (2) Formal Education (collegiate), Jeanne Narum of Project Kaleidoscope; (3) Informal Education/Youth Groups, Linda Fallo-Mitchell, then of the Girl Scouts USA (GSUSA); (4) Informal Education/Planetariums, Martin Ratcliffe and Jon Elvert of the International Planetarium Society (IPS); (5) Professional/Amateur Astronomical Organizations, Larry Lebofsky of American Astronomical Society Division of Planetary Sciences and Robert Havlin, then of the astronomical Society of the Pacific.

Expected Outcome

We are developing a series of white papers summarizing the synergistic needs and activities for NASA OSS E/PO and the general types of organizations we interviewed, to be supplemented by further research on other organizations in that community. The information will contain:

- Highlights of interaction opportunities recommended by the representative, including:
  - general information exchange and awareness raising throughout our respective communities
  - potential for partnerships
  - training opportunities
  - participation in conferences, publications, and product development and dissemination
- Specific information on external organizations within the community, including contact information, web sites, audiences and numbers reached, and dates and lead times for conferences and publications.

Summary of Recommendations

Several common themes emerged from these interviews regarding ways the communities would like to interact with NASA OSS E/PO.

- All representatives stressed the desire for and importance of sustaining relationships with their community, and ensuring on-going communication.
- Most expressed interest in a single-point-of-contact within NASA OSS for the relationship.
- Providing access to OSS material and experts through one-stop shopping also would facilitate smooth interaction.

- All were interested in the infusion into their communities of current and scientifically accurate space science content, put into appropriate context. All appreciated the inspiration that space science content can provide their audiences.

- Presentation of information, and therefore programs, are best done at a thematic level.

- In cases where it helped forward the community's mission, the organizations expressed a common need for training and professional development of staff and associates on space science content.

Specific recommendations for implementing the interactions will be detailed in the white papers. Presented here are general types of interaction opportunities and selected implementation examples.

There are various levels of involvement and commitment when working with external organizations. The sustaining and nurturing a long-term relationship is exemplified through the joint, strategic planning of activities between IPS and OSS NASA E/PO entities. Working with the leadership of IPS, OSS Forums and integrated OSS Broker/facilitator efforts, a long range plan is being formulated that includes: 1) Publication in the IPS quarterly newsletter, 2) Electronic notification to IPS members of relevant OSS news, to allow local events around OSS mission activities, 3) Attendance at IPS International Conferences, with exhibits and workshops for planetarium educators, and 4) Connecting scientists to regional planetariums for expert information and speakers.

Training of the organization's staff and/or members on relevant space science topics inspires and enlightens them, and helps assure their understanding (and in some cases their comfort) with the topic. A core element of the maturing high leverage relationship with GSUSA includes presentation of OSS content at their annual international trainer workshop, where attendees learn content to take back to their regional trainers and leaders. Many organizations have training mechanisms in place into which OSS content can be directly infused, typically involving hands on activities designed for their audiences.

Almost universally, organizations trust NASA to make information available to them that is scientifically accurate. The involvement of OSS scientists themselves is the most highly desired way of assuring this and providing relevant content for the audiences. OSS researchers can present content, develop and review materials, and act as role models. An example is the science cadre review of selected NASA web sites for SciLinks. The SciLinks activity provides web-based updates to support external web page links, printed by code in selected textbooks under an agreement with publishers.

An untapped resource for university-level partnerships is Project Kaleidoscope (PKAL), which is a national alliance to build strong learning alliances for undergraduate students in science, math, engineering, and technology departments. They also work to improve the public understanding of the purpose of a strong undergraduate science community. NASA can help to bring a unique focus to PKAL through work with community colleges and interaction with education departments. PKAL has a ten-year goal to promote the immediate transfer for research that's being done to the learning being done. Space science content can help motivate non-science majors to study science (including education students).

Further examples and recommendations will be detailed in the white papers.

Future Directions

After peer review, white papers detailing information on interacting with these types of organizations will be made available to the NASA OSS E/PO community. We anticipate they will provide significant input to the OSS E/PO working groups concerned with strategic planning and relationships with external communities, in particular the Community Based Groups and Museums and Planetariums Working Groups. Further, the SSE Forum will assist in brokering and developing further relationships with these communities. We also plan further interviews with representatives of the public library community, and with interpretive societies that present relevant science concepts in outdoor settings.

References

Cohen, S. B., Gutbezahl, J., & Griffith, J. 2001, Office of Space Science Education/Public Outreach January 2000-May 2001 Evaluation Report (http://spacescience.nasa.gov/education/resources/evaluation/ NASA_Interim_2001_Report.pdf)

Using Scientific Results from the Hubble Space Telescope to Create Curriculum Support Tools

Dan McCallister, Bonnie Eisenhamer, and Jonathan Eisenhamer

Space Telescope Science Institute, 3700 San Martin Drive, Baltimore, MD 21218, mccallis@stsci.edu

Denise A. Smith

Space Telescope Science Institute/Computer Sciences Corporation, 3700 San Martin Drive, Baltimore, MD 21218

Abstract. *Amazing Space* (http://amazing-space.stsci.edu) is a program designed to enhance the teaching of mathematics, science and technology skills using recent data and results from NASA's Hubble Space Telescope mission. The program partners teachers with scientists, graphic artists, writers, and multi-media Web developers. An extensive production cycle including specification, project definition, frequent reviews and several testing periods has proven to generate robust but flexible resources for classroom use.

Project Philosophy

Interactive resources can be designed to effectively integrate into and supplement pre-college science and mathematics curricula (Gooden 1996; Wiesenmayer & Koul 1998). Such interactive materials have the potential to enhance students' skills, especially if they contain actual data used from scientific research (Salomon & Perkins 1996), and the data are used in an engaging context. *Amazing Space* is a unique program designed to create interactive modules based on results from NASA's Hubble Space Telescope (HST). Current HST discoveries and technological concepts are woven into online resources that address specific national educational standards for mathematics, science and technical literacy. *Amazing Space* is produced by the Office of Public Outreach (OPO) at Space Telescope Science Institute (STScI), the home of the science operations for the HST observatory.

Scientist-Educator Collaboration

The core of *Amazing Space* is a collaborative relationship between scientists and teachers, augmented by professional graphic artists, Web developers, programmers and technical writers. By building on the combined talents of the individuals, *Amazing Space* teams develop comprehensive, interactive multimedia educational activities. The resources, available online through the Web, incorporate real scientific data from NASA missions (primarily HST) that can be used in the classroom to develop specific skills articulated in the national educational standards. For example, the *Galaxy Hunter* activity uses HST's Hubble Deep Fields, two of the deepest optical images of the universe, to reinforce statistics concepts, such as simple random sampling, bias, variability, and sample size.

Development Guidelines

Some of the key guidelines for development of *Amazing Space* activities are:

- activities must address specific, well articulated K-12 educational standards
- modules must relate to specific HST data and research results
- activities must be interactive, modules that can be integrated together in different ways to suit the learning environment
- activities must assist in developing an appreciation and hopefully some skill in the scientific process
- modules must supplement or augment existing curriculum, not replace it

Activity Development

Activity development is based on the Production Pyramid, developed through testing and evaluation over several years of Amazing Space. In addition to being a successfully tested model for production, the pyramid is based on associative learning principles and follows the same principle as Maslow's Hierarchy of Human Needs (Maslow 1998), particularly familiar to teachers.

Resource Review and Testing

The review of OPO's educational resources progresses in a number of stages and includes both alpha and beta testing. The following is a list of criteria used to assess *Amazing Space* online resources.

- *Content:* Scientific and technical accuracy, educational content, curriculum content, etc.
- *Design:* readability, texture, conformance to OPO design standards, treatment of science threads, clarity, appropriate use of technology, colors, other material, etc.
- *Pedagogical approach:* educational goals, adherence to educational standards, science accuracy, appropriateness to grade level, clarification of preconceptions, clarity of goals, accuracy of performance instruments, language, etc.
- *Usability:* Training required, network performance, computer performance requirements, appropriateness to curriculum, etc.

The extensive review process including internal reviews, workshops and field-testing has resulted in improvements in the modules, based on in-situ teacher feedback. Though the development, production and review of *Amazing Space* requires substantial effort, the procedures and methods, developed and improved over several years, has resulted in a suite of modular resources for classroom use that are based on contemporary scientific data.

References

Gooden, A. 1996, Computers in the Classroom: How Teachers and Students Are Using Technology to Transform Learning (San Francisco: Jossey-Bass Publishers)

Maslow, A. H. 1998, Toward a Psychology of Being, 3rd ed. (New York: John Wiley & Sons)

Salomon, G., & Perkins, D. 1996, in Technology and the Future of Schooling, pp. 111-130, ed. S. T. Kerr (Chicago: University of Chicago Press)

Wiesenmayer, R. L., & Koul, R. 1998, J. Science Education & Technology, 7, 271.

Working with Scientists in the AAS Division for Planetary Sciences

Ellis D. Miner

NASA OSS Solar System Exploration E/PO Forum, Jet Propulsion Laboratory, m/s 230-260, 4800 Oak Grove Drive, Pasadena, CA 91109
ellis.d.miner@jpl.nasa.gov

Introduction

On page 15 of its 1996 publication, *Implementing the Office of Space Science Education/Public Outreach Strategy* (NASA 1996), the NASA Office of Space Science states:

"Implementation of the OSS Education Outreach Strategy must involve the scientific participants in ways that: preserve OSS research goals; make a genuine contribution to education and the public understanding of science; permits scientists to continue to function as scientists; and make optimum use of their time and expertise ..."

"The strength of the OSS research community and the unique contribution that it can make to education and public outreach is based on the fact that it is a source of continuing new knowledge and new discoveries that can inform teachers at all levels and excite students and the public about science.
    Space scientists:

- Are a living demonstration that science is a human endeavor carried out by real people of all kinds who have many different types of skill and who approach their science in many different ways;

- Possess an enthusiasm for science and a way of looking at the world to try to understand why things work the way they do;

- Have a presence in colleges and universities, research laboratories, and industry in communities across the country;

- Can communicate and provide information on the latest exciting research results.

Their contributions to education and outreach must draw on these attributes.
    But to be most effective, the limited time and resources of the scientific community must be highly leveraged and properly channeled. ..."

In its role as an agent for NASA's Office of Space Science, the Solar System Exploration Education and Public Outreach Forum (SSE Forum) has attempted to establish a working relationship with the Division for Planetary Sciences of the American Astronomical Society (DPS) to assist DPS scientists to use their limited time and resources as effectively as possible in their E/PO efforts. Most

DPS scientists already spend some fraction of their time in such efforts, often at their own expense, but frequently those efforts tend to be local in their reach and duplicative in their content.

The SSE Forum and seven regional 'Brokers/Facilitators' constitute a Support Network for planetary scientists to help make connections between them and E/PO specialists who need their expertise and their data. The SSE Forum is located at

- NASA-JPL in Pasadena, CA
  (818-354-4450 or 818-393-7334;
  ellis.d.miner@jpl.nasa.gov or leslie.l.lowes@jpl.nasa.gov).

The seven regional brokers are located at

- Museum of Science, Boston, MA
  (617-589-0359; csneider@mos.org),

- College of Charleston, Charleston, SC
  (843-953-8279; cass@cofc.edu),

- Wheeling Jesuit University, Wheeling, WV
  (304-243-2388; nitin@cet.edu),

- De Paul University, Chicago, IL
  (773-325-1854; cnarasim@condor.depaul.edu),

- Lunar and Planetary Institute, Houston, TX
  (281-486-2109; Shipp@lpi.usra.edu),

- Space Science Institute, Boulder, CO
  (303-492-7321; camorrow@colorado.edu), and

- University of Washington, Seattle, WA
  (206-543-0214; nasaerc@u.washington.edu).

DPS E/PO Survey

In order to effectively serve the planetary science community, it was important to get some measure of the current involvement of DPS scientists. A survey was designed to meet the needs of both the DPS Executive Committee and the OSS E/PO Support Network. Copies of the questionnaire were e-mailed to all DPS Members, posted on the DPS Web Site, and made available at DPS 2000. Results were collected between May and November 2000. Early results were presented as posters at DPS 2000 in Pasadena, CA.

There were 122 (of approximately 1200 members of the DPS) who responded to the questionnaire. The geographic distribution included 14 European scientists, 34 scientists from the eastern time zone, 11 from the central time zone, 34 from the mountain time zone, 26 from the Pacific time zone, and three from

Hawaii. The respondents covered a wide range of scientific expertise, and almost all were already involved in some type of E/PO. Most often named specialities were planetary atmospheres (50), comparative planetology (48), planetary surfaces (47), planetary satellites (42), spectroscopy (39), solar system origin/evolution (38), physical properties of comets (37), physical properties of asteroids (30), and spacecraft instrumentation (29).

Almost all respondents were already involved in multiple kinds of public communication efforts. The most frequently listed involvements included public lectures (87), media interviews (80), answering science questions for public or educators (77), informal education efforts (70), community clubs (64), writing articles for popular magazines (60), working with lawmakers to improve their science understanding (59), pre-college classroom activities or course work (58), post-high-school classroom activities or course work (53), and reviewing E/PO products for scientific accuracy (50). Interestingly, 49 responded that they would be interested in being part of a DPS Speakers' Bureau. Organization of a DPS Speakers' Bureau is now underway.

Of the 122 respondents, only 24 indicated that they did not want to have help from the OSS E/PO Support Network, 38 indicated a desire to have such assistance, and the remaining 60 were undecided. It is the SSE Forum's intent to work with the 38 and provide additional information to the 60. Eventually, we would like to provide the names of interested scientists to the relevant regional brokers, who will act as filters to assure that requests for scientist time and expertise are kept within whatever bounds are set by the scientist, but will provide many opportunities for E/PO partnering.

Presence at Annual Meetings

The SSE Forum set up a large booth adjacent to the DPS 2000 registration desk and provided information on the available services in the registration packets. There were also several education session posters by members of the Support Network. A video depicting DPS scientists' involvement in E/PO was distributed to all participants at DPS NASA Night. NASA OSS Associate Administrator Edward Weiler praised the Support Network and its efforts in his address at NASA Night. The video and other help and contact information were subsequently distributed via CD-ROM to all DPS members. Support Network personnel were also instrumental in helping to facilitate a luncheon to discuss how scientists can effectively work with planetaria and in a Planetary-Society-sponsored student press conference on the final day of the DPS meeting.

Although not as extensive, similar efforts were carried out at DPS 2001 and are planned for future DPS meetings. Efforts will be adjusted to meet the mutual needs of the DPS scientists and potential E/PO partners.

Working Behind the Scenes

The author, who presently serves as DPS Press Officer, has worked with Dr. Larry Lebofsky (DPS Education and Public Outreach Officer) and other members of the DPS Executive Committee to make it easier for DPS members to become more effectively involved in E/PO efforts. Several Support Network personnel who are also DPS members have joined with Lebofsky and others in the Solar System Exploration Decadel Study support. The DPS Committee has agreed to waive DPS membership requirements for E/PO poster presentations by Support Network personnel (and others) on a case-by-case basis as recommended by the Press Officer or the E/PO Officer. Education posters are exempt from the DPS rule limiting authors to one paper at each meeting so that research scientists can report on their scientific results and on their E/PO 'findings.' NASA has made substantial improvements in the NRA E/PO proposal process based on DPS member feedback through the SSE Forum, and attempts are being made to provide alternative means of funding good E/PO projects through NASA and other sources. Support Network personnel at the Space Science Institute in Boulder have constructed a 'Menu of Opportunities for Scientists In Education' (MOSIE) to make it simpler for planetary scientists and other space scientists to be effectively involved in E/PO. Very soon an on-line journal will permit members of the American Astronomical Society and others to publish refereed E/PO papers. The Support Network will continue to be committed to providing as much help as limited funding will permit.

Summary

Planetary scientists represent an invaluable resource for information on, and understanding of, the solar system for all aspects of public communication. Yet most scientists have little time, funding, or connections to enable them to use that expertise. The OSS E/PO Support Network, with its theme-based forums and regional brokers/facilitators, is already well equipped to assist scientists, non-intrusively, to establish partnerships with those who need their scientific expertise. Such partners also often possess the funding and know-how to build effective public communication programs, activities, and products, resources that scientists alone often cannot easily access. The result of such E/PO partnerships is often a mutually satisfying symbiosis of talents that enables scientists to make much more effective use of their limited time and resources to accomplish grand E/PO designs.

References

National Aeronautics and Space Administration 1996, Implementing the Office of Space Science Education/Public Outreach Strategy (Washington: NASA)
  (http://spacescience.nasa.gov/admin/pubs/edu/imp_plan.htm)

An Investigation Into Student Understanding of Astrobiology Concepts

E. G. Offerdahl, E. E. Prather, and T. F. Slater

University of Arizona Steward Observatory, Tucson, AZ 85721, eofferdahl@as.arizona.edu

Introduction

Astrobiology can be defined as the study of the origin, evolution, distribution, and destiny of life in the universe. It defines itself as an interdisciplinary science existing at the intersection of astronomy, biology, chemistry, geology, and physics. Discoveries from this field have dramatically changed our view of the potential for life in the universe. For example, at least ten times as many planets have been discovered outside our solar system as there are within it. Perhaps even more impressive is that life has been found to exist under conditions previously thought impossible. This includes organisms that thrive in temperatures above the boiling point and below the freezing point of water, in extreme acidic and basic conditions, thousands of feet below the Earth's surface and on the ocean's floor, and in the extreme radiation conditions of outer space. As a result, our understanding of the limits on life has forever been changed. These discoveries are being made simultaneously with discoveries that strongly suggest that liquid water oceans exist under the icy surface of Jupiter's moon Europa and that running water was likely present on the surface of Mars in the past.

As a result of its truly interdisciplinary nature, teachers are actively considering the inclusion of astrobiology in their courses (Prather & Slater 2001). Consequently, there exists a rapidly growing need to create classroom-ready astrobiology instructional materials. However, the development of meaningful activities requires a robust research base that identifies specific difficulties and pre-instructional beliefs students have concerning astrobiology related concepts. For our investigation of astrobiology, we focused our research around two main goals. First is the documentation of conceptual and reasoning difficulties students experience when learning about astrobiology. The second is to design innovative instructional materials that are sensitive to these identified student difficulties and that are aligned with the content and instructional goals of the *National Science Education Standards*.

Our investigation was focused on four key astrobiology related topics: sunlight, water, temperature, and limiting environments. Over two thousand students ranging from fifth grade students through seniors in college were surveyed to document pre-instructional beliefs at each age level as well as identify beliefs common to all ages. Students were surveyed through the use of open-response style questions designed specifically to elicit student beliefs concerning the importance of sunlight, water, and temperature for the existence of life as well as the limitations on life.

Through careful analysis of student responses we found that the majority of students correctly identified that liquid water is necessary for life and that life forms can exist without sunlight. However, many students incorrectly stated that life cannot survive without oxygen. Furthermore, when students were asked to reason about life in extreme environments, they most often cited complex organisms (such as plants, animals and humans) rather than microorganisms.

Results from this investigation were used to guide the development of an interactive lecture-tutorial structured around a directed-inquiry approach. Overall, students demonstrated a significant improvement in their conceptual and procedural knowledge about life in extreme environments after instruction using these researched-based instructional materials.

References

Prather E. E. & Slater T. F. 2001, The Physics Teacher, 39(2), 20

Additional Information: For information on the research conducted by the Conceptual Astronomy and Physics Education Research (CAPER) team at the University of Arizona, please contact the first author.

ASTRO-VENTURE: Using Astrobiology Missions and Interactive Technology to Engage Students in the Learning of Standards-Based Concepts

Christina O'Guinn

NASA Ames Educational Technology Team, NASA Ames Research Center, MS 204-14, Moffett Field, CA 94035,
christina.m.oguinn@nasa.gov

Introduction

A challenge we often face when developing curriculum based on NASA's missions, is the integration of concepts that are far too advanced for K-8 students to grasp and do not always align well with standards-based curriculum. Many of these concepts are not found in the national education standards indicating that they are not developmentally appropriate for students. Furthermore, teachers' instructional time is already so limited that they do not have time to cover the additional content covered in NASA curriculum supplements unless it teaches standards. Thus, the challenge to curriculum developers is to develop standards-based curriculum supplements that are truly useful to educators, while still integrating NASA's missions. Astro-Venture is one example of an approach that achieves this, while using the engaging topic of astrobiology to provide a context and motivation for learning national standards related to the theme of Earth and space as a system.

Description

Astro-Venture http://astroventure.arc.nasa.gov is an educational, interactive, multimedia Web environment in which students in grades 5-8 role play NASA occupations as they search for and design a planet that would be habitable to humans. *Astro-Venture* is designed to include modules in the areas of astronomy, geology, atmosphere and biology with a culminating assessment module in which students apply knowledge from all areas to design a planet.

For each core science area, students engage in an online training module in which they isolate variables associated with that area and observe the affects on Earth. They then draw conclusions about which characteristics allow Earth to remain habitable. Following this experience, students engage in several classroom, hands-on activities that teach them core standards-based concepts which help them to understand why the identified characteristics are vital to human habitability. These concepts include states of matter, flow of energy, conservation of matter, planetary geology, plate tectonics, human health and systems theory. With an understanding of the "whats" and the "whys," students then engage in a mission module for that core science area in which they simulate the methods scientists would use to go about finding a planet with these characteristics. This helps them to understand the "hows." It is important to note that students are given a very simplified version of these methods in which they are evaluated on the standards-based skills of comparing data and drawing conclusions.

The Astronomy section of *Astro-Venture*, when looked at in detail, shows how standard concepts are truly met and not merely mentioned. In this section, students focus on the conditions required for liquid water, as a requirement for life. In order to understand these conditions, students first engage in standards-based lessons that teach the different states of matter and factors that change states of matter. This knowledge helps them to understand the conditions necessary for water to be a liquid on Earth's surface and how star type, orbital distance and atmosphere all work together to determine these conditions. Geology, Atmosphere and Biology will explore similar requirements for human survival. This theme is an important one in helping students to understand the Earth as a system, which was identified as a need and recommendation in the recent report from the 2001 National Conference on the Revolution in Earth and Space Science Education (Barstow & Geary 2001). This document states: "Students should not experience Earth and space science as a series of topics, but rather as a whole system-the interconnected geosphere, hydrosphere, atmosphere and biosphere." *Astro-Venture* is designed to use core astrobiology ideas to provide a context and motivation for understanding Earth and space-related core concepts required in the national education standards.

Throughout all of the modules and activities, students are engaged in the inquiry process and instructional strategies identified by Project 2061 as most effective in science teaching (AAAS 2001). The lessons use the five "E" approach, by first "Engaging" students with a problem or question often connecting back to the overall purpose of designing a habitable planet. Students then "Explore" possible answers to this problem or question through a hands-on activity or research and come up with their own "Explanations" and conclusions guided in a discussion by the instructor. Finally, students are involved in an "Extension" of the new concept to a different situation and are "Evaluated" on their understanding of the Earth-space system through, often through rubrics. Throughout, the lessons connect to each other and to pre-requisite concepts in a cognitive approach to learning that helps students to develop a schema of the concepts. When relevant, the lessons identify commonly held misconceptions, give suggestions for how teachers can ascertain if their students possess these misconceptions and recommendations for how to overcome them. Questioning strategies are modeled for teachers so that they can guide higher order thinking and help students to form their own learning, instead of lecturing to them.

Astro-Venture uses the fascinating topic of Astrobiology and the search for habitable planets to engage students in an overall scenario that provides the overall purpose, motivation, context and goal for all of the online and off-line instructional activities. In this way, *Astro-Venture* demonstrates one successful method of teaching core science curriculum concepts while integrating NASA's mission.

References

American Association for the Advancement of Science Project 2061 2001, Middle Grades Science Textbooks Evaluation http://www.project2061.org/newsinfo/research/textbook/mgsci/criteria.htm

Barstow, D. & Geary, E. 2001, Blueprint for Change: Report from the National Conference on the Revolution in Earth and Space Science Education (Cambridge, MA: TERC.), p.13

Project FIRST and Eye on the Sky: Space Science in the Early Grades

Ruth Paglierani

UC Berkeley, Space Sciences Laboratory, MC 7450, Berkeley, CA 94720-7450, ruthp@ssl.berkeley.edu

Sally Feldman

West Contra Costa Unified School District, Washington Elementary School, 565 Wine Street, Richmond, CA 94801

Elementary educators typically have only limited opportunity to teach extensive science units. This is due in great part to the primary focus on literacy and mathematics instruction in the early grades. It is not surprising then, that the time and resources allocated to science teaching are significantly less than those allocated to language arts and mathematics. The integration of elementary science curriculum with language arts provides one means of addressing the challenge of keeping science education robust in the elementary classroom. Project FIRST's *Eye on the Sky* suggests a model for the successful integration of science instruction with language arts through inquiry-based learning. *Eye on the Sky* offers an exciting opportunity to explore the dynamic Sun and share research discoveries of NASA's Sun-Earth Connection with the elementary education community. We will present *Eye on the Sky: Our Star the Sun*, a suite of integrated, inquiry-based lessons designed specifically for K-4 students. The lessons were developed and tested by UC Berkeley educators and NASA scientists in partnership with classroom teachers. We will review the program components and examine the benefits and challenges inherent in implementing such a program in the elementary school setting.

Scientists Actively Involved in Education and Public Outreach: A Successful Experience at the MIT Center for Space Research

Irene L. Porro

MIT Center for Space Research, 77 Massachusetts Ave. NE80-6079, Cambridge, MA 02139, iporro@space.mit.edu

Abstract. The Center for Space Research (CSR) at the Massachusetts Institute of Technology is an interdepartmental center that supports research in space science, astronomy and engineering. The Center has been involved in several community outreach initiatives for many years and in 1999 Kathryn Flanagan, research scientist with the Chandra mission, started the CSR Education and Public Outreach Office to coordinate the outreach efforts of the Center. The local but genuine success of these programs prompted a rapid growth in the number and variety of commitments of the E/PO Office that required the development of new initiatives. Today, our main strength consists in the availability of a number of scientists whose overall expertise covers a wide range of space science themes. Our mission is to provide the interface between the CSR scientific community and the formal and informal education communities. Our goal is to offer audiences of any age the opportunity to share in the excitement of the scientific enterprise, and to augment the awareness, in particular for underrepresented groups, of the science resources available to the local community. In this paper we present a sample of initiatives that show our involvement in both formal/informal education and public outreach.

Our initiatives rely on the direct participation of CSR scientists, and on the partnership with institutions that provide the expertise in formal and informal education. The initiatives are designed to conform to NASA OSS guidelines for E/PO, but are also flexible enough to meet the needs and requests of the local community. Indeed, an E/PO office with limited personnel like ours is continuously dealing with the tradeoff between quality of the outreach and the number of people reached out to. Small-scale outreach is usually a very effective and successful option: it is also not so intimidating, for both the volunteer speaker and the audience, and allows for individual interaction and follow up. The personal interaction with a scientist is often an event with a long lasting impact on both children and adults.

As part of our partnership with the Cambridge Public Schools, we hosted one-day professional development workshops for middle and high school teachers in which several scientists from the HETE-2, Rossi X-Ray Timing Explorer (RXTE), and Chandra mission participated. The main goal of the program was to address the concept of science as inquiry by showing "how scientists know what they know." After forming small groups the teachers were engaged in hands-on activities facilitated by the scientists. The activities involved using the scientists' everyday tools, from computers to photographic plates, to electronic equipment, to the 1:1 mockup model of the HETE-2 satellite. A questionnaire filled out by 15 teachers who attended the March 2001 workshop provided the following results: 60% of the teachers found that the workshop was very effective, and 40% fairly effective, in addressing the teachers' professional needs and interests; 90% found that the workshop was very good in providing new knowledge in astronomy and space science; 60% said that the workshop was fairly useful for direct application to the classroom; 70% suggested that future workshops should devote more time to each activity, and more informal time with the scientists. Encouraged by this feedback, we are now organizing a new series of teacher's workshops that will be five days long and open to about thirty middle and high school teachers from all of Massachusetts.

Our ongoing collaboration with the Current Science and Technology Center at the Boston Museum of Science featured a bilingual presentation on Chandra for Astronomy Day 2001. The positive feedback from that initiative motivated us to provide multi-lingual options of the same presentation. We took advantage of the multi-cultural environment of the CSR to ask scientists from different countries to translate the presentation into their own languages. The presentation is now available on our web site in English and in four other languages, with more to come. In a tentative attempt to introduce an interdisciplinary element to our effort, the presentation is always provided simultaneously in English and in another language. The purpose is twofold: to help the people who speak English as a second language to learn about Chandra and to get familiar with the English technical vocabulary; and to expose those who speak English as first language to a variety of languages and stimulate their interest in some of them.

To specifically address underrepresented groups, we are involved in several initiatives to offer opportunities for middle and high school girls to interact with CSR women scientists. We have been working with the MIT based "KEYs to Empowering Youth," a motivational program for 11 to 13 year old girls, and with "Eyes to the Future," a mentoring program directed by TERC, that links girls of all abilities with women professionals in science and technology fields. The feedback we received shows that these two initiatives had a remarkable impact. They strengthened the teenage girls' interest in science and opened them to a scientific view of the world to which they can contribute with their own creativity.

Finally, we recently started a collaboration with the Boston 2:00-to-6:00 After-School Initiative that is intended to enhance the academic support of inner-city students in middle and high school. The initiative will offer 6 visits to the Boston Museum of Science, plus one to the control center of the Chandra satellite in Cambridge, to 90 kids in grades 7-9 from a number of after-school programs. The effort that brings together CSR personnel, after-school programs providers, and Museum of Science volunteers, is indeed aimed at exposing the kids to a variety of scientific topics involving physics, space science, astronomy, etc. and widen their intellectual and, eventually, professional horizon.

In these and other E/PO initiatives, the role of the CSR E/PO Office is to provide the necessary interface to assure communication between the CSR scientific community, the local education community, and the beneficiary of the program. This task requires an understanding of both the science carried out at the CSR and the issues concerning science education. For the latter we rely on the expertise offered by our partners: teachers, school administrators, and museum coordinators, while our main task is to facilitate the partnership between scientists and educators and to support the integration of their different skills. It is not uncommon in fact to have to deal with an underlying diffidence and with concerns about competence, and interference in each other field of expertise, that affects professionals from both backgrounds. Most concerns however can be prevented, or properly addressed, by precisely identifying expertise and roles: scientists are asked to provide science content and to communicate the excitement of the scientific enterprise by showing what their work consists of and by sharing the passion in their research. The education professionals are asked to provide the framework and the guidelines according to which the science content has to be delivered. We make sure that the communication flows and provide all the management support to make things happen.

This strategy has been quite successful for the programs we ran in the last two years and encourages us to keep pursuing partnerships between science and education professionals by assuring a proper integration of their different skills.

The Search for Life in the Universe Online: A Distance Learning Course on Astrobiology for Teachers

Edward E. Prather and Tim F. Slater

University of Arizona Steward Observatory, Tucson, AZ 85721, eprather@as.arizona.edu

Introduction

Effective inquiry-based instruction requires that teachers possess a detailed understanding of the target concepts to be taught. Because astrobiology, the search for the origins and evolution of life in the universe, is a new field of science bringing together many fields of study, it presents a formidable challenge for most K-12 science teachers. The NASA-funded Center for Educational Resources (CERES) Project (URL: http://btc.montana.edu/ceres/) at Montana State University has designed a set of classroom activities and an accompanying Internet course for teaching astrobiology. These activities have been designed to combine on-line data resources from NASA with the student-centered inquiry instructional strategy emphasized in the *National Science Education Standards*. The activities have been developed and field-tested by pre-college science teachers and university faculty. The accompanying asynchronous Internet course is a 15-week, graduate-level course in astrobiology for teachers. The course integrates the *NASA Astrobiology Roadmap*, the NRC *National Science Education Standards*, and the astrobiology curriculum supplements available online at URL: http://btc.montana.edu/ceres/astrobiology/LabActivities/.

The two main goals of this course are: (1) To provide information on the central concepts related to the field of Astrobiology, and (2) To provide experiences with using and creating student-centered and inquiry-based curriculum materials for teaching astrobiology. Specifically, each of the teacher-participants, who are in-service teachers: identify NASA's science goals for the study of Astrobiology; complete lab activities designed to develop knowledge and skills in Astrobiology; create an original reading activity that synthesizes Astrobiology information; develop inquiry-based curriculum in Astrobiology education that are aligned with the *National Science Education Standards*; field-test an Astrobiology lab activity with middle or high school students; and write an implementation plan for teaching an existing lab activity with students. Course evaluations, conducted externally by Horizon Research in North Carolina, overwhelmingly suggest that the course meets its goals of improving teachers' astrobiology content knowledge and skills at implementing these topics in their classrooms. In total, 18 credit-hours are now available in a series of 'astronomy for teachers courses' which, when combined with 14 graduate credit-hours in curriculum and instruction, can constitute a foundation for a new astronomy education option in the Montana State University Master of Science – Science Education (MSSE) online degree program.

When NASA E/PO funds are available to subsidize this course, this course allows in-service teachers the opportunity to earn three graduate credits at a total cost of $100. When the course is run as a self-supporting course on tuition alone, the course costs each teacher-participant approximately $450. This is one of five astronomy for teachers courses currently offered by the CERES Project at Montana State University.

References

Prather E. E. & Slater T. F. 2001, The Physics Teacher, 39(2), 20

Slater T. F. 2000, The Physics Teacher, 38(9), 538

Slater T. F., Beaudrie, B., Cadtiz, D. M., Governor, D., Roettger, E. R., Stevenson, S., & Tuthill, G. F. 2001, Journal of Computers in Mathematics and Science Teaching, 20(2), 163

Additional Information: Please direct correspondence to the first author. Teachers can enroll in these online courses through the distance learning program at Montana State University which can be found online at: http://btc.montana.edu/nten/.

MarsQuest Online

Chris Randall

TERC, 2067 Massachusetts Ave. Cambridge, MA 02140
chris_randall@terc.edu

James B. Harold

Space Science Institute, 3100 Marine Street, Suite A353 Boulder, CO 80303-1058

Paul Andres

Jet Propulsion Laboratory

Overview of MarsQuest Online

The *MarsQuest Online* Web site extends the scope of MarsQuest, a traveling museum exhibit about Mars. *MarsQuest Online* creates an interactive, inquiry-based web environment that provides visitors opportunities to delve into the science related to the topics of life and water on Mars. The project is funded through a grant from the National Science Foundation.

Key Questions Addressed by MarsQuest Online

- Is there life on Mars?
- Where is there water?
- What information do we need to understand more about life and water on Mars and how do we get it?

MarsQuest Online Concept

- offers a variety of interactive, short duration experiences to engage the web-going public
- enhances users' understandings of Mars and makes Mars more of a real place
- provides users with the tools to obtain and analyze Mars data and images
- gives users practice in defining and carrying out their own investigations
- uses a learning-cycle model to help people learn about a topic

How MarsQuest Online extends the museum exhibit

- Uses the *MarsQuest* exhibit topics as starting point.

- Extends the reach of the exhibit to new audiences, such as those unable to attend the exhibit.

- Extends the scope of the exhibit by providing pathways for users to delve into Mars science.

- Provides an on-going resource for teachers and families that will be available after the exhibit has moved to another city.

The Partners

- TERC, an educational research and development firm based in Cambridge, Massachusetts, brings expertise in designing K-12 math and science curriculum and in developing inquiry-based materials and web sites

- The Space Science Institute of Boulder, Colorado brings an expertise in informal education, interactive web experiences, and museum exhibit design.

- NASA's Jet Propulsion Laboratory brings expertise in developing web-based tools, maintains the Mars data and image catalog, and is home to the Mars Exploration Program

Balancing Relevancy and Accuracy: Presenting Gravity Probe B's Science Meaningfully

Shannon K'doah Range

Gravity Probe B at Stanford University, Hansen Experimental Physics Lab, Stanford University MC4085, Stanford, CA 94305, kdoah@stanford.edu

A critical challenge of any science education outreach is to fulfill two mutually-important missions: 1) present science concepts, theories and information accurately, and 2) connect any "new" science concepts with science schemas that are familiar to students, so that they can understand the relationships between "old" and "new" science ideas.

Accomplishing the first part of the mission is usually straightforward, as scientists and producers of outreach materials possess a strong understanding of the scientific concepts and technology that is the foundation of their mission. Accomplishing the second part is a little trickier. The science educator must sense what science schemas are familiar to students and what science concepts they use in their daily lives. When she has a sense of where students are coming from, she must then build an intellectual bridge from their current scientific understanding to the "new" scientific concepts that she is attempting to communicate.

Building that bridge is the challenge for all science educators. Ideally, our work as science educators should reach a balance between accuracy and meaningfulness – outreach products that communicate central science concepts without presenting the science inaccurately or incompletely.

At Gravity Probe B (GP-B), we are presented with a challenging situation that is a case study in managing this balancing act. GP-B is a sophisticated and complicated mission, attempting to test the theory of curved spacetime using the world's most sophisticated, most spherical gyroscopes. The technology relies on complex science and the mission's goal is to test an abstract theory. For the most part, our chosen audience (high school students) knows little about the theory of general relativity or the technology of nearly-perfect gyroscopes.

The process of developing three particular GP-B educational products has highlighted different aspects of this balancing act. Below, I go into the detail of each educational product and the tension that arose between presenting the science content both accurately and meaningfully.

Teacher's Guide & Classroom Lesson Plan

This lesson plan guides the teacher and students through recognizing and understanding what is meant by curved spacetime through the technical aspects of the GP-B experiment. One of the major challenges is to introduce curved spacetime as something more than a science-fiction-like alternative to the theory of gravity. This is especially difficult when the students have only a simplistic notion of the theory of gravity. We present a model of spacetime that is used frequently in textbooks and classes – the model of a mass deforming a sheet. The reaction of the sheet to deform around the mass represents how spacetime is deformed by the presence of planets, stars, and galaxies. While this demonstration gives students a rudimentary sense of curved spacetime, it runs the risk of misleading students. This is a two-dimensional model of a four-dimensional effect. Students may be left with the impression that spacetime is curved "above" and "below" the planet, but not along the sides.

Instructional Classroom Video

York University and GP-B designed and produced a 15-minute animated video, focused on describing the GP-B experiment. The video relies on images, text, narration, and music to explain the concept behind the experiment and the technology involved. As with all images, this animation was interesting to look at and informative but struggled with a main issue: presenting images of too much complexity with unfamiliar objects and language that over-simplified the science behind the technology.

An example of this is when the video presents an image of a SQUID, a superconducting quantum interference device, that GP-B will use to monitor the alignment of the gyroscope axis. This image is simple enough for one to understand if they are familiar with the quantum-mechanical London moment and monitoring the movement of magnetic fields. However, the central idea of this technology is lost – the idea that when a spinning sphere is coated with a superconducting metal, and is cooled to temperatures near absolute zero, it creates a magnetic field that perfectly aligns with the rotational axis of the spinning sphere.

Classroom Wallsheet (Poster Image & Text)

In producing a standard NASA wallsheet, we were challenged to create a recognizable image that included many aspects of GP-B. The image was difficult to balance because the individual parts of GP-B are unfamiliar without some explanation. In general, describing GP-B to the layperson or the average high school physics student requires careful use of language and images. Otherwise, we will present a technically correct description which reaches no one.

Conclusion

The above critique is meant to make explicit a challenge that all producers of educational products must face – the challenge to balance accurate content with engaging the students on their level. The conclusion of this paper is that we, along with other outreach people, must work together to continue to push the bounds of our products. We must push them to be accurate, and we must push them to be accessible to our audiences.

Space Update: A Fun Way to Teach Space Science

Patricia Reiff

Rice Space Institute, 6100 Main St. MS 108, Houston TX 77005, reiff@rice.edu

Carolyn Sumners

Houston Museum of Natural Science, 1 Hermann Circle Drive, Houston, TX 77030

Abstract. In just a few short years the internet has gone from having too few sites of interest in science to too many – in both cases it can be difficult to find the information or activity that is needed to learn (or teach) a specific concept. Museum exhibits have gone from "mausoleums" of dusty science artifacts to vibrant, interactive, hands-on, real-time safe sources of up-to-date information and images. Space Update, developed with NASA DLT and OSS resources, can provide just the starting point for many a lifetime of exploration, and many activities and games along the way to make the journey fun.

The first online interactive realtime computerized exhibit of earth and space science was developed with resources from the Digital Library Technology (DLT) program: "Space Update," funded by IITA under the "Public Use of the Internet" program. First on display in late 1994 at the Houston Museum of Natural Science, this software was safer than a browser (since it did not allow unlimited access to the web) but yet allowed one-click updating of earth and space science imagery – views of the sun in various wavelengths, weather maps and NOAA satellite images, etc. Our first module "Shoemaker Levy 9" was on display only 6 weeks after the historic comet hit Jupiter; a full space module opened in October 1994. The earth section, "Earth Today," opened in summer of 1995.

The software was the first to attempt to bring the power of the internet to the citizens on the far side of the "digital divide." At that time very few of the public had access to the internet at the office, much less at home, and fear of computers was common. Yet kids were attracted to a display that dynamically changed in response to a touch, and a very popular kiosk was born and has been on display in some version at the museum ever since. This early software (developed prior to Netscape or Internet Explorer) nevertheless was instrumental in bringing earth and space hands-on information to over a million people. (It still doesn't need a keyboard in order to run!)

Although the software was developed for use as a museum exhibit, many teachers whose classes came through the museum or who participated in our workshops asked for their own copy of the software, and requested that we develop activities to go with the software. Responding to that plea, we created our first "Space Update" disk, called "Connected," in early 1996.

In 1999, NASA's first Cooperative Agreement Notice for innovative products in Earth science was released, and we were one of the first awardees for the ESIP (Federation of Earth Science Information Partners). As a part of that project, we developed a separate "Earth Update" software, a full interactive museum exhibit, "Earth Forum," and a series of immersive Earth science shows, including "Powers of Time" (about the cycles of Earth); "Force 5" (about the greatest storms on earth and in space, and "Night of the Titanic," a show that teaches about the conditions on earth and in space that, coupled with human error, caused that tragic sinking. At that time the earth part was removed from Space Update, leaving room to expand Space Update to include a simplified planetarium module plus a "space events" section. Earth Update became its own piece of software, with a consistent look and feel among the Atmosphere, Biosphere, Cryosphere, Geosphere, and Hydrosphere sections.

Space Update is a favorite among museums and schools for its robust design. Its five modules, Astronomy, Solar System, Sky Tonight, Space Weather, and Space Events, can run as a single linked exhibit, or all except Space Events can run as a stand-alone exhibit. Space Weather even expanded to become its own CD-ROM, with extra activities, movies and sounds highlighting the results from Space Weather missions like IMAGE. Over 11,000 copies are in the hands of museum and school educators and the public, with the software on permanent display in many fine museums. The software passed the NASA product review, and was recently reviewed in the December 2003 "Sky and Telescope": ... "I was amazed at Space Update's intuitive interface" ... "a must-have for science teachers" ... "You can bet that Space Update will be on display at our local observatory and my club's future Astronomy Day activities" ...

Space Update to Teach Space Science

- Making the study of space science more "real" for the students by using real images

- Giving many graphic examples for hard to understand concepts to help make information understandable to many different age groups

- The variety of presentation types helps reach all learners

- Real world applications to "Why do we need to know this?"

- Giving students new and interesting career ideas

Activities

- Gives relevant application of information presented in Earth/Space Update

- Shows how information would be used in the "real world"

- Activities may be presented in a variety of ways, not just lower level *Bloom's Taxonomy*

In Ms. Furitisch's classroom, the most exciting and most rewarding result of using new technology is to turn on the students who are just "marking time" through science. She has several examples of students who routinely failed science every six weeks, who are now actively engaged and considering a science career, coming back to her with their new knowledge of science even after they have gone on to high school. It is examples like those that keep us making interactive exhibits that are educational and fun.

For more info: http://earth.rice.edu

Acknowledgments

This project is supported by NASA's Office of Earth Science under the ESIP Federation NCC5-311 and by NASA's Office of Space Science under the IMAGE mission. Many federal agencies and individuals have participated in providing images and movies for our CD-ROMs, and many others in the review and improvement process. As a result of federal funding of the development, we are able to provide this software at a price that just allows us to continue the program. We also thank the many teachers who have field tested the software and the activities with many levels of students, and particularly thank the teachers in our experimental "Masters of Science Teaching" degree, who have given many hours late at night and on weekends to improve their skills and this software. Kudos to Katty Furitsch, Lollie Garay, Kevin Robedee, Katrina Miguez and Amy Jackson! ... PR

Issues in Formal Education: Building and Delivering Standards-Based E/PO Products and Activities

John Ristvey and Jacinta Behne

McREL, 2550 S. Parker Rd., Suite 500, Aurora, CO 80014, jristvey@mcrel.org

When Mid-continent Research for Education and Learning (McREL) was asked to serve as the Education and Public Outreach (E/PO) partner in NASA's Discovery mission, Genesis, an unprecedented partnership evolved between a U. S. Department of Education standards-designated research laboratory and NASA. The immediate charge was to take the content of a space science mission and turn it into standards-aligned classroom learning opportunities. Building E/PO standards-based products and activities that are instructionally viable and engaging ensures that the best of mission science and technology is delivered to classrooms in a meaningful way. The Genesis mission is one example of McREL E/PO work, with additional contract work for the Deep Impact mission, the Hubble Space Telescope office, and NASA's upcoming Discovery mission, Dawn. McREL is committed to building and delivering standards-based E/PO products and activities that include formal education materials and public learning materials and activities.

Module Design

Formal education materials centered on the science of a mission generally are presented as curriculum modules. An education module contains a module planning guide, teacher guides, student texts, student activities, formative and summative assessments, and teacher resources. An education module is science-specific, and often includes interdisciplinary applications. Depending on the discipline/s upon which it is based, an education module is aligned to education standards, including: *Content Knowledge: A Compendium of Standards and Benchmarks for K-12 Education*, (the McREL Compendium), the *National Science Education Standards* (NSES), the *International Society for Technology in Education* (ISTE) standards, and the *National Council of Teachers of Mathematics* (NCTM) standards.

Building a standards-aligned education module is a five-step process. It begins with a scientist-informed topic selection. Criteria include examining mission science and identifying a best-fit with a formal education audience. After topic selection, researchers investigate the standards addressed by the identified topic, from which the relevant grade level range is determined. Next, education researchers construct and apply the appropriate learning cycle, after which the module layout is created.

A McREL E/PO education module may also feature engaging technology applications and interactive learning exercises that enhance student learning. Because it serves as a curriculum resource, an education module supplements

rather than supplants existing curriculum. All education modules are print-optimized and are available to educators online on the respective mission Web site. Examples of education modules may be found at:
http://www.genesismission.org/educate/scimodule/moduleoverview.html and at: http://deepimpact.jpl.nasa.gov/collaborative_ed_module/index.html.

Delivery and Evaluation

Module evaluation is an important step in product development and is critical as the education module is implemented in the classroom. McREL E/PO facilitates a network of classroom teachers nationwide that is committed to using education modules and activities in the classrooms. These teachers may serve as individual field testers, or as part of a district effort. All teachers who field test McREL E/PO education modules in their classrooms agree to provide feedback, informing module developers of successes, and identifying where improvements can be made. Among the issues that field testers are asked to consider are

- How well do the materials work for students and teachers in a classroom setting?
- How adaptable are the materials in meeting the needs of local and state curricula?

All participants can network with field test colleagues nationwide by joining in an online education module discussion forum. In this venue they share activity adaptations that contributed to particularly successful classroom results. Development networks function as an integral part of education product development, in that they result in impact on classroom teaching and learning through classroom teachers via teacher feedback and input.

Learning Products and Activities

Opportunities for learning don't stop with curriculum modules. McREL E/PO has sponsored several contests and distributed a number of learning products to raise awareness and promote learning around mission information. In addition to the standard mission fact sheets, posters, and bookmarks, there are a number of multi-media items featuring the science of the Genesis mission. Many of these items feature accompanying standards-aligned classroom activities. Mission learning products and activities are a result of collaboration with scientists at partner institutions. Complementing the Genesis cleanroom education module, there is a learning video and an interactive electronic field trip that takes the participant into the cleanroom at NASA Johnson Space Center. A newly released video dedicated to testing of mission instrumentation work done at Los Alamos National Laboratory, and a Genesis mission screensaver, computer desktop wallpaper, electronic newsletter all result from scientist collaboration. Additionally, an online childrens' stickerbook, CD-ROMs, public modules, and multiple interviews with Genesis mission project members all serve to raise mission awareness and heighten career interest in space science.

All multi-media items and activities are featured on the Genesis mission Web site at: http://genesismission.jpl.nasa.gov

Conclusion

In the 21st century, NASA mission education and public outreach activities transcend the traditional print materials to encompass new learning opportunities for the formal classroom as well as the interested public audience. As the face of space science is ever-changing, so too are the formats and venues that can be used to reach the space science enthusiasts, from pre-school age to the adult learner. Creating products and activities that are aligned with today's education standards ensure a quality content-into-learning environment.

NASA Space Science Diversity Initiatives

Philip J. Sakimoto

Space Science Education and Public Outreach Program, NASA Headquarters, 300 E Street, SW Washington, DC 20546, phil.sakimoto@hq.nasa.gov

The NASA space science missions currently envisioned span the next several decades, during which time substantial turnover in the scientific workforce and changes in the Nation's demographics will occur. To prepare for this future, the NASA Office of Space Science (OSS) has been conducting targeted efforts to engage underrepresented minorities in space science activities.

Minority Universities

The first thrust in these efforts is the NASA Minority University and College Education and Research Partnership Initiative (MUCERPI) in Space Science. This grants program, carried out in collaboration with the NASA Office of Education, offers minority universities opportunities to develop academic and/or research capabilities in space science. The hallmark of this program-and perhaps the most important key to its success-is that OSS plays an active role in providing guidance and in engaging the community of OSS-sponsored researchers to serve as active partners in collaborations with the minority institutions involved in the program.

The first set of MUCERPI grants were awarded to fifteen minority universities in the year 2000. These grantees have already reported a remarkable set of success stories. They are engaged in research collaborations with 10 NASA space science missions or suborbital projects and in nearly 50 working partnerships with major space science research groups. In academic programs, they have established on their campuses 25 new or redirected space science faculty positions, 12 new or revised space science degree programs for which nearly 100 students have signed up, and 68 new or revised space science courses with a total enrollment to date of nearly 1,800 students. They are also engaged in a wide variety of teacher training, precollege outreach, and public outreach programs.

These successes clearly demonstrate that vibrant academic and research programs in astronomy and space science can be built at minority institutions provided that sponsoring Agencies offer serious opportunities to do so. Further information on these projects may be found in the 2002 OSS Education and Public Outreach Annual Report at http://ossim.hq.nasa.gov/ossepo/.

After a competitive solicitation conducted during 2003, a second round of MUCERPI awards was announced at the Congressional Hispanic Caucus and Congressional Black Caucus meetings in September 2003. These 16 awards were made to eight Historically Black Colleges and Universities (HBCU), five Hispanic-Serving Institutions (HSI), two Tribal Colleges and Universities (TCU), and one Minority-Predominant Institution (MPI), as listed below:

- Alabama A&M University [HBCU], Dr. Arjun Tan

- California State University at Los Angeles [HSI], Dr. Charles W. Liu

- California State University at San Bernadino [HSI], Dr. Susan Lederer

- Fisk University [HBCU], Dr. Arnold Burger

- Hampton University [HBCU], Dr. Patrick McCormick

- Medgar Evers College [MPI], Dr. Leon P. Johnson (see page 312)

- Norfolk State University [HBCU], Dr. Carlos W. Salgado

- North Carolina A&T St. Univ. [HBCU], Dr. Abebe Kebede

- Salish Kootenai College [TCU], Dr. Timothy S. Olson

- South Carolina State University [HBCU], Dr. Donald K. Walter

- Southern University, Baton Rouge [HBCU], Dr. J. Gregory Stacy

- Southwestern Indian Polytechnic Institute [TCU], Mr. Kirby Gchachu

- Univ. of the District of Columbia [HBCU], Dr. Abiose O. Adebayo

- University of Houston-Downtown [HSI], Dr. Penny Morris-Smith (see page 175)

- University of Puerto Rico at Mayagüez [HSI], Mr. Rafael Fernandez

- University of Texas at El Paso [HSI], Dr. Ramon E. Lopez (see page 207)

Eight of these institutions will be developing research capabilities in various areas of space science through partnerships with major space science research institutions. In addition, all 16 of them will be improving their academic capabilities in space science at various levels in the educational system, including 14 who are developing undergraduate courses or degree programs, three who are developing graduate courses or degree programs, 13 who are developing precollege outreach or teacher training programs, and six who are engaging in Public Outreach activities.

Each of the MUCERPI-2003 awards is a three-year grant offering up to $275,000 per year over the period from January 1, 2004, to December 31, 2006. More than 50 major OSS-sponsored research or educational institutions will be active partners in these projects. Synopsis of each project and a full list of the partners may be found at http://spacescience.nasa.gov/education/news/index.htm in the January 2004 issue of the OSS *Voyages* newsletter.

Professional Societies of Minority Scientists

The second thrust of OSS's diversity efforts is a response to discussions held over the past several years with leaders of various professional societies of minority scientists and also to a strong recommendation of the NASA Space Science Advisory Committee's Task Force on Education and Public Outreach (E/PO) that OSS "expand and intensify" it's "pioneering efforts to attract and better integrate minorities into E/PO projects and into the mainstream of OSS science programs."

An early meeting in May, 2001, at Western Kentucky University brought the officers of 11 professional societies of minority scientists together with members of the OSS E/PO Support network to explore potential collaborative projects. The societies involved in this effort, including one that joined during a follow-up teleconference, are the:

- American Indian Science and Engineering Society;
- Coalition to Diversify Computing;
- Council for African-American Researchers in the Mathematical Sciences;
- Institute for African-American e-Culture;
- National Association for Black Geologists and Geophysicists;
- National Association of Mathematicians;
- National Institutes of Health Black Scientists Association;
- National Organization for the Professional Advancement of Black Chemists and Chemical Engineers (see page 270);
- National Society of Black Physicists;
- National Society of Hispanic Physicists; and
- Society for the Advancement of Chicanos & Native Americans in Science (see page 307).

One example of a successful collaboration was the partnering of the National Society of Black Physicists (NSBP) with the NASA Sun-Earth Connection Education Forum (SECEF) in planning and implementing a national effort to establish local sites in African American communities for viewing a live broadcast/Webcast of the June 21, 2001, total solar eclipse from Africa. More that 20 NSBP members served as hosts and science experts at the sites.

OSS support for activities of the professional societies is also being enhanced through new relationships emerging from this initiative. As an example, OSS representatives have led workshops for teachers and/or provided space science exhibits at the last three annual meetings of the National Organization for the Professional Advancement of Black Chemists and Chemical Engineers (NOBC-ChE).

As a future initiative, OSS is sponsoring Chicago 2004: A Workshop to Foster Broader Participation in NASA Space Science Missions and Research Programs. This workshop is aimed at bringing together NASA personnel, current OSS-funded scientists and educators, and a diverse array of scientists and educators who are interested in participating in future OSS missions and research programs. A specific goal of the workshop is to seed personal contacts among a much more diverse community of investigators than has traditionally been active in NASA space science missions. In addition, all participants are expected to gain insights and contacts leading to a better understanding of how the NASA space science program is organized, planned, and conducted; how missions and research programs are conceived; how mission and research teams are formed; and how successful proposals are constructed.

Additional information on these initiatives may be obtained from the OSS Diversity Coordinating Committee by contacting: Dr. Philip Sakimoto, Space Science Education and Public Outreach, NASA Headquarters
(E-mail: Philip.J.Sakimoto@nasa.gov), Dr. Charles McGruder, Western Kentucky University and Past President, National Society of Black Physicists (E-mail: Charles.McGruder@wku.edu), or Dr. Carolyn Narasimhan, NASA/OSS Broker/Facilitator, DePaul University (E-mail: cnarasim@depaul.edu).

NASA Office of Space Science Education and Public Outreach Conference 2002
ASP Conference Series, Vol. 319, 2004
Narasimhan, Beck-Winchatz, Hawkins & Runyon

Why Things Fall: What College Students Don't Understand About Gravity and How To Teach It Better

H. L. Shipman, N. W. Brickhouse, Z. R. Dagher, and W. J. Letts

University of Delaware, Sharp Laboratory, Newark DE 19716, harrys@udel.edu

Abstract. An in-depth interview study of students in a college level astronomy course revealed that 19 of the 20 students only understood gravity as a phenomenon, not as a universal force or as an explanation for a phenomenon. We report the results of our in-depth study. As a result of this study, one of us (HS) developed a teaching sequence which, as our preliminary data shows, leads to an improved understanding of gravity.

In our astronomy courses, we not only teach facts, but we also teach explanations. Many people come into an astronomy course wanting to know how the Universe began. But an important part of understanding our origins is understanding why the scientific community accepts the Big Bang theory as a good explanation for cosmic evolution. In other astronomical areas we want to address "why" questions. Why do we believe that there are planets around other stars? And, at the most basic level, why do apples fall to the earth, and why does the moon stay in its orbit?

Twenty students in a college astronomy course were interviewed three times each by a research team member other than the instructor using a semi-structured protocol. In addition to interview transcripts, we examined student responses to assignments and exams. The interview data demonstrate that few students understand gravity as anything more than a phenomenon. Dionne's (students are given pseudonyms) understanding of gravity is typical:

> Dionne: Gravity I don't think is a theory.
> Interviewer: Why don't you think it's a theory?
> Dionne: Because how else would we be planted here? You know? Otherwise we'd be floating around and stuff. I think it's been proven. Why else would stuff fall towards the earth and like with the planets?

As a result of these findings, one of us (HS) has revised the course, introducing an activity where students compare alternative explanations of gravity, suggesting experiments for the instructor to do (in front of a class of nearly 300 students). Exam data, supplemented by a few interviews, show that students' understandings of gravity has considerably improved. (Only the first author of this paper takes credit or blame for the results described in this paragraph.)

Interdisciplinary Approaches to Teaching Astronomy: Special Topics and Team-Taught Courses

Harry L. Shipman

Physics and Astronomy Department, University of Delaware, Newark, DE 19716-2570, harrys@udel.edu

I teach a special topics course which has worked very well for 20 years: "Black Holes and Cosmic Evolution" (http://www.udel.edu/physics/phys145.) I have taught astronomically-related interdisciplinary courses on "Science and Religion" (taught with a philosopher). "Extraterrestrial Life" (taught with a biologist), and "Popular Writing About Science" (taught with an English faculty member). I have also taught "Ethical Issues in Scientific Research" with a biologist and a philosopher, but this course had no astronomical content. I know this area and could serve on a panel, but a limitation is that many of the interdisciplinary courses I teach go beyond astronomy. I could also do a workshop in this area; while I would focus my attention in this workshop on astronomy, there may be others at the conference who have not wandered off the reservation as far as I have.

A Systemic Approach to Improving K-12 Astronomy Education Using the Internet

Tim F. Slater and Edward E. Prather

University of Arizona Steward Observatory, Tucson, AZ 85721, tslater@as.arizona.edu

G. T. Tuthil

Montana State University Department of Physics, Bozeman, MT 58717-3840

Introduction

The scientific community's understanding of astronomy and space science is rapidly growing and public outreach efforts via the Internet provide up-to-date data and images. Nevertheless, text-based classroom materials largely remain behind this rapidly advancing knowledge front. Moreover, many pre-college teachers report that their lack of training and understanding of contemporary content makes it difficult for them to provide inquiry-driven instruction in astronomy and space science.

This project was initiated in 1997 to systematically implement the astronomy concepts in the National Science Education Standards (NRC 1996) by using NASA resources – both primary data and educational activities – on the Internet. The Center for Educational Resources (CERES) has had dual missions. The first was to create a series of exemplary WWW-based classroom lessons that would integrate the student-centered inquiry emphases of the NSES and on-line data resources from NASA. This was accomplished by bringing together classroom teachers, university professors, and research scientists to develop, field-test, and distribute classroom-ready lessons. In short, this part of the project was intended to provide a structure for students to do science, and help teachers to implement the NSES. To date, more than 30 innovative, classroomc-field tested lessons exist. The unique aspects of these lessons are that they were created specifically to meet the NSES and use the most recent research results from astronomy education research.

CERES's second task was to develop three graduate-level courses in space science and astronomy for K-12 teachers, to be delivered over the Internet. The first course, *"Comparative Planetology: Establishing a Virtual Presence in the Solar System,"* focused on the NSES content standards in astronomy for grades K-8, while the second, *"Studying the Universe with Space Observatories,"* addressed the standards for students in grades 9-12. The third course is an interdisciplinary approach to *"Astrobiology for Teachers"* and emphasizes exobiology. These asynchronous computer-mediated courses use a robust combination of WWW resources and conferencing software for extended participant interactions over a 16 week course. In total, now 18 credit-hours are available in a series of astronomy for teacher's courses which, when combined with 14 graduate credit-hours in curriculum and instruction, can constitute a foundation for a new astronomy education option in the Montana State University *Master of Science – Science Education* (MSSE) online degree program.

References

National Research Council 1996, National Science Education Standards (Washington: National Academies Press)

Additional Information: The CERES WWW site is freely available at URL: http://btc.montana.edu/ceres

Partnering Scientists and Educators: A Model for Professional and Product Development

Denise A. Smith, Bonnie Eisenhamer, Terry J. Teays, and Dan McCallister

Space Telescope Science Institute/Computer Sciences Corporation, 3700 San Martin Drive, Baltimore, MD 21218, dsmith@stsci.edu

Abstract. The Hubble Space Telescope (HST) Formal Education Group and the Origins Education Forum have been actively researching effective methods to communicate with and develop products for the K-12 audience. Through evaluation at the national and local level, we have found that the most effective efforts involve a strong partnership between scientists and educators.

Bridging the Education and Science Communities

To effectively work together, scientists and educators must understand each other's needs and backgrounds. Educators generally face a tremendous burden in meeting an increasing number of requirements with little time for developing new skills and incorporating new classroom materials. Educators wish to build upon their science and technology skills, and identify materials and teaching methods that can be integrated into the classroom. These goals must be accomplished in the face of competing professional development demands from state and local districts, however. Scientists wish to communicate the excitement and importance of their work, and to increase the scientific literacy and support of the public that ultimately funds their missions and research.

How can educators and scientists help each other? Educators can help scientists by providing experience with education standards, an understanding of student capabilities and learning styles, knowledge of the classroom environment and available resources, and insight into the structure of the education system. Scientists provide access to scientific data, a deep understanding of current research and its significance, scientific expertise and knowledge, and insight into career paths, but often do not fully understand the needs and capabilities of the K-12 audience. The HST Formal Education team has found that a facilitator can help bridge the culture gap between scientists and educators, helping them to assess the needs of the audience, to share information, to learn to work together and communicate effectively, and to establish a dialog between the two communities (Christian et al. 2001).

Professional and Product Development

Our most successful experiences occur when a scientist and educator actively work together, using information from a needs assessment of the target audience. These strong partnerships enable the team to bring science into the classroom in a meaningful way, by providing products and professional development that meet the needs of the audience and that can be easily integrated into the educational system. The partnerships also provide support to the scientific and educational communities, fostering an understanding between the two cultures that carries with it the potential for long-term partnerships.

HST Formal Education/Origins professional development workshops provide one application of this model. HST/Origins surveys of educators indicate that they are very interested in using NASA images in the classroom, but often do not know how to use them effectively. To address this issue, we draw on the skills of both scientists and educators. In tag-team format presentations, scientists provide recent HST images and outline the associated science and its importance. Educators show how this material can be used as an engagement tool to spark interest in science and our current understanding of the universe. Most recently, our scientists and educators have worked together to develop a Venn diagram to illustrate how the science in HST images of the Eagle and Eskimo nebulae can be used in a compare and contrast activity. These materials and presentations have been well received by educators, who indicate they appreciate hearing accurate science information from an authoritative source, and having an experienced educator place this information into a relevant educational context.

The partnership model is also central to HST's *Amazing Space* on-line curriculum support activities. HST surveys of educators indicated the need for activities discussing gravity, opportunities to develop inquiry process skills, and science background in a question and answer format. In response, scientists and educators worked together during a summer workshop to design *Planet Impact* (http://amazing-space.stsci.edu), an interactive activity based on HST data that gives students an opportunity to use inquiry process skills and explore gravitational effects. To assist educators, the development process is based upon familiar associative learning principles. For example, we use a production pyramid (Christian et al. 2001) reminiscent of Maslow's Hierarchy of Human Needs (Maslow 1998). This technique places unfamiliar content in a familiar context, facilitating the development process.

References

Christian, C. A., Eisenhamer, B., Eisenhamer, J., & Teays, T. 2001, Journal of Science Education and Technology, 10, 31

Maslow, A. H. 1998, Toward a Psychology of Being, 3rd ed. (New York: John Wiley & Sons)

The Art and Science of Storytelling in Presenting Complex Information to the Public, or, Give 'em More than Just the Facts

Anita M. Sohus and Alice S. Wessen

Jet Propulsion Laboratory/California Institute of Technology, 4800 Oak Grove Drive, Pasadena, CA 91109. Anita.M.Sohus@jpl.nasa.gov

Introduction

"All we want are the facts, ma'am," Sergeant Joe Friday appealed weekly in the popular 1950's cop show *Dragnet*. Unfortunately, in communicating science to the public, just the facts can leave the public baffled, bewildered, and bored. In communicating science to the public, we need to learn to tell the story, not just the facts.

Science and engineering is serious business, requiring precise language and rigorous reporting of "just the facts." Yet, we believe this very code of integrity has contributed to a public image, at best, of scientists as eccentrics and engineers as geeks, and at worst, as elitist snobs who speak in secret codes. The very heart of the science process – open discussion and disagreement – often leaves the public with the impression that scientists don't know which way is up.

Unfortunately, the competitive nature of modern science contributes to the problem. One of the best known science communicators, Dr. Carl Sagan, suffered the slings and arrows of his scientific peers who disdained his "popularization" of science. Yet a whole generation of new scientists and engineers cite Dr. Sagan as their inspiration for pursuing these careers. They are asking where is today's Carl Sagan?

And, Russ Rymer, a former editor for the American Association for the Advancement of Science and the New York Academy of Sciences, acknowledges, "science is easy to get wrong, even by the best-intentioned layman" (Rymer 2001). Or, we might add, even by a well-educated professional in another field. Each branch and speciality in science and engineering has evolved (and is evolving) its own language, which only the initiated understand. Translation is needed for outsiders – anybody outside the field.

Beginning in 2001, all Caltech juniors must prepare a feature length article on a scientific topic, suitable for publication in a lay magazine, as a requirement for graduation. The students are mentored by faculty, and the results are published in the online Caltech Undergraduate Research Journal. The official reason for the requirement is to improve their writing skills for the competitive world of grant writing. Editor Rymer believes, however, that since these high-achievers are in a better position than most of us to change our world, they had better be able to explain their work clearly (Rymer 2001). Says Rymer, "As it brings new knowledge to bear about our universe, science does more than provide new product; it establishes and reestablishes the metaphor of ourselves and society, the prevailing paradigms by which we live. It tells us, in ways other disciplines can't, who we are and what we mean ... this calls for intelligent interpretation" (Rymer 2001).

Storytelling

It is story that captivates people's attention and imagination: a Gallup poll in 1998 showed that 67% of Americans prefer to spend their evenings reading or watching TV, movies, or theater performances – in other words, being immersed in a story (Gallup Organization 1998).

Good stories have drama, suspense, and engaging characters – and so do science and engineering. Storyteller Susan Strauss asserts in her book *The Passionate Fact: Storytelling in Natural History and Cultural Interpretation*, "In literary terms, the scientific method is solving a mystery." She goes on to point out that scientists and historians are observers and recorders of phenomena, from which they extract data, which in turn they synthesize into amazing and often very beautiful stories. Yet, she believes, scientists and historians are often in great denial of their work as storytellers.

Strauss observes that content given as information does not invite us to question or wonder, while content expressed in a "story way" creates relationship, translates information into imagination and excites our imagining – our sense of wonder (Strauss 1996). This is especially important in education: Instilling a sense of wonder during the first eight years of schooling is more important than any content (R. Steiner).

In Jonesboro, Tennessee, the Storytelling International Foundation has been nourishing storytelling and storytellers for over 25 years. Besides sponsoring a legendary annual storytellers festival, the Foundation has established an International Storytelling Center, which conducts professional development workshops in storytelling for educators, corporations, and institutions. Their primary focus is on health, conflict resolution, leadership and management, and children and youth. "Stories constitute the single most powerful weapon in a leader's arsenal," they quote Dr. Howard Gardner, Professor, Harvard University, and author of *Leading Minds* on their website.

We believe that as science communicators and educators, we have much to learn from the world's great storytellers. The education/public outreach staff of several space science missions at JPL are forming a partnership with the International Storytelling Center to develop storytelling mentors and training.

Returning to Dragnet, a good deal of its success came from its basis in real-life, its attention to the details of the investigations, and its human portrayal of the cops – their frustrations, the interruptions of their private lives, and the outcome of their investigations (The Network & Cable TV Guide 2003). We can't help but see some parallels with the stories of many scientists and engineers!

Storymining

So where does someone who is unpracticed in storytelling start? One method is storymining, a way of looking at the facts to develop a story. Storymining enables storytelling. The concept of storymining is well-developed in NASA's Earth Science Enterprise, especially by such award-winning websites as Earth Observatory. Questions to ask when storymining include: Who are the protagonists? What are the objectives? What is compelling? What are the obstacles? What is the path to overcome the obstacles? Is there a crisis conflict? What is the path to understanding? What is the story of the solution? What are the connections? What will be the legacy? The next step? Then, tell the story.

Interpretation

Naturalist Freeman Tilden is credited with starting the interpretive movement in the 1950s. Interpreters try to connect you – your heart and mind – with time and place through storytelling. As defined by the Board of Directors of the National Association for Interpretation (NAI), "Interpretation is a communication process that forges emotional and intellectual connections between the interests of the audience and the inherent meanings in the resource." Although the interpretive profession focuses on nature and culture, there is much to learn from their methods, techniques, practices, perspectives, and skill sets for oral, written, and visual interpretation and storytelling. NAI sponsors a National Interpreters Workshop (this year in Virginia Beach, VA from Nov. 12 to 16, 2002), professional certification, interpretive research, regional chapters and regional workshops, and special interest sections ranging from living history to visual communications.

If you have ever visited a national park or a historical site, you have probably listened to an interpreter. The National Park Service's definition of informal education is interpretation. NPS has developed an Interpretive Development Program to develop the interpretive competencies of its staff, including presenting effective interpretive talks, conducting effective activities, presenting effective demonstrations, writing effective interpretive material, developing curriculum and interpretive media, coaching, and research. NASA's Earth Science Education Implementation Office is investigating the possibility of tailoring some of these courses for professional development of NASA Earth Science staff.

Summary

Storytelling is as old as humankind. Storytelling applied to communicating science and engineering content, however, may be a new concept to a generation of scientists and engineers trained to present "just the facts." While we have barely scratched the surface here, a world-wide renaissance of storytelling has many resources, including the fields of journalism and drama.

Acknowledgements

The authors acknowledge stimulating conversations with Maura Rountree-Brown, Art Hammon, Richard Alvidrez, Michelle Viotti, Anita Davis, and Tom Nolan.

References

Gallup Organization 1999, American's Favorite Evening Recreational Activity is Still Television
http://www.gallup.com/content/login.aspx?ci=4048
(Subscribers only)

The Network & Cable TV Guide 2003
http://www.geocities.com/TelevisionCity/9348/dragnet.htm

Rymer, R. 2001, Science and the Art of Storytelling, Los Angeles Times, June 17

Strauss, S. 1996, The Passionate Fact: Storytelling in Natural History and Cultural Interpretation (Golden, CO: Fulcrom Publishing)

Further Reading:

Beck, L. & Cable, T. 1998, Fifteen Guiding Principles for Interpreting Nature and Culture (Champaign, IL: Sagamore Publishing)

Hartz, J. & Chappell, R. 1997, Worlds Apart: How the Distance Between Science and Journalism Threatens America's Future (Nashville, TN: First Amendment Center)
http://www.freedomforum.org/publications/first/worldsapart/worldsapart.pdf

Tilden, F. 1977, Interpreting our Heritage (Chapel Hill, NC: University of North Carolina Press)

Websites

Caltech Undergraduate Research Journal
http://www.curj.caltech.edu

International Storytelling Foundation
http://www.storytellingfoundation.net

NASA's Earth Observatory
http://earthobservatory.nasa.gov

National Association for Interpretation
http://www.interpnet.com

National Parks Service Interpretive Development Program
http://www.nps.gov/idp/interp/

Using Sloan Digital Sky Survey Data in the Classroom

Robert Sparks and Chris Stoughton

Fermi National Accelerator Laboratory, Wilson and Kirk Road, MS #127, Batavia, IL 60510, rspark@prairieschool.com

M. Jordan Raddick

Department of Physics and Astronomy, The Johns Hopkins University, 10513 Demilo Pl. #301, Orlando, FL 32836

Abstract. The Sloan Digital Sky Survey (SDSS) will map 25% of the night sky down to 23rd magnitude, cataloging more than 100 million objects and taking spectra of over 1 million objects. All data from the SDSS will be publicly available on the Internet. These data include exact positions of stars, galaxies and quasars in the sky; magnitudes in five wavelengths; and spectra. From these data, astronomers will create a detailed map of the universe. With this map, they will better understand the universe's large-scale structure, yielding information on its evolution and ultimate fate.

SDSS data will give students a unique opportunity to conduct astronomical research using the same data that professional astronomers use. Students will create Hubble Diagrams to illustrate the expansion of the universe, and will create Hertzsprung-Russell diagrams, which will allow them to find the ages of and distances to stars and star clusters. Because these activities use real data, students will be led through some the difficulties that professional astronomers experience in analyzing data.

All SDSS data will be available on the *SkyServer* web site, http://skyserver.sdss.org. The vast amount of data – 13 million objects so far – can be browsed and searched using a variety of tools. The data range from tri-color images and processed spectra to raw image files to magnitude data in 5 different wavelengths. Student lessons available on the *SkyServer* web site range from the elementary to the introductory college level. Each lesson is designed to meet national standards for science education. Each lesson also has an extensive section of teacher notes available, providing appropriate background information and ideas for student evaluation.

Introduction

SkyServer is the education and outreach web site of the Sloan Digital Sky Survey (SDSS). The SDSS's early data release contains images of approximately 13 million objects and spectra of over 50,000 objects. All data in the early data release are available online to astronomers. *SkyServer* makes exactly the same data available to students and to the general public.

Student Projects Using SDSS Data

SkyServer features a variety of projects suitable for students from elementary school through the college introductory level. All projects use SDSS data extensively, and all come with teacher's notes that contain sample solutions and correlations to national education standards.

Old Time Astronomy

Old Time Astronomy (http://skyserver.sdss.org/en/proj/kids/oldtime/) is designed for younger students. The project teaches them about the history of astronomy, and about the importance of careful scientific observation.

Until photographic plates were developed, the only way for astronomers to record what they saw through their telescopes was to make a sketch. Instead of looking through telescopes, students will practice sketching using images from *SkyServer*. After getting some advice from an amateur astronomer on how to sketch astronomical objects, the students make their own sketches and compare them to sketches made by other students. Students then try to identify which objects their partners sketched.

From these exercises, students will gain an appreciation of how difficult astronomical research was before modern photography. They also will learn the importance of making good observations and recording them carefully.

The Hubble Diagram

In 1929, Edwin Hubble measured the distances to numerous galaxies, then used redshift measurements made by Vesto Slipher to create his famous "Hubble Diagram," which suggested that the universe was expanding. In *SkyServer*'s Hubble Diagram project (http://skyserver.pha.jhu.edu/en/proj/advanced/hubble/), students use SDSS data to create their own version of this famous diagram, and to see the expansion of the universe.

The SDSS will measure the spectra of hundreds of thousands of galaxies. All spectra in the *SkyServer* database have their important emission and absorption lines clearly marked. Students use these spectral lines to find redshifts of the galaxies.

Finding accurate distances to galaxies has always been a challenge to astronomers. In the Hubble Diagram project, students estimate the distances to galaxies by analyzing galaxy clusters. Once they have estimated relative distances to several galaxies, they plot galaxies' distances against their redshifts. Students will see an overall linear relationship, but they may also see some noise in the plot due to difficulties in estimating the distances to galaxies.

Hertzsprung-Russell Diagram

The Hertzsprung-Russell (or H-R) diagram is one of the fundamental tools for teaching stellar evolution. Making a plot of luminosity vs. temperature divides stars into several distinct groups, including main sequence stars, red giants, white dwarfs, and several types of variable stars.

In *SkyServer*'s H-R diagram project (http://skyserver.pha.jhu.edu/en/proj/advanced/hr/), students make H-R diagrams using data from a variety of sources, including parallax data from the Hipparcos satellite.

Hipparcos data only is available for relatively nearby stars. Students can make H-R diagrams for distant globular clusters using SDSS data. Using a simple search tool, students collect data for hundreds of stars in the globular cluster Palomar 5 and use this data to create an H-R diagram.

Other Projects Using SDSS Data

These three projects are just a sample of how SDSS data can be used in the classroom. Other projects include *Spectral Types of Stars*, *Colors of Stars*, *Asteroid Searches*, *Quasars*, *Image Processing*, and the *Constellation Game*. New projects are being created and will be added later this year.

Integrating Radio Physics Projects into Education: INSPIRE and Radio JOVE

William W. L. Taylor

NASA/Goddard and Raytheon, ITSS, Code 630, GSFC, Greenbelt, MD 20771

James R. Thieman

NASA/GSFC, Code 630, Greenbelt, MD 20771

Abstract. INSPIRE and Radio JOVE are scientific and educational projects whose objectives are to bring the excitement of building kits and observing radio waves to students. INSPIRE radios observe natural and manmade radio waves in the Very Low Frequency (VLF, audio) region and Radio JOVE radios observe 20 MHz radio waves from Jupiter and the Sun. Underlying these objectives is the conviction that science, mathematics and technology are the underpinnings of our modern society and that only with an understanding of science, mathematics and technology can people make correct decisions in their lives, public, professional, and private. Stimulating students to learn and understand science, mathematics and technology is key to them fulfilling their potential in the best interests of our society.

INSPIRE

INSPIRE began with a test bed project in 1989, ACTIVE/HSGS (High School Ground Station), which involved 100 high schools, making observations of transmissions from the Soviet ACTIVE satellite. The second major project was support to a NASA Shuttle/Spacelab mission, ATLAS 1 investigation, SEPAC, in which 1,200 schools participated. The third project was focused around the annular solar eclipse on May 10, 1994. Participants (students, teachers, etc.) observed radio waves before, during, and after the eclipse to study the effects of reduced solar UV on the ionosphere and its ability to propagate audio frequency radio waves. The most recent manned space flight project was in cooperation with space station MIR, which had a set of accelerators on board, similar to SEPAC's. Through an INSPIRE agreement with IKI (Space Research Institute, Moscow), the electron accelerator and plasma generator on MIR were fired over the US and Europe during weekends in November and April each year, starting in 1995 and ending with MIR's deorbiting in 2001. INSPIRE observers made observations to see if the radio waves generated by the accelerators propagate to the surface of the earth. Starting in 1998, INSPIRE observers have made observations during the Leonids meteor showers. In 1999, the first of annual Leonid's Balloons were flown by NASA/MSFC carrying INSPIRE receivers, real time telemetry and live streaming of the data over the internet. In 2000, INSPIRE data began to be streamed from MSFC, 24x7, and in 2002, INSPIRE data began streaming from the University of Florida Radio Observatory, 24x7.

Radio JOVE

Radio JOVE began offering kits to observe Jupiter and the sun at 20.1 MHz in the fall of 1998. A high sensitivity, low noise level receiver had been designed, which when used with a dual dipole antenna, is able to observe both sources. The radio telescope kits can be assembled by science classes and used to collect planetary or solar radio astronomy data. Schools may opt to use other equipment to collect this data, but use of the Radio JOVE kit is highly recommended and provides additional educational value to the students. The radio telescope kit is intended for high school level classes, but may be appropriate for introductory college courses or advanced middle school students. The students build the receiver kit using basic electronic tools under the supervision of the teacher. They also construct the special antenna needed to receive the planetary or solar emissions. The antenna requires construction of a basic structure using wood or pipe, ropes, stakes, etc. Once the kit is completely assembled and tested the students determine a good time to observe Jupiter based on predictions supplied on the Radio JOVE website. Note that Jupiter radio signals can only be received at night and the conditions are often best in the hours just before dawn. Also, the antenna needs to be set up in a location that is as free from electrical interference as possible. This may be possible near some schools, but it is recommended that observing be done in nighttime field trips to locations away from power lines and other sources of interference. If nighttime viewing or field trips are a problem, daytime viewing of the Sun at an outdoor location near the school may provide the equivalent observing experience

How Students and Teachers use INSPIRE and Radio JOVE

Participating students get hands-on experience in gathering and working with space science data. They obtain the data by either building a radio receiver and antenna and making observations with their equipment or by remotely using live data streamed on the web. They can then compare their results with other schools who had also observed and come to conclusions concerning the nature of the radio sources and how the radio waves propagate. Thus, they fully follow the method of scientific inquiry.

Web Sites

The associated web sites, http://image.gsfc.nasa.gov/image/inspire and http://radiojove.gsfc.nasa.gov contain science information, copies of the construction and instruction manuals, observing guides, and education resources for students and teachers. Software is also available on the web sites, as well as on a CD for Radio JOVE. The INSPIRE and Radio JOVE websites also provide links to live streams of data. The data from these streams can also be analyzed by students. This is useful for those schools that are not able to find suitable observing sites, but we believe that kit building is an important hands-on part of both projects.

Getting Started

To order an INSPIRE kit, print out the order form at:
http://image.gsfc.nasa.gov/poetry/inspire/orderform.html. To order a Radio JOVE kit, print out the order form at:
http://radiojove.gsfc.nasa.gov/office/order_form.txt. To allow the Radio JOVE project to keep in contact with the participating schools, we ask that you fill out and return an application (by mail or FAX). The application form is available on the Web on the Radio JOVE website at:
http://radiojove.gsfc.nasa.gov/office/rj_applications.htm

Educational Challenges

The issues associated with doing projects like these are:

1. It is often difficult for the teacher to incorporate the learning required to make the projects understandable and meaningful to the students into the curriculum.

2. The hands-on aspects are also time consuming. So much needs to be covered in a standard physics or earth/space science course that INSPIRE and Radio JOVE are often done as extra-curricular activities and so becomes an informal science activity.

3. For the teacher to be able to be comfortable using INSPIRE and Radio JOVE, the projects offer workshops to teachers (and sometimes students) to provide them with the required background. To arrange workshops, contact the authors (Taylor – INSPIRE and Thieman – Radio JOVE).

How Does an Educator Find NASA Space Science Resources?

Terry J. Teays, Carole Rest, Denise A. Smith, and Bonnie Eisenhamer

Space Telescope Science Institute/Computer Sciences Corporation, 3700 San Martin Drive, Baltimore, MD 21218

Abstract. The Origins Education Forum at the Space Telescope Science Institute operates and maintains the Space Science Education Resource Directory (SSERD) for NASA's Office of Space Science Education Support Network. Through a collaborative development effort between members of the Education Support Network, educators can now easily locate NASA space science resources for use in their classrooms.

Introduction

Many NASA space science missions and programs have developed materials for use in education and public outreach. Educators looking for NASA education resources need to know how to find and acquire these materials. The Education Support Network of NASA's Office of Space Science has created a tool to do this, the Space Science Education Resource Directory (SSERD).

What is the Directory?

The Directory is a Web-based tool that allows educators to browse for resources, find out key information about the resources they find, and go to the resource. Users can search by a variety of methods based on grade and subject, general topics, and/or key words. Currently the Directory only has electronically available materials (i.e. Web sites and PDF files), but in the future will allow "one stop shopping" for CDs and hard copy materials. All of the resources in the Directory have been verified for scientific accuracy. Users can also submit reviews of the products for others to read. The Directory also exports the information it contains to other education related databases, such as the Gateway to Education Materials, so that educators can locate the resources through other channels with which they may be already familiar. The Directory development team was awarded a NASA Group Achievement Award this past year for the effort to create the Directory. The Directory can be found at http://teachspacescience.org

The Past and the Future

A Working Group of the Office of Space Science's Education Support Network created the Directory. The current prototype was introduced October 2, 2002.

During this past year we have been working to evaluate how well it meets educators' needs and to improve its efficiency and usability. Extensive testing has been done with many teachers, and through formal usability laboratory work. We are currently working to increase the number of resources available. We are also testing procedures to do more extensive review of the products, and provide a "seal of approval" for those that pass. The ability to request products that aren't electronically deliverable is high on our list of capabilities to add in the future.

Tell Us What You Think: Everyone is invited to check out the Directory and provide feedback.

Schmidt Crater: Making Data from the Mars Global Surveyor Accessible to Introductory Astronomy Students

Frederick J. Thomas

Sinclair Community College, 444 West Third Street, Dayton, OH 45402-1460, fred.thomas@sinclair.edu

Introduction

Like 200,000 other college students nationally, a typical astronomy student at Sinclair Community College is a non-science major taking astronomy to satisfy a general-education science requirement for the Bachelor's degree. Images from the Mars Orbiter Camera (MOC) and data from the Mars Orbiter Laser Altimeter (MOLA) are the basis for a series of lab activities that engage these students in investigating the Schmidt Crater region, near the South Pole of Mars.

Goals of the Project

In cooperation with faculty from mathematics, geography, and sociology, the author developed a set of new lab activities that (a) enhance students' understanding of contemporary explorations of the solar system, (b) increase students' ability to use scientific reasoning and basic math skills to answer realistic questions, and (c) encourage students to make more detailed comparisons between Earth and Mars. The activities are also intended to be low-cost, to be suitable for either on-campus or distance-learning environments, and to be fun for both students and instructors.

Methodology

Schmidt Crater was selected as a region on Mars about which students have little or no advanced knowledge but which is similar in size to a familiar object – the state of Ohio. Altimeter data for the region were extracted from the MOLA PEDRs, Volumes 2010 through 2054 (Smith et al. 1999). Wide and narrow angle photos of the region were obtained from the Planetary Data System web site (Malin et al. 2003). Using Perd2tab, ArcView, and other software, the data and images were converted to forms that introductory, non-science students can use without the need for expensive or highly specialized software. The activities were tested during the spring and fall of 2001 and are being revised to incorporate data from Mars Odyssey. Students use free software (Scion Image [Scion Corporation 2004] or ArcExplorer [ESRI 2004]) and print materials to interpret images and data both qualitatively and quantitatively. They complete a series of "authentic learning tasks" related to planning for a permanent base near Schmidt Crater.

Sample Activities

Maps and Coordinate Systems. Coordinate systems are an important topic in introductory astronomy, and one that many students find difficult. The astronomer's system of right ascension and declination is traditionally described as a projection onto the sky of latitude and longitude, and teaching Einstein's ideas about relativistic space-time warps also depends upon an analogy with latitude and longitude. Unfortunately, only a few students begin college astronomy classes familiar with map coordinate systems. The Schmidt Crater region provides an opportunity to build the students' skills in quantitative analysis of locations, elevations and displacements within an environment that is small enough to seem familiar but large enough to show distinct effects of the planet's curvature. Students use the GIS program, ArcExplorer, and printed maps to plan a pipeline to potential water sources in the high plateaus south of Schmidt Crater. They also investigate possible transportation links to a polar research station.

Interpreting Data. Assessing the quality of data is an essential skill for scientists and for everyone else. Unfortunately, introductory science students typically do little more than calculate a "percent error" for some of their own measurements. They frequently ascribe any shortcomings to "student error" or "faulty equipment," and assume that "real" scientists with "real" equipment never have to worry about such things. Even with the best possible equipment, of course, it is extremely difficult to collect, validate and disseminate over half a billion high-precision MOLA measurements. The data quality is amazingly good, but one incident provided an opportunity to engage students in a quantitative analysis of the MGS orbit and to give them a more realistic view of the difficulties faced by scientists. When preliminary data from December of 2000 were added to our set, a new "fault line" suddenly appeared. It turned out the data in question was collected on Christmas Day, when only one of the usual three tracking stations was monitoring the satellite. For 18 hours, MGS was on its own and the quality of the data suffered.

Where Do You Want to Live? Using MGS data and GIS software to answer Earth-like questions provides a way to engage introductory college students in testing the limits of our current knowledge about Mars. Most key locations (including that of "Schmidtville") were selected to make use of narrow-angle photos from the Mars Orbiter Camera. As a potential "homestead," students also investigate a 4-km^2 area for which there is no narrow-angle photo. The 38 MOLA hits provide important details, but many questions remain. Are there really 4 separate peaks in the lower left of the homestead, or is it an artifact of the cartographer's interpolation method? What do we really know about this small area of Mars, and what else would you like to know?

Frontiers of Knowledge. Even before Odyssey provided new evidence for the presence of subsurface water ice on Mars, students were looking at the "pitted terrain" south of Schmidt Crater as a potential source of water. In the activities, students made comparisons with water use on Earth and in the International Space Station to estimate the water needs of a colony on Mars. They also made comparisons with the permafrost and the glaciers of Alaska to estimate the quantity of water available. The very first Odyssey image provided important new information about the temperature variations associated with the pitted terrain. The early results from Odyssey's epithermal and high-energy neutron spectrometers provided evidence that water ice is likely to be present beneath the entire Schmidt Crater region. The use of GIS software made it very easy to integrate the new data with the old.

Conclusion

Data sets from the Mars Global Surveyor provide a wide variety of opportunities to engage general-education astronomy students in quantitative investigations near the frontiers of planetary exploration. The compilation and translation of data into forms appropriate for use with low-cost or free software (such as ArcExplorer and Scion Image) can greatly expand the educational uses of the data, both by allowing faculty to produce new print materials and by allowing students to interact directly with the data.

References

Environmental Systems Research Institute, Inc. 2004
   http://gis.esri.com/
Malin, M. C., Edgett, K. S., Carr, M. H., Danielson, G. E., Davies, M. E., Hartmann, W. K., Ingersoll, A. P., James, P. B., Masursky, H., McEwen, A. S., Soderblom, L. A., Thomas, P., Veverka, J., Caplinger, M. A., Ravine, M. A., Soulanille, T. A., & Warren, J. L. 2003, PDS Mars Global Surveyor Mars Orbiter Camera (MOC) Image Collection
   http://ida.wr.usgs.gov/
Scion Corporation 2004
   http://www.scioncorp.com
Smith, D., Neumann, G., Ford, P., Arvidson, R. E., Guinness, E. A., & Slavney, S. 1999, Mars Global Surveyor Laser Altimeter Precision Experiment Data Record, NASA Planetary Data System, MGS-M-MOLA-3-PEDR-L1A-V1.0

A Successful Formula for Teacher Retention and Renewal: The Teacher Leaders in Research-Based Science Education Program

Constance E. Walker and Stephen M. Pompea

National Optical Astronomy Observatory, 950 N. Cherry Ave Tucson, AZ 85719, cwalker@noao.edu

Teacher Leaders in Research-Based Science Education (TLRBSE) is a Teacher Enhancement Program funded by the National Science Foundation and hosted by the National Optical Astronomy Observatory in Tucson, AZ. Consistent with national priorities in education, TLRBSE seeks to retain and renew middle and high school teachers of science and mathematics by integrating the best practices of Research Based Science Education with the process of mentoring. Components of the program include 1) an on-line distance learning course to learn about astronomy, leadership skills, pedagogy, image processing and research-based science education; 2) an in-residence summer institute at Kitt Peak National Observatory and the National Solar Observatory engaging participants in an authentic research experience; 3) an extension of the research projects to the classroom with datasets and support provided by NOAO; 4) mentoring support for the Teacher Leaders and their Learning Colleagues (novice teachers they mentor); and 5) a professional community of RBSE educators linked by on-line discussion forums and face-to-face presence at professional meeting. For further information, please look on the web at http://www.noao.edu/outreach/tlrbse/.

NOAO: Supporting Astronomy Education Across the Spectrum

Constance E. Walker and Stephen M. Pompea

National Optical Astronomy Observatory, 950 N. Cherry Ave Tucson, AZ 85719, cwalker@noao.edu

The National Optical Astronomy Observatory (NOAO) supports science education at all levels, including the education of the next generation of astronomers. Programs for students, teachers and astronomers make use of NOAO's unique facilities to promote discovery and lifelong learning, returning to the public their investment in research and development. *Project ASTRO-Tucson* partners elementary and secondary school teachers with astronomers to enrich their astronomy and science teaching. The *Teacher Leaders in Research-Based Science Education* program is designed to develop master teachers in research-based science education and to prepare them as leaders to mentor learning colleagues. NOAO's *Research Experiences for Undergraduates* and *Research Experiences for Teachers* participants work in close collaboration with members of scientific staff on research programs. Graduate students are awarded telescope observing time through a competitive review process. Postdoctoral research appointments are awarded to those of outstanding promise. The most recent member of the repertoire is the *Astronomy Education Review*, a new on-line journal that provides a meeting place for all who are engaged in astronomy or space science education, in either a formal or informal setting. For further information on workshops, materials and opportunities, please look on the web at http://www.noao.edu/education/outreach.html.

PROJECT ASTRO-TUCSON: The Art Of Learning About The Cosmos Around Us

Constance E. Walker and Stephen M. Pompea

National Optical Astronomy Observatory, 950 N. Cherry Ave Tucson, AZ 85719, cwalker@noao.edu

Project ASTRO-Tucson matches professional and amateur astronomers with K-12 teachers and community educators who want to enrich their astronomy and science teaching. During a 2-day workshop, the partners focus on learning hands-on, inquiry-based activities that put students in the position of acting like scientists – as they come to learn and understand more about the cosmos around them. After the 2-day workshop, which includes an inspirational visit to Kitt Peak, each volunteer astronomer "adopts" a class and makes at least four visits during the year. Examples of activities the partners bring away from the workshop and into the classroom include interdisciplinary investigations of the Moon by students integrating art and writing within recordings of their scientific observations. In addition to partnerships, a training workshop, and hands-on activities, other major components of the program include continued staff support, effective educational materials, follow-up workshops, and connections to community resources like local amateur astronomy associations. For further information, please look on the web at http://www.noao.edu/education/astrotucson.html.

HOKU: An Online Astronomy Newsletter for Educators and Parents

Lisa Wells and Liz Bryson

Canada-France-Hawaii, Telescope Corporation, P.O. Box 1597, Kamuela, HI 96743, lwells@cfht.hawaii.edu

Abstract. HOKU is an online newsletter developed at CFHT for use by educators and parents. In past issues of HOKU we have compiled useful websites with exercises for the classroom, as well as pictorial, and informational sites. Here we are highlighting the May 2002 issue. HOKU is registered with the Library of Congress, ISSN 1538-5140.

Introduction

HOKU is a quarterly online newsletter developed at the Canada France Hawaii Telescope (CFHT) for use by educators (grades 4-8) and parents. Hoku is the Hawaiian word for star. It is a compilation of useful websites with exercises for the classroom. It includes pictorial and informational sites as well.

We have also included many text references at grade and high school levels for use in the classroom or for a parent with a child interested in astronomy.

CFHT has received many requests by local educators on the Big Island of Hawaii for information on astronomy and jobs related to the running of an astronomical observatory. At a recent open house, educators expressed a strong interest in having a resource available which would enhance astronomy and space science education at the grade school level – grades 4-8. HOKU was developed specifically for this purpose. Our presentation highlights the issue Volume 2, Number 2 from May 2002.

Design and Distribution

The website was designed in a split format to provide quick links on the left, and more detailed descriptions of the linked websites on the right. The quick links on the left are organized by the target audience; parent/student or teacher, and further categorized by type; picture, activity, or informational site. It is likely that a site highlighted in an issue is found in several of these categories. We list the contents of each issue for easy viewing of the topics covered. The descriptions of each site are taken directly from the site to which we have linked otherwise we summarize the content. When we find a good reference book, those are listed at the bottom on the right side.

HOKU is distributed to educators around the Big Island of Hawaii and to any and all requests for updates of new issues released. We have included distributions to others around the world such as ESO, Lowell Observatory, and the Lunar and Planetary Institute. The newsletter resides in the CFHT library page and would be easily found by web surfers looking for an educational site. The URL is: http://www.cfht.hawaii.edu/Reference/Library/hoku.html

If you wish to receive notice of new issues released, please advise us by email at Bryson@cfht.hawaii.edu or connect to our site and use the mail option at the top right side of the page. We also have a link to the HOKU archive on the left so teachers may review older issues for ideas and classroom materials.

Newsletter Content

The websites featured in HOKU come from a variety of sources. First we look through online sources such as The Scout Report, Yahoo's Picks of the Week, Blue Web'N, and the Librarian's Index to the Internet. If we decide to highlight a certain topic, the topic searches are done using various search engines such as Yahoo!, Google, and Excite.

Our May issue begins with research being conducted at CFHT in collaboration with others at the Space Telescope Science Institute. We have included many space science sites, including the Starshine and HESSI projects, and imaging from various space satellites in the Catalog of Spaceborne Imaging. The Planetary Fact Sheets is another great site for learning about the planets in our solar system. The Hands-On Universe is a site where students or teachers may request observations of a certain target in the sky. Understanding the Universe, and the Earth and Sky Radio Series sites all have great ideas for teachers, and students, lesson plans, and fun & games. The Atlas of the Universe is a great site for teaching distance scales. The Quiz Center of the Trivia Portal is a great site for testing a student's knowledge in all areas of science.

The last few issues have included a newer feature, Upcoming Astronomical Events. It gives the dates and times (in Hawaii Standard Time) for upcoming meteor showers, eclipses, comets, and planetary events. The books which have been highlighted in the Suggested References in past issues include a link to Amazon.com. We give the ISBN number for those who wish to order the book from their local retailers or use another online bookstore.

We welcome any and all comments and suggestions for future issues. When a 4-8th grade teacher finds or creates an interesting resource for their classroom, we encourage them to share it with us. All educators creating sites for educational purposes are strongly encouraged to send us the information for inclusion in future issues.

Astronomy Education Review: A New Journal/Magazine for Astronomy and Space Science Education

Sidney Wolff

Nat'l Optical Astron. Observatories, P.O. Box 267323, Tucson, AZ 85725, swolff@noao.edu

Andrew Fraknoi

Foothill College, 12345 El Monte Rd., Los Altos Hills, CA 94022

We are proud to introduce a lively new journal and magazine for those working in astronomy and space science education. Called *Astronomy Education Review (AER)*, the on-line journal will also become a database of information on educational topics. Our hope is to engender research, information sharing, discussion, networking and controversy, while providing a place to showcase what our colleagues in education and outreach are doing. We particularly want to invite everyone doing E/PO work at NASA to contribute to the new journal.

The project received the endorsement of both the American Astronomical Society and the Astronomical Society of the Pacific, and is currently being supported by the National Optical Astronomy Observatories.

Its web address is: http://aer.noao.edu

Five Distinct Sections

AER is both a journal and a more general magazine (just as *Science* and *Nature* are) and will include:

- Refereed papers on research in astronomy education (with commentary on how to apply the research in "real life")

- Short reports on innovative techniques, approaches, activities, and materials (even if they are still in the development phase)

- Annotated lists of useful resources for educators at all levels

- Announcements of opportunities (jobs, conferences, grants, material testing, funding sources, chances for cooperation on joint projects, etc.)

- Editorials, reviews, opinion pieces, and interactive discussion

More specific guidelines for each section (and some sample articles) can be found on the Web Site. Except for the announcements (which will be up only as long as they are relevant), all the other contributions will be archived and searchable in a variety of ways. We hope to build a permanent repository of the best ideas and practices in the field.

Rationale for Having a New Journal

Astronomy is the only major science field with no vehicle for communication for educators and public outreach professionals. Yet, especially with the growth of NASA OSS funding, the number of people involved in doing astronomy and space science education has increased dramatically in the last few years. Those doing E/PO are often working alone or in small groups and many such projects have only begun recently. Isolation means that we frequently reinvent the wheel and that good ideas or innovations do not always spread to the community.

Even when good research is being done on how learning and teaching can take place most effectively (and much of this research is just beginning in our field), the results of such research are rarely reaching practitioners in the classroom. All these were among the reasons that connecting those who work in education and the integration of education into professional review and reward structures are key recommendations of the new Decadal Survey. We agree that the time has come for a central place for educational resources and information to be easily available, and hope that *Astronomy Education Review* will become that place.

How It Will Work

The journal/magazine will serve all five of the main arenas where astronomy and space science education takes place:

1. K-12 classrooms and the professional development of K-12 teachers
2. undergraduate-level college classes (both for non-majors and science majors)
3. graduate-level training
4. informal education in museums, planetaria, youth groups, amateur clubs, community centers
5. public outreach via newspapers, magazines, radio, TV, the web, books, and other media

A strong Board of Editors with members from each arena has been appointed and they are joined by a larger Advisory Board to broaden the expertise available to the journal.

AER will be an electronic journal with free subscriptions. Eventually, we hope to publish quarterly issues, with an e-mail announcement to subscribers when an issue is ready. Articles are posted as soon as they are accepted (after refereeing), but each article will be part of a defined issue (so that the journal can be cited in CVs and in other publications.) Articles will be classified by topics and arena and become part of a searchable database on the Web site (so you can read back issues or search specifically for what you need.)

Many issues about the journal remain to be resolved, and we very much welcome ideas, support, practical advice, and contributions of articles from everyone in the education community as we move forward with the project. Please contact the editors at the addresses given in the heading of this paper.

Closing Remarks

Dear Meeting Participants,

I'm excited to have the last couple of words in this remarkable meeting. It's always so important for somebody who's been put in the spotlight to have some perspective at their own meeting. So I will be mindful of the fact that I'm really the last obstacle between the audience and lunch. That's the real perspective!

First of all, I think I have to start out with a series of thanks. It is easy to underestimate the huge amount of effort that was required to pull this off, and the number of people who put in countless, countless hours. First and foremost, I'd like to thank Carolyn Narasimhan, who had a vision about this conference. She had a vision about doing something that was very different from the typical kind of conference. We found that when we put this together there were many different views of what this conference could be. They were all valid but they were quite divergent. We really had to think strategically about this conference and Lynn had the strategic vision. She worked with the organizing committee and it was not only her vision but an act of extraordinary persistence to have maintained that vision and put this conference together. Lynn, if you could come up here. Thanks from your colleagues.

I'd like to also express thanks to Lynn's staff: Victoria, Karen, Bernhard, and to the others who really made this happen. This was a fabulous job. The accommodations were fabulous. The plenary speakers, the work of the panelists, the work of the moderators. Pulling off a conference like this, particularly in the unique way that it was arranged, was a true trick. We'd also like to thank our local museum host, the Adler Planetarium. Thank you, Paul Knappenberger, for hosting a wonderful evening at the Adler.

I just want to close with brief observations about this conference. First of all, the remarkable assemblage of people that we brought together, both from the education community and the space science community. We have a wonderful mix of people here today and for the last few days. Scientists, educators, people wearing both hats, all interesting people. One of the most interesting things when we looked at the registration was the number of people we didn't know, that weren't direct participants in the OSS E/PO program. That basically says there really was an intrinsic interest in this conference. I think that what happened during this conference is really a measure of how much people wanted to gather together, to bring these various communities together, to begin to intensify the nature of the interaction.

This has been a very active conference by design. It was not the typical showcase conference in which we would self-congratulate and say how important we all are and how proud we are of our own work. This was really a meeting where people interacted in an active way, encouraged by the structure of the meeting itself. You engaged. You didn't sit back. You rolled up your sleeves. You advanced your ideas. Your shared what you were doing, focusing on key issues in education and public outreach. We talked earlier that one of the purposes of this meeting was community-building and it really accomplished that. Perhaps some of the most important outcomes of this conference are the connections that were made, because I think many of them will evolve into long-lasting partnerships.

We realized early on that we needed to deepen, extend and intensify the dialogue between the space science and education communities. The education community is, after all, our ultimate customer in our efforts. Whatever the Office of Space Science does, whatever NASA does, if it really doesn't benefit the education community, we're doing the wrong thing. When we put out the original OSS Education Strategic Plan, it was called "Partners in Education." That was a title that was deliberately chosen. It was never our intention, still isn't our intention, to bring the scientific community charging in saying what's good for education. We think we do better than that. Instead, we felt that we could marshal the talents of the scientific community allowing them to do what they do best, namely, science research and insight into the process of science. We wanted to marry those talents with the insight and expertise of the education community, and we are succeeding. It's through those kinds of collaborations that we can help make a difference. The real trick was to figure out how to arrange those collaborations.

In a real sense the science community has historically been isolated from the education community. But we're in a society that is more and more dependent on science and technology. Which means that in all sorts of ways, science and technology permeate our society. Yet, the people who have to make decisions don't understand the role that science and technology play in their very daily existence.

Somehow that problem has to be solved, and one way to do it is to try and get your arms around the two communities and start to bring them back together. But there's a prerequisite to that happening, which involves facilitating dialogue between the two communities, enabling real understanding of each other's viewpoints, establishing a common language. So this is not the end of the story. This is really the beginning of what I hope is a long term dialogue between the two communities, deepening the interaction over the years, over the long term.

This meeting gave us lots of new ideas and lots of things to think about. One of the things in designing this meeting that I hoped would happen is that everyone would be pushed and stretched to think about new things. That we would have our ideas and our preconceptions challenged. That we would be forced to take a much more critical look about what we were doing in education. I think that actually happened at this conference. Again, this was due to the wisdom of the people who put this program together.

Now I think the most important thing of all is what happens next. Sometimes conferences take place, the food is good, sometimes the food is too good; this conference was a broadening experience in more ways than one, I suspect. But the real issue is that this conference is just a way to point along the path towards a bright future.

We are determined to take a critical look at the outcomes from this conference. What were people trying to tell us? What do the lessons learned at this conference tell us about how we're doing business? What kinds of changes do we need to make? We had some indications that we needed to do more professional development for E/PO professionals from missions and research programs, and for the people involved in the Support Network. Maybe we need to start to pay a lot more explicit attention to that and really develop a program, put together a program of professional development. We were offered a challenge by several experts at this conference. The issue of educational coherence came up, and perhaps we need to think about developing a space science curricular framework and perhaps strands, so we can serve our user community better. That would actually provide context for mission products people have been putting together. I think more coherence would also be useful to people in the informal education world, providing context and a way to begin to link materials.

A final thought, but an important thought, is to keep in mind the reality that the OSS E/PO program is less than six years old. If we had had this conference five or six years ago, if we had been foolish enough to make that attempt, maybe twenty people would have attended. People would have talked about what they were doing and they would all have been proud that they made those presentations and they would have gone home and nothing would have happened. That's not what's going to go on in this particular case. This has really given us an important basis of information for thinking about what needs to be changed; for example, how do we begin to think consciously about professional development? How do we think much more consciously about making resources fit together in a coherent fashion? So you've actually offered us, the OSS E/PO program, the Support Network, something invaluable – your expertise and your willingness to contribute towards improving our efforts.

So I'd like to conclude by thanking all of you for having given us this help, and I hope that you will find ways to continue this dialogue on a regular basis. Thank you.

Jeff Rosendhal
Chicago, June 14, 2002

Appendices

Appendix A –

The NASA Office of Space Science Education and Public Outreach Program

The NASA Office of Space Science Education and Public Outreach Program

Presented at the 2002 Conference of the Committee on Space Research (COSPAR) in Houston

Jeff D. Rosendhal, Philip J. Sakimoto, Rosalyn A. Pertzborn, and Larry Cooper

NASA Headquarters, 300 E Street, SW Washington, DC 20546, Jeffrey.D.Rosendhal@nasa.gov

Abstract. Over the past six years, NASA's Office of Space Science has implemented what may well be the largest single program in astronomy and space science education ever undertaken. The program goals include the public sharing of the excitement of space science discoveries, enhancement of the quality of science, mathematics and technology education – particularly at the precollege level – and supporting the creation of our 21st century scientific and technical workforce. This paper provides an overview of the program origins, policies and philosophies, and describes the development and growth of the program. Program accomplishments and the challenges that remain are discussed along with potential opportunities for international collaboration.

Introduction

Education and contributing to the public understanding of science have long been important components of NASA's mission dating to the 1958 Space ACT which created NASA. The current NASA Office of Space Science (OSS) education and public outreach (E/PO) program really began with work done by the Space Telescope Science Institute in the early 1990's to share the wonder of the Hubble Space Telescope scientific discoveries. More recently, education at NASA has gained significant additional prominence with its elevation to the level of an Agency Core Mission as expressed in the 2002 NASA Mission Statement.

- To understand and protect our home planet;
- To explore the Universe and search for life;
- To inspire the next generation of explorers

    ... as only NASA can.

The mission "To Inspire the Next Generation of Explorers" encompasses all of NASA's education activities to 1) inspire and motivate students to pursue careers in science, technology, engineering, and mathematics and 2) engage the public in shaping and sharing the experience of exploration and discovery. The OSS E/PO program is well aligned with these new Agency goals and expects to make major contributions toward achieving them.

The quality of the U.S. educational system has also long been an area of public policy debate. In recent years concerns over the interplay of education and our country's future economic and national security has received considerable attention. Excerpts from *Road Map for National Security: Imperative for Change* (The United States Commission on National Security/21st Century 2001) indicate:

- The capacity of America's education system to create a 21st century workforce second to none in the world is a national security issue of the first order. As things stand, this country is forfeiting that capacity.
- Education is the foundation of America's future. Education in science, mathematics, and engineering has special relevance for the future of U.S. national security, for America's ability to lead depends particularly on the depth and breadth of its scientific and technical communities.
- The health of the U.S. economy, therefore, will depend not only on professionals who can produce and direct innovation in a few key areas, but also on a populace that can effectively assimilate a wide range of new tools and new technologies.
- The American educational system does not appear ready for such challenges...

NASA recognizes its broader responsibilities to education and the unique contributions that it can make. *Science in Air and Space: NASA Science Policy Guide* (NASA 1996a) states

> "Both the public and the political system expect benefits broader than purely scientific ones to be derived from NASA research programs and missions... There are two particular areas in which NASA and the NASA-supported research community can make especially significant contributions:
>
> - Education and raising the general level of public understanding and appreciation of science and technology
> - Developing advanced technology in support of NASA's science missions and research programs which also has uses beyond the space program, thereby contributing to the country's technological base and its long-term economic competitiveness."

With respect to education the document then goes on to say:

> "It is NASA policy to use its space research missions and research programs and the talents and resources of its research and development communities to make significant and measurable contributions to meeting national goals in the reform of science, mathematics, and

technology education – particularly at the precollege level – and the general elevation of scientific and technological literacy throughout the country. NASA will implement this policy by formally making education and public outreach integral components of ground-based and space flight research conducted by NASA and NASA-supported scientists throughout the country."

"The economic vitality of our nation depends increasingly on new scientific knowledge and its application. For NASA, this means ensuring that the ideas and capabilities of the widest possible talent pool are brought to bear on its missions."

"The responsibilities of scientists must include explaining scientific results to the public and communicating the role and importance of science and technology in contemporary society."

As noted earlier, the NASA OSS Education and Public Outreach program is part of a larger NASA effort in education. As expressed in the FY 2000 OSS Strategic Plan (NASA 2000), OSS has sought to support the Agency through specific programs designed to use space science and the space science research community to share the excitement of space science discoveries with the public, enhance the quality of science, mathematics and technology education, particularly at the precollege level and help create our 21st century scientific and technical workforce. The program is designed around the unique contributions that OSS and NASA can bring to education the science and the scientists.

The body of this paper provides background and details on the approach used in creating the OSS Education and Outreach program an effort that is now believed to be the single largest program in astronomy and space science education ever undertaken.

Background

Development of OSS E/PO Policy, Strategy and Implementation Principles

The first critical steps in creating a formal OSS E/PO program were taken beginning in late 1993. Drawing upon a diverse cross section of national leaders representing scientists and science educators, science centers and planetariums, state education departments, universities, and NASA Centers, an education and public outreach strategy and set of implementation principles were collaboratively developed over the next two years. Guiding the entire planning effort were a number of observations about NASA OSS, the educational community it desired to impact, and the space science community it desired to engage that were fundamental to shaping the subsequent program. These basic operational principals, included the need for the OSS program to:

- Coordinate with NASA's overall effort to support the national education agenda;

- Support current national and state efforts of systemic education reform and the establishment of national standards and benchmarks;

- Emphasize, wherever possible, active, experiential involvement in NASA research programs and missions for both teachers and students;
- Use outside advice from the scientific, educational, and minority communities in the planning, development, implementation, and assessment of all our education and outreach activities;
- Tie into existing programs for women and minorities and, in doing so, contribute to the education and training of groups currently underrepresented in technical disciplines;
- Conduct regular tests and assessments of impact and effectiveness;
- Make educational materials and the results of our education and public outreach programs as widely available as possible using both existing dissemination mechanisms and new information technologies;
- Maintain a focus on excellence as the standard for performance in all education and public outreach activities that are undertaken.
- Base all of OSS's E/PO efforts on collaborations between the scientific and education communities, thereby drawing upon and marrying the appropriate expertise of the two communities.

The governing strategy (NASA 1995) and implementation plan (NASA 1996b) placed major emphasis on: helping scientists become involved in education/outreach through several actions:

- creating a network of brokers/facilitators (to be described later),
- providing opportunities for appropriate training, and
- removing contractual and other impediments to participation.

Attention was also given to the need to enhance the breadth and effectiveness of partnerships among scientists, educators, contractors, and professional organizations as the basis for education and outreach activities by:

- focusing on high leverage opportunities,
- building on existing programs, institutions, and infrastructure,
- emphasizing collaborations with planetariums and science museums,
- coordinating with other ongoing education and outreach efforts inside NASA and with other government agencies,
- involving the contractors in OSS's education/outreach programs,
- making materials widely available and easily accessible, using modern information and communication technologies where appropriate, and
- providing meaningful opportunities for underserved/underutilized groups.

Finally, evaluation for overall quality, impact, and effectiveness was highlighted as a crucial component of the overall strategy.

What is Different About the OSS Approach?

The NASA OSS approach to E/PO specifically recognized that the number of space scientists available to participate in E/PO would be completely inadequate to impact the national educational system. (There are only about 10,000 space scientists and over 100,000 schools across the county. Also the geographic distribution of space scientists is very nonuniform.) A departure from the traditional linear approach was needed. In the traditional approach the collective programmatic impact is simply the sum of a set of individual disconnected efforts. The approach OSS embarked on was to create a decentralized, highly networked system in which the impact of individual efforts could be greatly amplified. Amplification would be achieved through finding the key entry points into the education system and through archiving and disseminating nationally the best products and programs. A key to achieving these goals was to establish a Support Network of Broker/Facilitators and Educational Forums.

OSS E/PO Support Network

The first element of the OSS E/PO support network, "Broker/Facilitators," were conceived of as regional agents charged with helping to identify and catalyze high leverage E/PO opportunities at the local and regional level. This was a new concept developed to address a critical issue in realizing the OSS E/PO program. As described in the *Education/Public Outreach Implementation Plan Chapter VII: Help Scientists Become Involved in Education/Outreach*:

"OSS must do more than place a new requirement on the participants in the space science program. OSS must take active steps to help the scientific community become involved in education and public outreach – help in looking for high-leverage opportunities, help in arranging partnerships and alliances with educators, help in understanding what is now happening in education and what sorts of materials are appropriate for the classroom, help in removing impediments that get in the way of scientists participating in education and outreach even if an individual wants to do so. A number of approaches are possible to providing such help."

The Task Force viewed the Broker/Facilitator concept "as central to the systems approach being recommended for implementation of the OSS Education/Outreach Strategy. The job is going to be a difficult and demanding one requiring familiarity with the OSS program and scientific community, familiarity with the needs of the education community, links to the education system at many levels, and an aggressive approach to identifying high-leverage opportunities and arranging alliances."

The other portion of the Support Network was to be comprised of a set of "Education Forums" aligned with the four NASA OSS research themes. The Forums were to provide the necessary sustained effort and long-term continuity required to effectively work with the education system – a continuity that could not be provided by short duration missions or activities undertaken as a part of individual research grants. Forums were to provide a "home base" for smaller missions so that each mission did not have to create its own infrastructure; play the role of national broker/facilitator for missions and research programs associated with particular OSS themes; work with the education community to develop educational programs and products suitable for national distribution;

serve as a national archiver/disseminator for education/outreach programs and products; and create and maintain an accessible directory of education/outreach products and materials.

The integration and coordination of the activities of the Support Network and other participants in the OSS E/PO program was addressed through creation of the OSS Education Council. From the OSS E/PO Implementation Plan:

> "Based on the previous recommendations, it is clear that close coordination of all activities and a strong interaction among the various institutions and organizations participating in the OSS Education/Outreach program must be achieved if the proposed approach is to realize its full potential. To achieve such coordination, the Task Force recommends that an OSS Education/Outreach Council be set up to assure optimized performance across the entire "Ecosystem." Membership of such a group should include representatives from all of the key groups playing a role in the execution of this Implementation Plan – OSS, the NASA Education Division, the Office of Equal Opportunity Programs, the "Education Forums," the broker/facilitator groups, and other appropriate participating organizations."

Work undertaken through this group has addressed a number of critical OSS-wide issues that would not have been address through any individual program or mission. The Space Science Education Resource Directory, http://teachspacescience.org, which is intended to provide a single place where educators can look for materials, is an example of the work being undertaken through the OSS Education Council.

Funding

Adequate resources to conduct an effective national education and public outreach program were viewed as essential. The Space Science Advisory Committee Task Force that produced the OSS E/PO Implementation Plan, recommended that, as a long-term goal, OSS should plan to spend 1-2% of its total budget on education and the public understanding of science. Elements to be funded would include: 1) Education/outreach components of individual flight missions, 2) Education/outreach components of individual research grants, 3) An OSS-wide program of small education grants, 4) A small number of major, high-profile education programs and projects, 5) A network of Broker/Facilitators and "Education Forums," and 6) A program Evaluator. The predominant fraction of available funding was to be used to support individual or mission-oriented education/outreach programs and projects carried out across the country with the direct involvement of the OSS research community. All these recommendations have been implemented.

Implementation

Beginning in 1997, NASA OSS initiated its E/PO program. All new major missions research solicitations included requirements that E/PO programs be developed and funded at 1-2% of the mission cost. The E/PO program was specifically identified as a factor that would be considered in mission selection. In practice E/PO has actually been a deciding factor in a number of mission selections. A small grants program was begun to add supplemental funds for E/PO programs to individual research awards. Evaluation criteria for E/PO proposals were developed and an *Explanatory Guide to the Evaluation Criteria* (NASA 2002) was created. Groups were selected as Broker/Facilitators and Forums.

Over the next five years, the Support Network built important infrastructure for the E/PO program, including processes to increase coordination of E/PO activities across missions, providing staffing and other support for education conferences, and creating a valuable series of internal and external newsletters. An OSS E/PO Tracking and Reporting System was created to facilitate collection, analysis, and evaluation of E/PO program activities and products and compilation of an E/PO Annual Report:
http://ossim.hq.nasa.gov/ossepo. A process was developed and implemented to conduct reviews of OSS E/PO products. Finally, as noted above, an organized approach to product dissemination was formulated which led to creation of an online Space Science Education Resource Directory:
http://teachspacescience.stsci.edu and the "Space Science Access" Informal Education Web site:
http://cfa-www.harvard.edu/seuforum/wateringHole/.

Efforts to increase the involvement of minorities and underserved were initiated. Under the Minority University Initiative, 15 institutions were selected for multi-year funding to enhance their space science curriculum, faculty and outreach efforts. (Sakimoto 2001). Dialog with several Minority Professional Societies was begun to explore areas of common interest leading to their involvement in several mission E/PO programs.

Current Status

The OSS Education and Public Outreach Annual Report for 2001,
http://ossim.hq.nasa.gov/ossepo, details over 400 E/PO activities and new products; nearly 3000 discrete E/PO events, a presence in all 50 states, the District of Columbia, and Puerto Rico; a presence at 20 national and 36 regional E/PO conferences and more than 50 awards and other forms of public recognition received. The geographic reach of the OSS program is shown in Figure 1. Over 100 OSS missions and programs were involved with nearly 900 OSS-affiliated scientists, technologists, and support staff and nearly 500 institutional partners, including 180 science centers, museums, and planetariums, 40 precollege educational organizations, school districts and boards, and 24 minority colleges/universities. Preliminary analysis of the 2002 data shows substantial continued growth in the number and variety of activities and events being undertaken within the OSS E/PO program.

In 2001, NASA OSS invested nearly $35,000,000 in its E/PO program. Appropriately, one-third went to formal education programs and products which focus on educators and students in the school classroom, one-third to informal education programs such as exhibits and shows provided through science centers

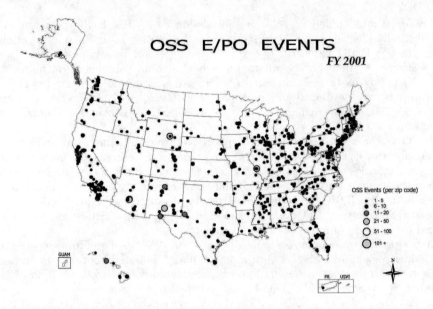

Figure 1. Geographic distribution of OSS E/PO events in 2001.

and planetariums, and one-third on public outreach activities such as public lectures, star parties and Web sites.

OSS's E/PO partners are also investing significant resources. As a consequence, the total scope of the OSS E/PO program is far larger that might be expected from the direct NASA funding amounts. As examples, the Sun-Earth Day 2001 attracted participation from hundreds of institutions around the country – each providing facilities and staff to host events at their locations. OSS missions and science are the focus of major science center exhibits and television programs funded primarily from non-NASA sources. Many scientists and educators are donating their efforts to develop and conduct educator workshops and public events. In general OSS has succeeded in developing products and programs that others want to use and/or participate in because it helps them to do their jobs better and not because OSS is providing funding.

Some especially significant accomplishments over the past five years have been the creation of a one to 10 billion scale model of the solar system, shown in Figure 2, on the national mall in Washington D.C. [Goldstein, 2002), development of educator guides on the "Reasons for the Seasons" (SECEF 2001), Astrobiology (Origins Forum 2001), and the "Invisible Universe (Cominsky & Plait 2003), creation of several traveling museum exhibits such as MarsQuest, shown in Figure 3, (Space Science Institute 2001), Hubble Space Telescope – New Views of the Universe (Space Telescope Science Institute 2001), and Cosmic Questions: Our Place in Space and Time (Dussault 2003), and development of materials for the special needs educators and students such as "Touch the Universe - A NASA Braille Book of Astronomy" (Beck-Winchatz 2001) and the Multi-sensory Space Science (MSS) Kit (SERCH 2001).

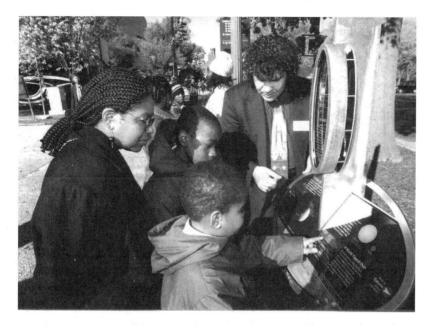

Figure 2. Jeff Goldstein, Voyage project director, visits Uranus with a class of Washington, DC middle school students. Photo credit; Challenger Center for Space Science Education.

Opportunities for foreign participation have primarily come about through international collaboration in the missions. For example, the European and Japanese partners on the Cassini mission are translating E/PO materials and planetarium show scripts into their native languages with their own funding. Most NASA E/PO products are now available in electronic format over the Internet, which has significantly increased access by parties outside the United States. We encourage the international community to use these materials and to help in providing translations.

Where Are We Now?

A substantial national education and outreach program is now underway and E/PO has strong backing from the top management of OSS. The quality of mission E/PO programs has improved significantly but is still uneven and needs additional attention. With the minimal staff available, emphasis to this point has had to be on large mission E/PO programs.

The Support Network has provided a broad array of services and identified a wide range of educational opportunities. Close connection to the grassroots of the education system has been important. There is increasing integration and coordination with NASA Education Division Programs that will be further enhanced through the new NASA Education Enterprise.

The Minority University Initiative (MUI) has broken new ground for NASA and provided a realistic pathway to move minority institutions into the mainstream of NASA research efforts. Under this initiative, 15 minority institutions

Figure 3. The MarQuest exhibit has been at seven venues across the county since opening in 2000. Photo credit; Space Science Institute.

are engaged in research collaborations with nine NASA space science missions or suborbital projects and in more than 30 working partnerships with major space science research groups. The start-up of the "systems approach" has been more difficult than expected. We tried to do too many new things simultaneously and ambition has exceeded the resources. We are now starting to take a more strategically focused approach to the definition and implementation of the core programmatic goals, while deliberately allowing for some aspects of the program to evolve and change. Over the next few years, we will pay critical attention to improving the coherence, quality, and availability of NASA space science educational materials; providing professional development for personnel engaged in OSS E/PO efforts; and more fully understanding the impact of the OSS E/PO program on the audiences it is designed to serve.

Some tactical considerations we have learned from the first five years include: 1) starting at home and then working outward from within NASA, the Support Network, or your own home town or state, 2) making sure that all the "obvious" connections to potential partners and supporters have really been made, 3) taking advantage of opportunities that have been handed to you, 4) working with the people who want to work with you, 5) focusing on making a modest number of things real, and 6) capitalizing on successes and publicizing them. OSS has discovered that it can not do everything. It has also discovered that the process of making strategic decisions based on well defined operating principles as to what programs should be undertaken is a very powerful approach.

Program Evaluation

Since 1998 the Program Evaluation and Research Group (PERG) at Lesley University has been conducting an evaluation of the OSS E/PO program to determine the effectiveness of OSS in carrying out its E/PO Implementation plan. The first two reports are online at
http://spacescience.nasa.gov/education/resources/evaluation/index.htm. Some of the major findings detailed in the second report (Cohen, Gutbezahl & Griffith 2002) are:

- OSS E/PO resources have been growing in number and diversity since the implementation of the current E/PO effort. This growth has been documented in the OSS Annual Report, regular newsletters, and through NASA's EDCATS system (Education Division Computer-Aided Tracking System – a NASA-wide database of education activity).

- The reach of the OSS E/PO effort has expanded to include groups of learners who have not previously benefited from NASA's educational resources. Resources have been developed to address the needs of audiences of varying ethnicities, of all ages, in all areas of the country, and at various levels of physical and intellectual development.

- OSS has also been building relationships with scientists who are members of Minority Professional Organizations. Minority scientists and researchers reported that their connection to OSS provides them with tools they need to act as role models in their communities and mentors to their students and colleagues.

The report noted that "in general, data from the contacted audience groups are extremely positive. Individuals consistently report that they (and those they are educating) find space science exciting, engaging, and complex; this indicates that OSS is succeeding in its first goal of sharing the excitement of space science." Report recommendations include:

- OSS should consider systematically including audiences in the creation and assessment of new resources in a variety of ways;

- OSS should provide scientists with increased access to educational expertise;

- OSS should work to improve communications and coordination both within the OSS E/PO program and the users and participants in OSS E/PO programs.

In addition to the Lesley University effort, an E/PO Task Force was set up in 2002, under the NASA Space Science Advisory Committee (SScAC), to assess the progress of implementing the OSS E/PO program, identify issues, and examine possible directions for program evolution. The Task Force conducted a series of program review meetings with a cross section of stakeholders in the OSS E/PO program including scientists, program managers, the Support Network leadership, and E/PO program leads. In addition surveys were made of educators, program participants, and other users. In March 2003, the Task Force presented of its preliminary findings to SScAC (Education and Public Outreach Task Force 2003). The Task Force found that the program's accomplishments and successes to date included:

- Direct engagement of OSS missions and the space science research community in education and in contributing to the public understanding of science;

- A rich harvest of educational programs and materials directed towards many types of audiences in diverse communities across the country;

- Significant steps towards involving minorities in the mainstream of OSS's scientific, technical, and educational programs and in developing educational materials directed towards audiences that have not previously been served by NASA; and

- Substantial leveraging of resources through collaboration with hundreds of educational institutions and organizations across the country.

The Task Force singled out a number of areas for particular attention, areas which the Task Force believed will yield especially rich rewards in taking the OSS E/PO program to even higher levels of maturity, effectiveness and accomplishment. Major findings and recommendations for OSS include:

- Make educational products more accessible and organize them in a more coherent way;

- Increase the inclusiveness of the program by involving new audiences, science topics, materials and partnerships;

- Expand and intensify pioneering efforts to attract and better integrate minorities into E/PO projects and into the mainstream of OSS science programs;

- Enhance efforts directed towards quality control and obtaining a better understanding of program impact;

- Increase the effectiveness of the OSS E/PO Support Network by focusing the activities of the Broker/Facilitators on their primary roles;

- Strengthen and expand professional development efforts for E/PO professionals, scientists, and the education community; and

- Enhance internal and external communications.

The Task Force Report noted that 1) the approach that OSS has taken in implementing its E/PO Program provides a model that is unique to NASA, to the government, and to science education in general, 2) based on its successes to date and the prospects for even greater successes in the future, the approach could well serve as a guide for future NASA educational efforts across the Agency, and 3) the program is already a credit both to the OSS and to all of the very talented people who have been involved in its planning and execution. OSS plans to pay close attention to and implement the recommendations received from the Task Force.

	Then **1995**	**Now** **2002**
Status:	A Plan	A Major National Program
Breadth:	Space Telescope Science Institute + a few other things (a handful of activities)	Every mission & research program (hundreds of activities)
Infrastructure:	Mostly focused on Space Telescope Science Institute	Extensive interactions with a national infrastructure involving 13 institutions
OSS $:	A few million/yr	More than $ 35 M/yr and growing
Non-OSS $	None	A significant fraction of the OSS investment
Minority Universities	Occasional Technical Advice to NASA Minority University Research and Education Division	Minority University Initiative ($ 3M/yr). OSS involvement in oversight and other solicitations
Peer Review:	A small number of proposals	Research grant supplements & Explorer & Discovery Class Mission Downselects
HQ Procurements:	A few $ 100K/yr	More than $ 10M/yr
Major HQ Sponsored Projects	None	Support Network, 2 television series, Scale Model Solar System, Training, Product Review, etc.
Coordination:	None	Extensive/OSS Education Council
Oversight:	Little required	Needed for large programs (and there are now many)
Conferences:	Limited — OSS presence only at National Science Teachers Association convention	Significant OSS presence at many science/education conferences
Interaction with Other Groups	Limited interactions with internal NASA education groups	Many interactions with internal NASA education groups, as well as, numerous external groups
Evaluation:	None	Major activity now underway

Table 1. Changes in the NASA OSS E/PO program over the past 7 years.

Concluding Remarks

Over the past 7 years, the OSS Education and Public Outreach program has undergone significant development, growth and change. Changes in program scope are summarized on the following page. The substantial progress of the NASA OSS E/PO program over a relatively short period of time can be attributed to the dedicated work of many individuals and organizations.

The process approach defined in the OSS E/PO Implementation Plan has shown significant adaptability and resilience. Changes have been made in the program on an ongoing basis in response to problems as they been encountered. As we move forward, we will continue to focus on increasing the quality of our program activities and products. The program is and will likely remain an ongoing and evolutionary process in keeping with the vision expressed at the program's inception:

> "The overall approach described in this Report is an experiment. The focus on process as the centerpiece of this experiment, rather than on the identification of a set of specific programs, represents a deliberate choice to depart from the practice of simply creating a collection of stand-alone activities having purely local impact. The proposed process offers the prospect of enormous amplification of OSS's education/outreach efforts. The only way to tell whether the experiment will work is to try it. Flexibility will be required, progress on the experiment will have to be monitored closely, and adjustments made on an ongoing basis."

> "Realistic expectations are important. No single education or outreach program undertaken or sponsored by OSS will, by itself, have a significant, long-term sustainable impact on the American educational system. Rather, it will be the total effect of a broad ensemble of high-leverage activities carried out over a long period of time that can make a difference."

Excerpts from Chapter XVI, Concluding Remarks
OSS E/PO Implementation Plan

Thus far, the experiment appears to be a success. The major challenge now facing the program is how to sustain and build upon that success in the face of an ever-changing political and programmatic environment.

References

Beck-Winchatz, B. 2001, Voyages in Education and Public Outreach, 3, 1

Cohen, S., Gutbezahl, J., & Griffith, J. 2002, Office of Space Science education and Public Outreach Interim Evaluation Report October 2001-June 2002 (http://spacescience.nasa.gov/education/resources/evaluation/NASA_Interim_2002_Report.pdf)

Cominsky, L., & Plait, P. 2003, Voyages in Education and Public Outreach, 7, 6

Dussault, M. 2003, Voyages in Education and Public Outreach, 7, 1

Education and Public Outreach Task Force 2003, Implementing the Office of Space Science Education/Public Outreach Strategy: A Critical Evaluation at the Six-Year Mark (Washington: NASA) (see p. 461 of this volume)

Goldstein, J. 2002, Voyages in Education and Public Outreach, 4, 1

National Aeronautics and Space Administration 1995, Partners in Education: A Strategy for Integrating Education and Public Outreach into NASA's Space Science Programs (Washington: NASA) (http://spacescience.nasa.gov/admin/pubs/edu/educov.htm)

National Aeronautics and Space Administration 1996a, Science in Air and Space: NASA's Science Policy Guide (Washington: NASA)

National Aeronautics and Space Administration 1996b, Implementing the Office of Space Science Education/Public Outreach Strategy (Washington: NASA) (http://spacescience.nasa.gov/admin/pubs/edu/imp_plan.htm)

National Aeronautics and Space Administration 2000, FY 2000 OSS Strategic Plan (Washington: NASA)

National Aeronautics and Space Administration 2002, Explanatory Guide to the NASA Office of Space Science Education & Public Outreach Evaluation Criteria (Washington: NASA) (http://ssibroker.colorado.edu/Broker/Eval_criteria/Guide/)

Origins Forum 2001, Voyages in Education and Public Outreach, 2, 4

Sakimoto, P. J. 2001, Voyages in Education and Public Outreach, 2, 5

Sun-Earth Connection Education Forum 2001, Voyages in Education and Public Outreach, 1, 6

Southeast Regional Clearinghouse 2001, Voyages in Education and Public Outreach, 1, 7

Space Science Institute 2001, Voyages in Education and Public Outreach, 1, 3

Space Telescope Science Institute 2001, Voyages in Education and Public Outreach, 1, 4

The United States Commission on National Security/21st Century 2001, Road Map for National Security: Imperative for Change (http://www.nssg.gov/PhaseIIIFR.pdf)

Appendix B –

The Office of Space Science E/PO Support Network

The Origins Education Forum

Ian Griffin, Denise A. Smith, and Carole Rest

Space Telescope Science Institute, 3700 San Martin Drive Baltimore, Md 21210, griffin@stsci.edu

The Origins Education Forum is one of the four national Education Forums established by NASA OSS to help organize the education and public outreach (E/PO) efforts of OSS missions and research programs. Led by the Space Telescope Science Institute (STScI), the Forum has specialized in evaluation and educational technology, offering formative evaluation services to Origins missions and playing a key role in the development of the NASA OSS Space Science Education Resource Directory (SSERD), an on-line directory of educational resources produced by OSS missions and programs. (Visit http://teachspacescience.org and see the September 2001 *Voyages in Education and Public Outreach OSS Newsletter*, issue 3).

The Forum provides Strategic leadership for Origins mission E/PO efforts and is an association of the E/PO leads of the Origins missions, with a Secretariat based at the Space Telescope Science Institute. Forum members work together to determine how the various mission E/PO efforts fit into the broader context of Origins research, which seeks answers to the fundamental questions: Where do we come from? Are we alone?

It seeks high-leverage opportunities to maximize the usefulness and effectiveness of E/PO materials and programs created by Origins missions. Forum members collaborate to enhance activities in formal education, informal education (e.g., exhibits), news, and online outreach, and to avoid duplication of effort and utilize limited resources for maximum effectiveness. The Forum coordinates distributed activities and serves as an information resource for measuring the impact of E/PO efforts conducted by the Origins missions and research programs.

The activities of the Origins Education Forum have been largely centered in the areas of sharing good ideas and developing recommended practices on such matters as product development and evaluation, and helping enhance the OSS Support Network infrastructure and services. Within these areas, the Origins Forum has specialized in evaluation, computer technology and informal science education, drawing on the experience of the Hubble Space Telescope E/PO programs and staff.

As Origins and OSS E/PO efforts have matured and evolved, the Forum has been re-evaluating the needs of the communities it serves in anticipation of needs for new services. At a recent strategic planning retreat in Baltimore, Maryland, forum members worked hard to develop a new charter, mission vision and goals for the next five years, and a new strategic plan resulting from these deliberations was published in March 2003. Our planning process will also include an implementation plan, an annual products and services delivery plan, quarterly reports and an annual review. The Origins Education Forum views the planning process as an inclusive and informed process, in which Forum plans build upon mission and Support Network plans. Through a robust annual planning process, the Forum aims to bring coherence to E/PO efforts and identify opportunities to leverage mission and forum resources. To this end, the Origins Education Forum has been actively involved in visiting forum and mission leads to develop the framework for coordinated planning and joint projects.

In our efforts to maximize the effectiveness of mission E/PO efforts and to share the story of the astronomical search for origins (ASO) with our audiences, the Origins Education Forum completed a joint poster in 2003. "The Electromagnetic Spectrum" poster brings together high quality classroom activities and content from individual missions under a unifying theme that is directly relevant to the education community. (The poster also updates and expands the content provided by an earlier out-of-print STScI poster on this theme, building on a proven and highly popular design.) As we believe our new product has wide appeal, Forum and mission staff worked with professional societies to deliver the poster to a wide audience of educators. "The Electromagnetic Spectrum" poster has appeared in journals of the National Science Teachers Association (NSTA; 41,000 copies), the American Association of Physics Teachers (10,000 copies), and the International Planetarium Society (750 copies). The NSTA journals also included articles to accompany the poster, authored by Forum and mission staff.

This effort provides one example of how Forum and mission staff are working together to identify common threads in their individual E/PO efforts and to synthesize their materials into effective products that meet the needs of their audiences. Other efforts include professional development workshops held in association with meetings of the National Science Teachers Association and the National Organization for the Professional Advancement of Black Chemists and Chemical Engineers (NOBCChE), and trainings for the Girl Scouts of the USA. The Forum is also pursuing strategic partnerships with the Lawrence Hall of Science (LHS) and the Pacific Science Center, in formal and informal science education, respectively. The Lawrence Hall of Science is building materials from their Great Explorations in Math & Science (GEMS) program into a Space Science Core Curriculum Sequence. In partnership with the Public Broadcasting Service and the National Science Foundation, the Pacific Science Center is leading a national outreach effort to accompany the 2004 Origins PBS series. This effort complements other initiatives supported by Origins missions, such as the Space Science Institute's Cosmic Origins traveling exhibit. These Forum activities and partnerships are designed to synthesize materials to tell the larger stories of Origins and to bring these stories to new and diverse audiences.

The observatories and programs comprising the Origins program include the Hubble Space Telescope, the Far-Ultraviolet Spectroscopic Explorer, the Space Infrared Telescope Facility, the Stratospheric Observatory for Infrared Astronomy, Kepler, the James Webb Space Telescope, the Navigator program, and the NASA Astrobiology Institute.

For more information on the Origins Education Forum, please email origins@stsci.edu or visit our website at http://origins.stsci.edu

NASA Office of Space Science Education and Public Outreach Conference 2002
ASP Conference Series, Vol. 319, 2004
Narasimhan, Beck-Winchatz, Hawkins & Runyon

NASA's Solar System Exploration Education and Public Outreach Forum

Ellis D. Miner, Leslie L. Lowes, and Shari E. Asplund

Jet Propulsion Laboratory, m/s 183-301, 4800 Oak Grove Drive, Pasadena, CA 91109, ellis.d.miner@jpl.nasa.gov

Introduction

In an attempt to coordinate education and public outreach (E/PO) efforts within NASA's Office of Space Science (OSS), the NASA OSS E/PO Support Network was established several years ago. As it is presently constituted, the Support Network includes four national theme-based forums, one of which is the Solar System Exploration (SSE) Forum, based at NASA's Jet Propulsion Laboratory in Pasadena, California. The four theme-based forums are supported and augmented by seven regional Broker-Facilitator groups, whose responsibility it is to assist in establishing partnerships between scientists and formal and informal education specialists in their regions. This paper will describe the scientific objectives which are addressed by Solar System Exploration, delineate the personnel who work for or with the SSE Forum, outline the present cast of NASA SSE missions, summarize some of the major SSE Forum activities to date, and highlight the uniqueness of the national networks that work with the SSE Forum.

NASA's Solar System Exploration Objectives

Among the eight science objectives outlined by NASA's OSS in the November 2000 version of *The Space Science Enterprise: Strategic Plan* are two that are addressed directly by the Solar System Exploration theme. These are to

- Understand the formation and evolution of the Solar System and Earth within it

- Probe the origin and evolution of life on Earth and determine if life exists elsewhere in our Solar System.

SSE also addresses limited portions of four other objectives: (3) "Learn how galaxies, stars, and planets form, interact, and evolve," (4) "Look for signs of life in other planetary systems," (7) "Understand our changing Sun and its effects throughout the Solar System," and (8) "Chart our destiny in the Solar System."

SSE Forum Organization

As indicated earlier, the SSE Forum is headquartered at NASA's Jet Propulsion Laboratory in Pasadena, California. Leslie Lowes and Ellis Miner co-direct the

SSE Forum, with Leslie handling the management responsibilities and Ellis handling the science responsibilities. During the months April through June 2002, Leslie was on maternity leave; Shari Asplund handled Leslie's responsibilities in the SSE Forum during that period. SSE Forum Staff members include Rosalie Betrue, Phil Davis, Eddie Gonzales, and Becky Knudsen. Other regular liaisons include: Discovery Program (Shari Asplund), Solar System Ambassadors (Kay Ferrari), Education Applications (Art Hammon), and Scientist Involvement (Rosaly Lopes).

Active SSE Missions

The SSE missions currently active include MESSENGER (Mercury orbiter), the Mars Program, Astromaterials (meteorites and Moon), Stardust (comet), Galileo (Jupiter orbiter), Europa Orbiter, Cassini-Huygens (Saturn orbiter and Titan probe), New Horizons (Pluto and KBO flybys), Genesis (Solar System origins), Voyager (Solar System boundary), and NASA's Deep Space Network.

Major Activities to Date

Specific activities of the SSE Forum include development of both internal (to the SSE Forum) and external relationships, support of on-going OSS Support Network activities, including working groups, development of science and math "quilts" (matrices) relating NASA products and national education standards, and developing methods to work more effectively with the planetary science community.

Unique National Networks

The SSE Forum works closely with a few national networks of volunteers and professional educators (both formal and informal) to enable wide dissemination of solar system products and information. These networks include the Solar System Ambassadors, the Solar System Educators, the Space Place web site and museum/library dissemination system, the Regional Planetary Imaging Facilities, and the volunteer networks of Girl Scouts and 4-H Clubs. The efforts of these groups and their leaders are greatly appreciated and valued.

Universe Education Forum: NASA – SAO Forum on the Structure and Evolution of the Universe

Roy Gould

Harvard-Smithsonian Center for Astrophysics, 60 Garden Street, MS 71, Cambridge, MA 02138, rgould@cfa.harvard.edu

The Universe Forum is a national center for teaching and learning about the scientific study of the structure and evolution of the universe (SEU). The Forum is headquartered at the Harvard-Smithsonian Center for Astrophysics in Cambridge, Massachusetts.

The Forum aims to foster partnerships that use the unique resources of NASA's space science research program to create exciting and substantive learning experiences for students, teachers and the public. Using questions about our universe, how it began, how it is evolving, and our place within it, we aim to inspire learning and promote understanding of the common principles underlying all science.

As a part of NASA's Office of Space Science (OSS) Support Network, we work in tandem with the three other Forums as well as NASA Brokers to bring scientists and educators together on projects which enhance space science learning with the latest NASA mission and research science.

The Forum maintains collaborations throughout the space science, education, and outreach communities. Examples of our partners are the Cambridge Public School System, the Girl Scouts of America, the Boston Museum of Science and centers of space science education, including Sonoma State University and the Goddard Space Flight Center. Along with our partners we develop dynamic programs in formal, informal, and community-based education environments. These include programs and resources for museums and planetaria, the development of classroom curricula and resources such as the MicroObservatory Telescope Network and activities, professional development experiences for teachers, and workshops for Girl Scout trainers.

The Universe Forum and our NASA SEU mission and research partners engage the public in the universe's deepest mysteries - such as the origin of time, the nature of black holes, the secret of dark energy, and the ecology of our unfolding universe.

The Forum is sponsored by NASA's Office of Space Science through a collaborative agreement with the Smithsonian Institution.

Visit our website http://cfa-www.harvard.edu/seuforum/ or contact seuforum@cfa.harvard.edu

The Sun-Earth Connection Education Forum

Jim R. Thieman

Goddard Space Flight Center, Code 633.2, Greenbelt, MD 20771, thieman@nssdc.gsfc.nasa.gov

Isabel Hawkins

Center for Science Education, Space Sciences Laboratory, MC 7450, University of California, Berkeley, Berkeley, CA 94720, isabelh@ssl.berkeley.edu

The Sun-Earth Connection Education Forum (SECEF), like the other three forums in the Office of Space Science Education and Public Outreach Support Network, has the responsibility of inspiring students and the public using the latest discoveries in the science of their theme area. In the Sun-Earth Connection (SEC) theme area, scientists study the Sun-Earth system, in particular: how the Sun as our nearest star works; how it varies; and how it affects the near-Earth environment as well as the other planets in the solar system. These scientists will excitedly tell you about their latest discoveries involving the ionosphere, magnetosphere, aurora, bow shock, solar wind, corona, flares, sunspots, solar storms, etc. SECEF works together with many dedicated SEC mission scientists and education professionals to spread excitement and research discoveries to a diverse national audience.

SECEF is a partnership, with teams working at both Goddard Space Flight Center and the University of California, Berkeley. Both of these locations have a large number of SEC scientists and involvement in many of the SEC missions. In addition to the SEC science community, SECEF enlists the partnership of leaders in the K-12 and informal education communities, including the Exploratorium, Lawrence Hall of Science, Maryland Science Center, and Ideum.

SECEF has taken a number of approaches to fulfill the role of promoting SEC science. In the formal classroom education area, example programs are the Student Observation Network (S.O.N.) for middle school classrooms, and the Eye on the Sky program, which integrates literacy and space science, aimed at the K-4 grade levels.

S.O.N. was created specifically as an interactive hands-on science program with activities that enable students to understand how events on the Sun can affect our environment on Earth. Through S.O.N. students observe the variability of the Sun through their own observations and by monitoring the observations made by both ground-based and space-based observatories. They also study the effects of the Sun's variability on the Earth's magnetosphere through direct observations of magnetic field variability and the generation of aurora. The S.O.N. is considered an exemplary program and recommended for wide usage by the NASA Explorer School program.

SECEF identifies specific needs of the formal education community. For example, there were very few products on SEC subjects in the K-4 grade levels.

Products like the "Auroras" book, "Our Very Own Star, the Sun" book, and a suite of fifteen integrated units from the "Eye on the Sky" program are filling this gap.

In the informal education arena, SECEF has worked with science centers, museums, planetariums, amateur astronomers, girl scouts, Solar System Ambassadors, etc., Partnerships such as these are especially useful in the context of our annual "big event" – Sun-Earth Day, sponsored by SECEF in partnership with SEC missions and scientists. Each year, SECEF uses some type of event to promote SEC science on a national level. Participating groups are given educational materials and training on SEC-focused E/PO programs conducted in their localities. In the first few years (and coming up in 2006) these events were solar eclipses. In 2004 we will focus on the transit of Venus across the Sun on June 8. For 2005, we will explore the theme of "Ancient Observatories" for Sun-Earth Day. Big events are promoted for use in formal classroom settings as well. Teachers can sign up at a SECEF website to receive a packet of educational materials relevant to the event, as well as training and support through NASA Explorers Schools and NASA Centers.

The SECEF team is led by co-Directors Isabel Hawkins and Richard Vondrak, and consists of Troy Cline, Karin Hauck, Nate James, Elaine Lewis, Carol Lunsford, Lou Mayo, Karen Meyer, Carolyn Ng, Sten Odenwald, Ruth Paglierani, Darlene Park, Lori Persichitti, Don Robinson-Boonstra, Igor Ruderman, Greg Schultz, Bryan Stephenson and Jim Thieman. The dedicated efforts of this group working together with SEC missions, scientists, educators, volunteers and many others has made the excitement of NASA Sun-Earth Connection discovery an inspiration to both the explorers of tomorrow and the citizens of today.

DePaul University Space Science Center for Education and Outreach

Carolyn Narasimhan, Jim Sweitzer, and Bernhard Beck-Winchatz

DePaul University, Space Science Center for Education and Outreach, 990 W Fullerton, Suite 4400, Chicago, IL 60614, cnarasim@depaul.edu

The DePaul Space Science Center for Education and Outreach is one of seven regional Broker/Facilitators funded by OSS to encourage and facilitate the development of partnerships between the space science and education communities. DePaul University is located in the heart of Chicago and the Space Science Center serves the states of Illinois, Indiana, Iowa, Michigan, Minnesota, Missouri, and Wisconsin.

The DePaul Broker/Facilitator team consists of Carolyn Narasimhan, a mathematician and currently the Executive Director of the Center, astrophysicists Jim Sweitzer and Bernhard Beck-Winchatz, Director and Associate Director of the Center, Tamra Gentry, Assistant Director, and Victoria Simek, Special Projects Manager.

Our general approach as Broker/Facilitators is to identify communities of potential partners for space science education initiatives, create opportunities to assess the interests and needs of the communities, and collectively develop action plans to address the needs. In the process, we recruit small groups of advisors or consultants from the communities who act as our partners in developing plans that are educationally or operationally sound while being accessible to or appropriate for the targeted community. Thus, we have established a Chicago Teachers' Advisory whose purpose is to develop ways to bring space science to the students of Chicago Public Schools. Events initiated by the Advisory have reached close to 400 teachers over the past four years and have resulted in a number of self-sustaining projects in space science.

In Wisconsin, we are facilitating the development of a different sort of network: a partnership among individuals, schools and school districts, higher education, and organizations, associations, and agencies who have some interest and stake in earth and space science education. The partnership will make use of a growing electronic network of science educators in that state and will encourage, connect, broker, and conduct earth and space science activities.

A third example is our work with an already existing informal education network, the Great Lakes Planetarium Association (GLPA). Over half of planetarium attendees in the U.S. visit modest sized planetariums like those of the GLPA. These planetariums primarily serve K-12 audiences. We have taken advantage of this special opportunity by offering small grants programs and brokering professional development in space science to GLPA members.

The DePaul Broker group took the lead in organizing and hosting the NASA OSS Education and Public Outreach Conference in June 2002 upon which these proceedings are based, and we are playing a similar role for the Workshop to Foster Broader Participation in NASA Space Science Missions and Research Programs to take place in Chicago in June 2004.

DePaul's Space Science Center sees itself as a learning organization committed to understanding, serving and adapting to the unique needs of educational communities in our region. The Midwest has traditionally been underserved by NASA's educational programs. We have now fostered partnerships between scientists and educators in all seven of our states. Yet even with this broader perspective, we remain deeply committed to serving smaller groups with special needs. A good example is the book we brokered: *Touch the Universe*, which featured Hubble images made tactile for the visually impaired. For more information about the DePaul Broker/Facilitator, visit our website at http://analyzer.depaul.edu/NASABroker/.

Mid-Atlantic Region Space Science Broker

Nitin Naik
Center for Educational Technologies, Wheeling Jesuit University, 316 Washington Av., Wheeling, WV 26003, ossbroker@cet.edu

A Regional Point of Contact

The Mid-Atlantic Region Space Science Broker (MARSSB) serves as a regional point of contact for scientists and educators seeking information or involvement in the Office of Space Science (OSS) Education and Public Outreach program. The mid-Atlantic region includes Delaware, Kentucky, Maryland, New Jersey, New York, Ohio, Pennsylvania, Virginia, West Virginia, and the District of Columbia.

Through the development of its web site, activity tracking support tools, visibility at regional conferences, and personal communications with scientists and educators throughout its region, MARSSB reaches out to those seeking information about NASA Space Science education. The MARSSB team responds to requests from research institutions, colleges and universities, museums, community organizations, and NASA centers seeking space science resources, expertise, and funding information.

The mid-Atlantic region includes diverse communities and needs from rural Appalachia to urban New York, Philadelphia, and Washington, DC. Building relationships with educational partners such as Medgar Evers College, Franklin Institute, DC-area schools, and the Appalachian Rural Systemic Initiative helps the MARSSB team match Office of Space Science events and activities with regional interests and needs. Administrative support from a web-based database helps MARSSB track individual and organization contacts such as space scientists, educators, informal science centers/programs, NASA affiliates, and commercial/technical contacts within the MARSSB region.

Science Center Shows and Exhibits

The MARSSB team works with science centers, museums, and planetariums in its region to assist informal science educators in developing and presenting education outreach programs that feature current space science events and discoveries. One of the highlights of the Mid-Atlantic team's activities in this area includes collaboration with Maryland Science Center SpaceLink Teacher Thursdays program. This is a distance learning program offered once a month that features a timely NASA mission and guest NASA-affiliated scientist who has been involved with that mission. When the featured mission is space science related, the MARSSB team facilitates offering this distance learning opportunity to sites outside the Maryland Distance Learning Network.

Targeted Outreach

Targeted outreach projects arc those that emphasize meaningful participation in OSS activities by individuals from groups that are currently under-served by NASA. Here is one example of a targeted outreach activity. During "Mars-Watch 2003," MARSSB provided broker/facilitator services to support a five-day series of Mars Watch Outreach activities presented at the Howard University Planetarium. Each night of this event featured a guest space scientist. This series of evening presentations was open to the public and was co-sponsored with the S.M.A.R.T. Inc. organization designed to promote minority participation in science and technology. Dr. George Carruthers, a space science researcher at the Naval Research Institute, directs this non-profit organization.

Educational Products

OSS mission scientists work in collaboration with professional educators to develop space science products and materials. The mid-Atlantic broker team is available to help OSS product developers get their materials evaluated and placed into the Space Science Education Resource Directory as well as other NASA-affiliated regional and national distribution systems. MARSSB promotes effective product development among OSS forums and missions by referring product developers to the Virtual Design Center developed by the Center for Educational Technologies Educational Research Team. Through its on-line workshop, product developers are guided to follow research-based pedagogical techniques and effective use of technology in their development of space science materials.

Educational Activities

The MARSSB team provides broker/facilitator services to facilitate partnerships between space scientists and educators for planning and carrying out OSS education outreach activities in the mid-Atlantic region. Our goals for the 2003 year have been to build relationships that lead to partnerships in the promotion and offering of professional development and curriculum support for teachers and scientists involved in space science outreach activities. For example, the solar system workshops offered in conjunction with Space Explorers, Inc., and the Mid-Atlantic Region Space Grant Consortium, introduced teachers to inquiry-based space science activities that could readily be implemented in their existing curriculum.

Support for Scientists Seeking Education Outreach Funding

Another important dimension to the broker/facilitator role is to assist space scientists in identifying and formulating education outreach projects and programs that are consistent with OSS guidelines. The nature of MARSSB broker support is to encourage members of the space science community to be involved in education and public outreach activities and to improve the effectiveness of scientists' participation in outreach activities. In 2003 the MARSSB team provided email, telephone, and face-to-face support to scientists in Kentucky, New York, Pennsylvania, New Jersey, Maryland, Delaware, Virginia, and Washington DC who were pursuing education outreach funding.

Evaluation

The MARSSB team participated in the OSS Support Network process of fine-tuning the scope and role of the broker/facilitator program during the 2003 year. On a regional level, MARSSB staff use qualitative evaluation techniques as well as summative evaluation survey tools to examine the quality and scale of each outreach event that it offers in the mid-Atlantic area. MARSSB works closely with the pre-service faculty and rural educators in its region to help them develop long-range plans for professional development. MARSSB also provides rural teachers and pre-service faculty opportunities to explore ways to use space science as a context to achieve goals for curriculum improvement and to promote scientific literacy among all students.

NESSIE: New England Space Science Initiative in Education

Cary Sneider, Cathy Clemens, and William Waller

Museum of Science, Science Park, Boston, MA 02114-1099, NESSIE@mos.org

The New England Space Science Initiative in Education (NESSIE) is one of seven regional broker/facilitators within the NASA/OSS Education and Public Outreach Support Network. NESSIE is charged with fostering and nurturing productive collaborations among space scientists and educators within both the formal and informal education communities. Its region of operation spans the six New England states of Connecticut, Maine, Massachusetts, New Hampshire, Rhode Island, and Vermont. NESSIE itself is a collaboration among space scientists and educators at the Boston Museum of Science, Harvard-Smithsonian Center for Astrophysics, and Tufts University. Its office is located at the Museum of Science, next to the Charles Hayden Planetarium. Electronic access to NESSIE is provided through its website http://www.mos.org/nessie, e-mail address nessie@mos.org, and telephone number (617) 589-0227.

In its second year of operation, NESSIE has matured from an infrastructure-building phase to a more programmatic phase. Our website now serves as an effective clearinghouse of space science education resources and opportunities. For example, educators, media professionals, and the general public are accessing the website's extensive listings of observatories, planetaria, and amateur astronomy clubs. We are continuing to enhance the website's capabilities by compiling information on collegiate departments that offer astronomy courses and science education programs. We have recognized a need to provide timely information on student internships, professional development workshops for space scientists and educators, and funding opportunities for scientist/educator partners.

Besides consulting with scientists who seek us out, we have begun to reach out to the scientists via presentations at the annual AAS meetings, visits to astronomy departments, and introductory workshops for the curious but uncommitted. These awareness-building activities have led to scientist/educator partnerships and fruitful educational programs. We are compiling a master list of scientists with E/PO experience and plan to award certificates of commendation to those who have been most active. We are also reaching out to educators through a mix of broad-but-shallow and narrow-but-deep engagements.

We continue to be leaders in partnering space scientists and educators by coordinating professional development workshops – wherein the scientists and educators learn together – and by facilitating subsequent collaborations in classrooms and other community venues. We are using our experience in running scientist/educator workshops to help others carry out their own workshops in space science education. We have contributed to workshops run by the NASA Aerospace Education Service Providers, New England Space Grant Consortia, Harvard-Smithsonian Science Education Department, Structure and Evolution of the Universe Educational Forum, Chandra X-ray Observatory, Wright Center for Excellence in Science Education, Maine Mathematics and Science Alliance, and WestEd – a nationwide educational research, development, and service agency.

New and emerging programs that we are brokering and facilitating include after-school programs for middle-school students in the Boston/Cambridge area, NASA-wide professional development workshops, peer-reviewed research on basic astronomy education, curricula on space science education in colleges and universities, and a major symposium on teaching introductory astronomy that will occur July 2004 at Tufts University (see http://www.astrosociety.org/events/cosmos.html).

SCORE – South Central Organization for Researchers and Educators

Stephanie Shipp

Lunar and Planetary Institute, 3600 Bay Area Boulevard, Houston, Texas 77058, Shipp@lpi.usra.edu

The South Central Organization for Researchers and Educators (SCORE) is one of seven Broker/Facilitators within NASA's Office of Space Science (OSS) Support Network. SCORE serves the six-state region of Arizona[1], Kansas, Louisiana[2], New Mexico[1], Oklahoma, and Texas. SCORE works to broker partnerships among OSS space scientists, formal and informal educators, and educational organizations and to help facilitate high-leverage education and public outreach (E/PO) activities undertaken within this community.

The recent activities of the SCORE program have focused on establishing a community of space science researchers and educators who are interested in developing partnerships. The SCORE Web site (http://www.lpi.usra.edu/score), currently in development, provides OSS news and information, opportunities, and access to resources for researchers and educators. Visitors to the site are invited to join the SCORE community and to receive the SCORE electronic newsletter. The SCORE program hosts workshops and exhibits at science educator meetings in the south central region, such as the annual conferences of the Louisiana Science Teachers Association and the Kansas Association of Teachers of Science. These meetings offer opportunities to expand the community, share space science content and resources from the Forums, connect local OSS researchers with teachers, and learn more about local needs.

SCORE also supports an experimental "traveling planetarium" program developed to establish a network of educators and researchers that partner to share space science missions and information through traveling digital theaters in portable planetarium domes. Four digital portable planetarium theaters, hosted through the Houston Museum of Natural Science, Rice University, Lodestar Astronomy Center in Albuquerque, and the Louisiana Arts and Science Center in Baton Rouge, will be available to travel throughout the south central region. These small traveling planetariums can visit communities that traditionally do not have access to planetariums and the space science experiences. Through the small dome presentations, children, educators, and the general public can explore Mars, the Moon, findings from the Hubble Space Telescope, and other topics. Show content will be enhanced by NASA and OSS resources and hands-on activities. Central to the program is the training of educators and researchers in the use of the planetarium equipment and content. Rice University and

---

[1]The Space Science Institute (SSI) also serves Arizona and New Mexico (http://www.spacescience.org/Products/Brokering/)

[2]SouthEast Regional ClearingHouse (SERCH) also serves Louisiana (http://serch.cofc.edu/serch)

the Houston Museum of Natural Science will train educators and researchers through formal classes and through informal workshops at science and education conferences. These trained facilitators will, in turn, train others. Once trained, participants can bring the traveling planetariums into their local communities.

In an effort to facilitate partnerships among researchers, educators, and educational organizations, SCORE assists interested individuals and groups in locating partners and identifying potential support structures to leverage their programs. SCORE also supports two programs to help initiate partnerships, the Educator/Researcher Workshops and the Educator/Researcher Collaborations programs. The Educator/Researcher Workshops program encourages educators to partner with OSS researchers to host local content- and resource-rich workshops for educators that are defined by local needs and interests. SCORE also offers a limited number of grants to initiate Researcher/Educator Collaborations. These collaborations are intended to result in a product or program that increases student or public understanding of space science content. The collaborations are planned to occur over approximately a year so that the researcher(s) and educator(s) have the opportunity to build a longer-term partnership. Like the educator/researcher workshops, the collaboration design and outcomes are flexible and intended to be based on local needs and resources.

Another primary goal of the SCORE program is to develop understanding and expertise in issues of translating space science content into the classroom and to the public. What are the needs? Challenges? Examples of success? SCORE and the Solar System Exploration Forum will co-host the spring 2004 workshop "Learning from the Frontier: Getting Planetary Data in the Hands of Educators," a workshop that explores the challenges of accessing electronic planetary data for educational use. Education specialists, curriculum developers, scientists, OSS Mission E/PO leads, and others from the space science and education community are invited to attend and participate. Future workshops may focus on the role of pre-service teacher programs, out-of-school-time programs, library programs, and planetarium programs in connecting the public with current space science research.

As the SCORE program matures, it will focus on developing stronger connections to state-based educational organizations across the south central region to assist in the implementation of space science programs to address educational needs. SCORE will continue discussions to understand how space science can be made accessible to pre- and in-service educators, to assess what topics can be incorporated into the curriculum, and to assist in bringing space science content and materials into these programs. Insights gained from SCORE's other activities will help build a foundation of understanding of the areas for growth and development, and for facilitating partnerships within the community of researchers, administrators, and educators.

We invite interested space science researchers and educators to join the SCORE community. More information can be found on the SCORE Web site (http://www.lpi.usra.edu/score) or by contacting the SCORE office at the Lunar and Planetary Institute in Houston, Texas at score@lpi.usra.edu.

Southeast Regional Clearinghouse (SERCH): a NASA Office of Space Science Broker/Facilitator

Cassandra Runyon

College of Charleston, Lowcountry Hall of Science and Math, 66 George Street, Charleston, SC 29424, cass@cofc.edu

SERCH works closely with 14 Space Grant consortia (AL, AR, DC, FL, GA, KY, LA, MD, MS, NC, PR, SC/VI, TN, & VA) throughout the southeastern United States. SERCH serves a population of more than 78 million, nearly one-third of which is minority. Our region also includes seven large urban centers with a population greater than 1 million.

Our staff at the College of Charleston includes:
Dr. Cassandra Runyon, Director; Kathryn Guimond, Program Manager; and Craig Anthony, Web / GIS Developer. Dr. Donald Walter and Dr. Linda Payne, our Minority Program Coordinators, are located at South Carolina State University.

Our goals are to:

- develop a network of educators and researchers interested in space science

- be an effective interface between researchers and educators in the area of space science

- be a primary information and resource clearinghouse for space science data information and educational products

- support OSS mission scientists in their educational outreach activities

- facilitate the modification of OSS materials to meet the needs of diverse educational environments

- be a leader in serving exceptional students and the general public

- enhance minority involvement across NASA OSS programs

- develop an accessible nationwide GIS (Geographic Information System) database that provides spatially related information of targeted NASA educational resources

To accomplish these goals SERCH focuses on five areas of emphasis /expertise:

- GIS Internet Mapping Solutions – an online OSS Support Network web-based Resource Locator Tool that offers database management with query tools and data that includes NASA OSS E/PO venues, partnerships, and scientists. Users can make custom maps for specific purposes. The tool provides the ability to print and save custom maps.

- SERCH Education/Outreach Initiatives – Educator training and mentoring; Mini-awards support scientist-educator partnerships; Conference Support, Girl Scouts

- Web Site [Space Science Materials and Information Resource] – http://serch.cofc.edu/serch

- Exceptional Needs [Partnership Workshops and Mini-awards; MSS Kit; Braille Map] – partnering scientists and OSS product developers with exceptional needs educators

- Minority Initiatives with our partners at SCSU – Extensive network of partnerships throughout the Southeast and across the country with Historically Black Colleges and Universities (HBCU), Hispanic Serving Institutions (HSI), Other Minority Universities and predominately-minority attended K-12 schools.

Our strategies include working WITH our partners, finding out what THEIR needs and goals are and working to address THEIR needs, understanding the resources and limitations of the project and respecting the diversity and culture of our audiences. Lastly, we work closely with both the product developers and end-users to assure that the materials and resources are effective, scientifically correct and fun to use.

The NASA Office of Space Education and Public Outreach Broker/Facilitator Program at the Space Science Institute of Boulder, CO

Cherilynn A. Morrow, James B. Harold, and Christy L. Edwards

Space Science Institute, 4750 Walnut Street, Suite 205, Boulder, CO 80301, camorrow@spacescience.org

The Space Science Institute (SSI) of Boulder, Colorado is home to one of seven regional Broker/Facilitator programs that support the education and public outreach (E/PO) efforts of NASA's Office of Space Science (OSS). The core mission of Broker/Facilitators is to cultivate opportunities and partnerships between the education and space science communities that can address important educational needs in their respective regions. Each Broker program also represents unique capabilities and expertise depending on the nature of the organizations in which they are embedded. SSI provides an especially supportive environment for the Broker/Facilitator role.

SSI is a non-profit organization founded in 1994 on the principle of integrating scientific research with E/PO in the space and earth sciences. The Institute leads the development of traveling science exhibits, innovative instructional materials, and professional development experiences for scientists in education, and for educators in science.

The SSI Broker program now serves a large swath of the US, starting in the upper midwest in North Dakota and extending to California (AZ, CA, CO, NB, ND, NM, NV, SD, UT). Four of the states in this region (AZ, CA, CO, NM) have significant populations of OSS-supported space scientists. Five states (NB, ND, NV, SD, UT) have very few space scientists and vast rural regions that include significantly underserved areas with Hispanic and Indigenous populations.

The goals of our Broker program are to provide strategically valuable support for: 1) space scientists' effective E/PO involvement, 2) formal education (emphasizing state-based agendas), 3) informal education (emphasizing planetarium associations, Girl Scouts, and traveling science exhibits), and 4) underserved populations (emphasizing Indigenous and Latino educators). This strategic support includes providing professional development opportunities; facilitating access to and use of exemplary materials; and facilitating E/PO participation and/or partnership.

To address the vastness of our region, we have developed new electronic resources (E-Brokering), that include two monthly bulletins called ROSIE (Regional Opportunities for Scientists in Education), and BESS (Bulletin for Educators in Space Science), as well as a web-based "Menu of Opportunities for Scientists in Education" (MOSIE).

For more information, please visit us at: http://ssibroker.colorado.edu/broker/.

Space Science Network Northwest

Julie Lutz
*University of Washington, Box 351310, Seattle, WA 98195-1310,
nasaerc@u.washington.edu*

Space Science Network Northwest (S2N2) is one of seven NASA Office of Space Science (OSS) Education Support Network Broker/Facilitators (B/F). The states that are covered by S2N2 are Washington, Oregon, Idaho, Montana, Wyoming, Alaska and Hawai'i. Headquarters for the organization is the Department of Earth and Space Sciences at the University of Washington.

The goal of S2N2 is to increase awareness of and participation in OSS missions and Education/Public Outreach (E/PO) programs by connecting various constituencies (formal and informal education, space scientists, general public, clubs, service organizations, etc.) and providing opportunities for them to work together. S2N2 will also be a clearinghouse for OSS mission information and related education products throughout the northwest. The OSS is the branch of NASA that is responsible for solar system missions, astrobiology and space science research satellites such as the Hubble Space Telescope and the Chandra X-ray Observatory.

The S2N2 approach is to have a strong B/F presence in each of the partner states. The Space Grant Consortium program in each of these states is host to S2N2 operations and activities. Thus it is possible to match up the needs and constraints in each state with the opportunities available from OSS E/PO.

The types of services S2N2 provides are quite diverse. Space scientists interested in E/PO can contact S2N2 to learn about opportunities and possible partners. Educators can use S2N2 to make contacts and learn about possible sources of funds for curriculum development and reform, special events, and professional development opportunities in space science. Museums, science centers, libraries, planetariums, clubs, community groups and other organizations that desire to disseminate the excitement of space science to a wide audience can come to S2N2 for help in defining and implementing projects and finding resources to carry them out.

For further information about S2N2, see the Web site (www.s2n2.org) or contact Dr. Julie Lutz, Space Science Network Northwest, University of Washington, Box 351310, Seattle, WA 98195-1310; Phone: (206) 616-1084; E-mail: nasaerc@u.washington.edu

Appendix C –
Excerpts from:

Implementing the Office of Space Science Education/Public Outreach Strategy: A Critical Evaluation at the Six-Year Mark

A Report by the Space Science Advisory Committee Education and Public Outreach Task Force

Full Document can be found at:
http://spacescience.nasa.gov/education/resources/evaluation/OSS_EPO_Task_Force_Report.pdf

March 21, 2003

Executive Summary

Over the past 6 years the Office of Space Science (OSS) has used the Plan developed by the 1996 Space Science Advisory Committee (SScAC) Education/Public Outreach (E/PO) Task Force to implement a bold and innovative approach to the planning and execution of its education and public outreach program. The program now underway has enabled the science and education communities to work together to bring scientific research, discoveries in space science, and educational products and activities associated with OSS missions and research programs to students and the public alike. The OSS E/PO program is based on goals most recently expressed in *The Space Science Enterprise Strategic Plan 2000* and has a clear approach to implementation as set forth in the 1996 Report *Implementing the Office of Space Science Education/Public Outreach Strategy*. The Program operates on the premise that achieving genuine success in affecting the quality of science, technology, engineering, and mathematics education in America will not be won through short-term activities with immediate results, but rather through a long-term commitment requiring a sustained effort in education and public outreach.

To assess progress at this stage of implementing the OSS E/PO program, identify issues that need to be addressed in the next stage, and examine possible new directions for program evolution, an E/PO Task Force was set up under the SScAC in 2001. The Task Force has found that the program's numerous accomplishments and successes to date include:

- Direct engagement of OSS missions and the space science research community in education and in contributing to the public understanding of science;

- A rich harvest of educational programs and materials directed towards many types of audiences in diverse communities across the country;

- Significant steps towards involving minorities in the mainstream of OSS's scientific, technical, and educational programs and in developing educational materials directed towards audiences that have not previously been served by NASA; and

- Substantial leveraging of resources through collaboration with hundreds of educational institutions and organizations across the country.

It is clear that significant progress has been made to date. The ensemble of activities that are now being undertaken are, by design, highly leveraged and broadly distributed throughout the nation. The approach that OSS has taken in implementing its E/PO Program provides a model that is unique to NASA, to the government, and to science education in general. The program is already a credit both to the OSS and to all of the very talented people who have been involved in its planning and execution.

To build on this progress, the Task Force has singled out a number of areas for particular attention, areas which the Task Force believes will yield especially rich rewards in taking the OSS E/PO program to even higher levels of maturity, effectiveness and accomplishment. Findings and recommendations have been grouped into eight broad subject areas.

- Make educational products more accessible and organize them in a more coherent way;

- Increase the inclusiveness of the program by involving new audiences, science topics, materials and partnerships;

- Expand and intensify pioneering efforts to attract and better integrate minorities into E/PO projects and into the mainstream of OSS science programs;

- Enhance efforts directed towards quality control and obtaining a better understanding of program impact;

- Increase the effectiveness of the OSS E/PO Support Network by focusing the activities of the Broker/Facilitators on their primary roles;

- Strengthen and expand professional development efforts for E/PO professionals, scientists, and the education community;

- Enhance internal and external communications; and

- Identify and acquire critical resources required for long-term sustainability.

The OSS E/PO program, with its established productive partnership between the space science and education communities and its use of a national network to identify and sustain high-leverage opportunities, has made remarkable progress in a relatively short period of time. Based on its successes to date and the prospects for even greater successes in the future, the approach that has been taken could well serve as a guide for future NASA educational efforts across the Agency.

Recommendations: Taking the OSS E/PO Program to the Next Level of Accomplishment

To build on the substantial progress that has been made over the past six years, the Task Force has singled out a number of areas for particular attention, areas which the Task Force believes will yield particularly rich rewards in taking the OSS E/PO Program to even higher levels of maturity, effectiveness and accomplishment. Some of these recommendations can be responded to straightforwardly by sharpening the focus of or reorienting efforts that are already underway. Implementing other recommendations will necessarily lead to the establishment of major new activities requiring careful planning and new resources.

1. Make educational products more accessible and organize them in a more coherent way

    Many educators aren't certain of how space science topics can be used in their classrooms. To reach the next level of effectiveness, the OSS Education Program should create a Space Science Education Framework to develop a bridge between the science and mathematics of OSS missions and research and the needs of the educational system. This Framework should be aligned with the National Science, Mathematics and Technology Standards. It should provide an appropriate, standards-aligned, sequencing of space science topics throughout the K-12 years, and give overall direction and context for the materials being produced by each mission and research project. The existence of such a Framework will make it easier for educators to use the space science materials in their classrooms. The development of the Framework should be a cooperative effort between NASA and leaders of educational organizations like the American Association for the Advancement of Science, National Science Teachers Association, Lawrence Hall of Science (LHS), and TERC. Involvement of such organizations is crucial if the Framework is to have credibility in the education community.

The Framework would provide recommendations for a sequence of content that would help students meet appropriate standards, such as the NRC content standards, that specify what students should have achieved by the end of 4th grade, or 8th grade or 12th grade. OSS could then encourage the development of curriculum sequences and materials that address concepts and topics in the NASA Space Science Framework using coherent and high quality approaches. It could also foster the development of assessment instruments correlated to the concepts in the Framework to enable K-12 teachers to accurately measure student understanding of key space science concepts leading to the increased rigor and effectiveness of space science education nationwide.

Good programs that have already been developed and fit within the Framework should be sustained. New missions and research projects would then develop additional E/PO programs and materials that fit within this Framework. When new programs are proposed they should identify the science, mathematics and technology education gaps in the Framework and address these in ways appropriate to the mission or research project. This will allow OSS missions and research to avoid duplicating efforts and to focus on developing fewer, better products. Adopting this approach will require greater coordination throughout the Support Network. If the products are logical in flow and connectivity, they will be easier for teachers, students and the general public to locate and use.

The OSS E/PO program must also take actions to make educational materials more accessible. The NASA Space Science Education Resource Directory (SSERD) is a good first step in providing accessibility to some of the products of the E/PO program. It has not yet gone far enough, however, in organizing and evaluating the materials it includes; and the Framework can provide a useful context for this. The Directory could then be expanded to include exemplary materials inspired by NASA science, but developed by other reliable organizations, that also align with the Framework. While having a web-based database of materials is useful, it does not address the wider needs and the larger issues of distribution, particularly for under-served and non-technology oriented groups. Other means of cataloging and distribution should be pursued to better serve these groups.

2. Increase the inclusiveness of the program through reaching out to, involving, and incorporating new audiences, science topics, materials and partnerships

The E/PO program has done a good job in forming partnerships outside of NASA, particularly the involvement of scientists and some scientific and educational institutions in E/PO work. Nevertheless, additional efforts should be made to include other important organizations and audience groups. A more explicit collaboration with the National Science Foundation's educational programs should be established. The OSS E/PO program can by necessity be only a small, but important, part of the nation's educational structure, but to further leverage its impact it should find more ways to partner with educational associations such as NSTA, ASP, NEA, and AFT.

Programs incorporating community colleges, colleges of education and undergraduate students are needed and could produce substantive results. Stronger and continuing partnerships could be developed with the informal education community, especially planetaria and museums, through a peer-reviewed grant program to support key projects aligned with OSS E/PO goals. Initial efforts to work with community groups could be further expanded to include additional amateur astronomers, scouts and youth groups, civic organizations, and especially minority religious and community groups. Partnerships with associations, businesses, and distributors could be pursued to help reproduce and distribute educational materials.

3. Expand and intensify pioneering efforts to attract and better integrate minorities into E/PO projects and into the mainstream of OSS science programs

OSS E/PO has already taken a number of significant steps towards increasing minority engagement in NASA space science research and education programs. The Space Science Minority University Initiative and OSS's work with minority scientific professional societies have broken new ground for NASA and have helped to establish new relationships between OSS and the minority research and education communities. Such activities should continue and be expanded.

The demographics of America are changing in a dramatic way. Currently, about a quarter of the U.S. population consists of minority group members, but during this century, the minority will become the majority. The E/PO participant community (and the research community as well) should reflect these changing national demographics. One of the explicit goals of the OSS E/PO Program is to help create the technical workforce of the future. This goal cannot be achieved without actively reaching out to minorities. The fact that, currently, 55% of the graduate students in physics and 60% in engineering are non-Americans raises a number of important issues. The future technological workforce needs of America cannot be met solely through the non-minority portion of the American population. It is vital that minorities become more involved in science and engineering. So the fundamental unanswered question is not whether the OSS E/PO Program should strive for more inclusion of minorities, rather it is: how is this inclusion is to be achieved.

One part of the answer to this question is to make more effective use of the Support Network in working with minorities and minority institutions. Another is to encourage mission E/PO leads to include minorities and minority institutions in the planning of their programs from the beginning. Both groups require help in order to be effective in acting in this role. Our discussions with Forum staff, Broker/Facilitators and E/PO leads revealed a lack of familiarity with minority communities. All of them require additional training on strategies for approaching minority communities and on how to best plan and implement exemplary programs involving minorities.

One fundamental principal of the OSS Education program is the direct involvement of space scientists in E/PO. The implementation of this principle has been responsible for much of the resounding success of the program. However, this principle may need to be considered and applied more broadly as there are too few minority space scientists available to work with OSS to have an impact on a national scale. There are, for example, only about 25 African-American astronomers and astrophysicists in the United States. This means that the OSS E/PO program needs to broaden its sphere of influence to achieve its objectives vis-à-vis minorities. A good first step in this direction was the Workshop, held at Western Kentucky University in May 2001, which involved the Broker/Facilitators and Forums and the leadership of ten minority professional scientific organizations. The dialogue was valuable, new connections were made, and, subsequently, one of these organizations (the National Organization of Black Chemists and Chemical Engineers) became a major participant in a mission E/PO Program. OSS had not previously thought about working with chemists and chemical engineers. Efforts like this need to be continued and expanded. The Task Force also recommends that such efforts be expanded in the future to include mission E/PO leads.

Because of the long history of exclusion of minorities from science (for instance between 1920 and 1962 only 17 African-Americans obtained a Ph.D. in physics or astronomy) role models and intense interaction of minority scientists with their communities is required to attract more minorities into technical fields. The models for such interaction have yet to be worked out, and the effort required to develop viable models and foster interactions between minority scientists and their communities needs to be increased. The infusion of minority cultural contributions to space science missions in OSS science/space missions will be important.

4. Enhance efforts directed towards quality control and obtaining a better understanding of program impact

    As emphasized in this report, the Task Force commends the OSS E/PO for notable successes in engaging astronomers, space scientists, and their professional communities in effective education and public outreach activities. These successes have been documented in the OSS E/PO annual reports in the form of (1) narratives describing individual activities and (2) data showing the numbers of several types of activities carried out each year. During the initial, formative years of the E/PO program, these types of performance measures were understandably the most readily available means of tracking progress. While these descriptions convey the excitement, breadth, and continued potential of the program, more complete measures of the actual impact of the program are needed.

As mentioned earlier in this report, the quality control and evaluation efforts that are built into the program's design have provided useful feedback that is being used to continually improve operations. As the program matures, further assessment of quality and effectiveness of E/PO activities and products is desirable. The Task Force readily acknowledges that indicators of quality and effectiveness are difficult to specify, let alone evaluate; but we nonetheless urge OSS to give them specific attention during the coming years. We recommend increasing efforts to enhance evaluation by continuing to use the standards of good educational practice already employed and by engaging those who know education well. To this end, we suggest

- Longitudinal studies to follow cohorts of students to determine significant influences on their college choices and career paths. Such studies are routinely undertaken by the National Center for Education Statistics (NCES), the Higher Education Research Institute (HERI), and the National Science Foundation (NSF). Examples include the NCES National Education Longitudinal Study (NELS) of students who are tracked from eighth grade through enrollment in undergraduate science and engineering programs; the HERI Freshman Norms Survey of intended majors by first-year students in four year colleges and universities; and the NSF's Scientists and Engineers Statistical Data System (SESTAT) that collects information on the employment, education, and demographic characteristics of individuals with science and engineering degrees in the U.S. The Task Force acknowledges that the complexity and expense of longitudinal studies preclude OSS E/PO from undertaking their own study, but OSS E/PO could explore the possibility of joining ongoing longitudinal studies by adding to them questions related specifically to the role of space science E/PO activities.
- Annual peer-reviews and selection of outstanding individuals for their contribution to E/PO efforts could provide models to help others evaluate their own efforts, and could go far in encouraging effective participation and dynamic innovation in the program.
- Thorough and critical reviews of products for quality and eventually, for alignment to the recommended Space Science Education Framework. OSS could also present awards for meritorious E/PO products. The nomination process might require documentation of the effect/impact of the product, thus lessening the burden on OSS to document success and encouraging the community itself to design and implement impact measures.

5. Increase the effectiveness of the OSS E/PO Support Network by focusing the activities of the Broker/Facilitators on their primary roles

As noted earlier, the Support Network has played a large role in the overall success of the OSS E/PO Program. The Support Network consists of four Education Forums aligned with the OSS themes (Sun-Earth Connection, Solar System Exploration, the Astronomical Search for Origins, and the Structure and Evolution of the Universe) and seven Broker/Facilitators distributed regionally throughout the United States. Our discussions with members of the Support Network suggested that the Forums had a very clear idea of what their role is in the OSS Program and are now operating effectively. However, the role of the Broker/Facilitators is far more ambiguous (even as seen by the participants), and their performance (and effectiveness) has been far more uneven. Core roles for the Broker/Facilitators were clearly spelled out in the 2001 Broker/Facilitator Cooperative Agreement Notice (CAN) and the Task Force believes these core roles are appropriate. The Broker/Facilitators should spend more of their time on brokering as defined in the CAN (searching out good opportunities and connecting space scientists and educators) and less on developing and implementing their own programs.

Broker/Facilitators (and Forum leaders) could also be more proactive in working with research and mission proposers to ensure that there is greater cooperation among scientists and educators in planning and preparing E/PO Proposals right from the beginning. Specific suggestions are given in recommendation 6 for raising awareness and skill levels of Support Network members in educational expertise - a crucial step if the Support Network is going to be more effective in the future. Specific steps are also given in section 3 for raising the skill levels of the Support Network members concerning diversity. Finally, a clear set of metrics needs to be developed for the Broker/Facilitators and Forums against which they can judge their success. While some indicators are already in place, the time has now come for a more rigorous approach to judging the effectiveness of the Support Network.

6. Strengthen and expand professional development efforts for E/PO professionals, scientists, and the education community

OSS has created a new profession – space science education and public outreach specialists – and it is now necessary to establish the standards for the profession and to put into place professional development programs. E/PO specialists include the Broker/Facilitators, Forum Directors, mission E/PO leads and others engaged in the E/PO program. The institutions that manage space science missions and conduct space science research programs employ many of these professionals. These specialists plan and implement the E/PO efforts of the missions and programs. Some are scientists or technologists who have adopted E/PO as a new career, while others have come from the educational community. All need to have a better understanding of the practical aspects of developing E/PO activities and products that are based on the best knowledge and practices of the professional education community.

The creation of the refereed journal the *Astronomy Education Review*, which is endorsed by both the AAS and the ASP and has received start-up funding from NASA, is one step in the direction of developing an archival record of best practices. Professional development opportunities in the form of workshops, short courses, and special sessions at scientific meetings are also needed for current members and new entrants into space science education. The third *Lesley University Report* (available on-line through the OSS Education homepage) notes that many in the E/PO community are up to the limits of their knowledge of educational practices. There is certainly a lot of talent, enthusiasm and creativity available, but there is a requirement to deepen the OSS E/PO community's understanding of the needs of educational community with special attention paid to best practices for creating products relevant to the age and learning needs of students.

Professional development is also needed for members of the education community, particularly classroom teachers and developers of curricular materials. Approaches to providing sustained help for educators must be found, not just a single workshop or conversation. As an example, programs that engage teachers in space science research projects over several summers would go a long way towards this goal.

7. Enhance internal and external communications

In a system as diverse and geographically distributed as the OSS E/PO effort has become, clear and efficient communication is essential. We note and applaud steps already taken to maintain communication within the system, and between elements of the system and external communities of space scientists and educators. These include the OSS Education Council, monthly OSS E/PO newsletters, and presentations at professional meetings of astronomers and educators. Given the rapid increase in the size and complexity of the system, however, we find that additional efforts to enhance communications are advisable.

Communications within OSS at Headquarters must be improved. For example OSS E/PO managers are not always informed of successful research grants to which E/PO components may be added, nor are the science program officers within OSS always informed about E/PO awards that have been made. This situation may be a result of too many demands on the time of existing staff, but it is still a problem that must be solved. We have also noted barriers to communication between OSS education programs and other education efforts underway at NASA. We recognize that the new organizational arrangements for education within NASA may help alleviate this situation. However, such coordination efforts will require additional staff time. These gaps in communication both within the Support Network and within NASA Headquarters can, and should, be fairly easily closed.

Another substantial need is for a clear, consistent and sustained dialogue with three external communities: space scientists and astronomers, educators and their professional societies, and the emerging E/PO professional community. We urge increased attention to communication with these three groups, especially with professional societies like the AAS, AGU and NSTA.

Clearer and more efficient communication is also crucial for fostering better cooperation between scientists and members of the formal and informal education communities when they join together to prepare E/PO proposals. The Task Force notes that more effective communication at earlier stages in the preparation of proposals is likely to produce better proposals and more substantive programs.

8. Identify and acquire critical resources required for long-term sustainability

The OSS E/PO Program has made remarkable progress in a relatively short period of time in large part because of the unstinting efforts of a small number of people at NASA Headquarters, and because of the willingness of OSS management to devote substantial financial resources to E/PO. Missions and research programs have allocated 1 to 2 percent of their baseline budgets for E/PO, and OSS has also provided additional resources for the Support Network and for funding a small number of individual E/PO efforts such as the competitively selected IDEAS Program. Such actions are clear indicators of OSS's genuine commitment to E/PO. However, the sheer scale of the program and the level of activity now underway have reached the point where a number of issues must be faced if the program is to be sustained over the long term.

The size of the OSS E/PO staff is presently inadequate to support a program of this scope, while at the same time substantial new demands are being put on the current staff. There simply aren't enough people to do the job well! The scale of the OSS E/PO Program can be expected to continue to grow substantially along with the growth in the overall OSS Program. Large new scientific and technology programs within OSS such as Living with a Star, Beyond Einstein, and Project Prometheus offer exciting prospects for major new E/PO efforts. To realize their potential, such efforts will require strong leadership and careful planning necessitating a major commitment of staff time. Interactions with the new NASA Office of Education are also likely to have a substantial impact on the activities of the current staff, as will the planning of new activities recommended by the Task Force, such as developing the Space Science Education Framework and planning and implementing a comprehensive Professional Development Program.

Taking the OSS E/PO Program to the next level of accomplishment will also require additional resources in a few carefully selected areas. New funding will be needed to develop the Framework, implement the recommended Professional Development Program, and undertake new activities in a few key areas such as programs focused on preservice teachers, expanding successful activities targeted towards minorities, and enhancing efforts to evaluate programs and assess their impact.

OSS must now proceed to analyze its real staffing needs, work to obtain new positions where needed, assess the resources required to carry out critical new activities, and identify how such resources can be acquired. Requirements are likely to be modest, but meeting those requirements will be critical if the OSS E/PO Program is to be sustained and enhanced.

Author Index

Adams, M., 235, 237
Alexander, C. J., 178
Allen, J., 175
Andres, P., 370
Arvidson, R. E., 239
Asplund, S. E., 241, 443
Atkins, R., 111
Austin, S., 312

Babin, E., 120
Baguio, M., 244
Barber, J., 95
Barstow, D., 107
Batt, C. A., 333
Beaman, B., 254
Beck-Winchatz, B., xviii, 231, 301, 448
Behne, J., 377
Bennett, M., 77, 246, 286
Bergman, J. J., 178
Betrue, R., 249
Bishop, J. E., 196
Bobrowsky, M., 200
Bothun, G., 114
Bowman, C. D., 239, 252
Boyer, F., 252
Bradbury, H., 281
Brettman, O. H., 254
Brickhouse, N. W., 384
Briggs, M., 111
Brinza, D. E., 333
Brunsell, E., 115
Bryson, L., 408
Burton, K., 292

Callahan, C., 284
Carruthers, G. R., 187, 257
Chippindale, S., 286, 289
Christensen, P. R., 323
Clemens, C., 453
Cominsky, L. R., 260

Cooley, J., 111
Cooper, L., 423
Craig, N., 262, 263
Croft, S. K., 264
Crowe, R., 102, 267

Dagher, Z. R., 384
Davis, D. L., 270
Davis, L., 85
Deardorff, C. R., 178
DeVore, E. K., 77
Dingus, B., 111
Dobinson, E., 337
Doxas, I., 272
Dusenbery, P. B., 275, 278

Edwards, C. L., 459
Eisenhamer, B., 281, 352, 388, 400
Eisenhamer, J., 352
Ellis, L., 284

Feldman, J. E., 333
Feldman, S., 364
Fischer, M. W., 244
Fortson, L. F., 221
Fowler, W. T., 244
Fraknoi, A., 214, 286, 289, 292, 410
Friedman, L., 333
Fuerstenau, S. D., 333

Gallagher, D. L., 235, 237
Gardiner, L., 178
Garvin-Doxas, K., 295
Garza, O. G., 175
Gelderman, R., 298
Genyuk, J., 178
Goldstein, J., 200
Gould, R., 445
Grice, N., 185, 301
Griffin, I., 441
Guzik, T. G., 120

Gyulai, C., 333

Halzen, F., 111
Hammon, A., 305, 326
Haro, L., 307
Harold, J. B., 370, 459
Hawkins, I., xviii, 147, 263, 446
Hecht, M. H., 333
Hemenway, M. K., 202, 309
Henderson, S., 178
Hering, J, 252
Herrick, R. R., 130
Hoette, V., 109
Howson, E., 289
Hughes, D., 337
Hundt, L., 307

Jaffe, D. T., 309
Jew, G., 348
Johnson, L. P., 312
Johnson, R. M., 178

Kawakami, A. J., 102, 267
Keller, J., 315
Kelly, L., 333
Kiefer, W. S., 130, 318
Klein, M. J., 321
Klug, S. L., 323
Knappenberger, P., 139
Knudsen, R., 326, 329
Koczor, R., 235
Kuhlman, R. R., 333

Laatsch, S., 174
Lacy, J. H., 309
LaGrave, M., 178
Larson, M. B., 262
León, M. J., 252
Lee, S. A., 44
Letts, W. J., 384
Leung, K., 318
Lewis, C., 333
Liggett, P., 337
Limaye, S. S., 341
Lindstrom, M., 175
Livengood, T., 200
Lochner, J. C., 345
Lopez Freeman, M. A., 17
Lopez, R. E., 207
Lowes, L. L., 348, 443

Lutz, J., 460

Möller, L. E., 333
Madsen, J., 111
Mahootian, F., 87
Martin, D., 337
Martin, M., 337
Martinez, C., 284
Mastie, D., 178
Mayo, L., 278
McCallister, D., 352, 388
McGruder, C. H., 220
McNeil, R., 120
Meloy, T. P., 333
Mendez, F. J., 194
Millar, S., 111
Millar, T., 111
Miller, J. D., 26
Miller-Bagwell, A., 262, 263
Miner, E. D., 355, 443
Moroney, L., 160
Morris, P. A., 175
Morrow, C. A., 275, 459
Morse, D., 292

Naik, N., 450
Narasimhan, U., xvi, xviii, 440
Nelson, G. D., 55

O'Guinn, C., 361
O'Leary, J. P., 194
Obot, V. D., 175
Offerdahl, E. G., 359
Oslick, J., 333

Paglierani, R., 364
Pertzborn, R. A., 341, 423
Plait, P., 260
Polk, K., 333
Pollock, G., 133
Pompea, S. M., 405–407
Porro, I. L., 365
Powell, G., 333
Prather, E. E., 125, 315, 359, 368, 386
Preston, S., 202

Raddick, M. J., 394
Randall, C., 370
Range, S. K., 372

Ratcliffe, M., 182
Rawlins, K., 111
Reiff, P., 155, 374
Rest, C., 400, 441
Richter, M. J., 309
Riddle, B., 200
Ristvey, J., 377
Roller, J. P., 321
Rosendhal, J. D., 3, 415, 423
Runyon, C., xviii, 231, 457
Russell, R., 178

Sadler, P. M., 35
Sakimoto, P. J., 380, 423
Salgado, J. F., 191
Schatz, D., 289
Semper, R., 163
Sherman, D. M., 239
Sherman, J., 333
Shipman, H. L., 88, 384, 385
Shipp, S., 455
Slater, T. F., 125, 315, 359, 368, 386
Smith, D. A., 352, 388, 400, 441
Smith, M., 284
Smith, S., 200
Sneider, C., 453
Sohus, A. M., 390
Sparks, R., 394
Squyres, S. W., 239
Steele, D., 111
Stevenoski, S., 111
Stoughton, C., 394
Sumners, C., 155, 374
Sweitzer, J., 165, 448
Sword, B., 337

Takamura, E., 187
Taylor, W. W. L., 397
Teays, T. J., 73, 388, 400
Thieman, J. R., 397, 446
Thomas, F. J., 402
Thomas, V. L., 187, 257
Thompson, P. B., 130
Towner, M. C., 333
Treiman, A. H., 130
Trowbridge, K., 333
Tuthill, G. T., 386

Valderrama, P., 323

Waldron, A. M., 333
Walker, C. E., 405–407
Waller, W., 453
Watt, K., 323
Weiler, E., 5
Wells, E. L., 408
Wentworth, B., 301
Wessen, A. S., 390
Whitt, A., 237
Winchatz, M. R., 301
Wolff, S., 410
Wooten, J., 175
Wyatt, R., 169

York, D. G., 94

Zirbel, E., 312

Acronym Index

4-H: Head, Heart, Hands and Health, 147, 249

AAAS: American Association for the Advancement of Science, 35, 55, 88, 95, 207, 326
AACTE: American Association of Colleges of Teacher Education, 44
AAPT: American Association of Physics Teachers, 207
AAS: American Astronomical Society, 88, 301, 355, 453, 463
AAVSO: American Association of Variable Star Observers, 246
ACE: Advanced Composition Explorer, 178, 278
ACTS: Advanced Communication Technology Satellite, 77
AEL: Aeronautic Education Laboratory, 312
AER: Astronomy Education Review, 410
AES: Advanced Education Services, 133
AFT: American Federation of Teachers, 463
AGI: American Geological Institute, 284
AGN: Active Galactic Nuclei, 221, 315
AGU: American Geophysical Union, 207, 463
AIM: Aeronomy of Ice in the Mesosphere, 178
AMANDA: Antarctic Muon and Neutrino Detection Array, 111
AMNH: American Museum of Natural History, 169

ASCD: Association for Supervision and Curriculum Development, 44
ASL: American Sign Language, 185
ASO: Astronomical Search for Origins, 441
ASP: Astronomical Society of the Pacific, 214, 246, 286, 463
AST: Amateur Space Telescope, 254
ASTC: Association of Science-Technology Centers, 139, 275, 278
ASU: Arizona State University, 249, 323
ATIC: Advanced Thin Ionization Calorimeter, 120
ATLAS: ATmospheric Laboratory for Applications and Science, 397
AURA: Associated Universities for Research in Astronomy, 182

BESS: Bulletin for Educators in Space Science, 459
BOCES: Board of Cooperative Educational Services, 196
BRAS: Baton Rouge Astronomical Society, 120
BREC: Baton Rouge Recreation Commission, 120
BSCS: Biological Sciences Curriculum Study, 17

CAN: Cooperative Agreement Notice, 463
CAPER: Conceptual Astronomy and Physics Education Research team (Univ. of Arizona), 125, 359
CAPS: Committee for Action Program Services, 85

CASDE: Consortium for the Application of Spacedata for Digital Earth, 337
Cassini-JMOC: Cassini-Jupiter Microwave Observing Campaign, 321
CCD: Charge Coupled Device, 109, 120, 241, 309, 312, 315, 341
CERES: Center for Educational Resources Project (Montana State Univ.), 368, 386
CERN: Conseil European pour la Recherché Nucleaire (European Laboratory for Particle Physics), 163
CfA: Center for Astrophysics, 35
CfAO: Center for Adaptive Optics, 109
CFHT: Canada France Hawaii Telescope, 408
CHIPS: Cosmic Hot Interstellar Plasma Spectrometer, 260
CISM: Center for Integrated Space Weather Modeling, 207
CME: Coronal Mass Ejection, 278
CONTOUR: Comet Nucleus Tour, 115, 241
CORE: Central Operation of Resources for Educators, 345
COSPAR: Committee on Space Research, 207
CPS: Chicago Public Schools, 94
CSR: Center for Space Research, 365
CSSS: Council of State Science Supervisors, 133
CUIP: Chicago Public Schools / University of Chicago Internet Project, 94
CUNY: City University of New York, 312
CWD: Chicago Web Docent, 94

DCPS: District of Columbia Public Schools, 257
DEA: US Drug Enforcement Administration, 85
DISD: Dallas Independent School District, 85

DLR: Deutsches Zentrum für Luft- und Raumfahrt (German Aerospace Center), 77
DLT: Digital Library Technology, 374
DPI: Wisconsin Department of Public Instruction, 44
DPS: Division for Planetary Sciences of the American Astronomical Society, 355
DSN: Deep Space Network, 115, 321

EA: Education Ambassador (GLAST mission), 260
EDCATS: Education Division Computer-Aided Tracking System, 423
ELL: English Language Learners, 17
ERIC: Educational Resources Information Center, 35, 125
ESIP: Federation of Earth Science Information Partners, 374
ESW: Earth Science Week, 284
EXES: Echelon Cross Echelle Spectrograph, 309

FIDO: Field Integrated Design and Operations Rover, 239
FIRST: For Inspiration and Recognition of Science and Technology, 252
FIRST: Foundations in Reading Through Science and Technology, 364
FOSS: Full Option Science System, 44, 88
FOSTER: Flight Opportunities for Science Teacher EnRichment (KAO), 77

GAVRT: Goldstone Apple-Valley Radio Telescope, 321
GEMS: Great Explorations in Math and Science, 95, 423, 441
GIS: Geographic Information System, 402, 457
GISS: NASA Goddard Institute for Space Studies, 312
GLAST: Gamma Ray Large Area Space Telescope, 260

GLOBE: Global Learning and Observations to Benefit the Environment, 341
GLPA: Great Lakes Planetarium Association, 196, 448
GP-B: Gravity Probe B, 260, 372
GSFC: NASA Goddard Space Flight Center, 187, 278, 312, 326, 397
GSUSA: Girl Scouts of the USA, 249
GTN: GLAST Telescope Network, 260

H-R diagram: Hertzsprung-Russell diagram, 394
HAWC: High-resolution Airborne Widebandwidth Camera (SOFIA), 109
HBCU: Historically Black Colleges and Universities, 380, 457
HERI: Higher Education Research Institute, 463
HESSI: High Energy Solar Spectroscopic Imager, 262, 278, 408
HETE-2: The High Energy Transient Explorer, 365
HISD: Houston Independent School District, 155
HMNS: Houston Museum of Natural Science, 175
HOU: Hands-On Universe, 109, 298
HRPO: Highland Road Park Observatory (Baton Rouge, LA), 120
HSGS: High School Ground Station, 397
HSI: Hispanic-Serving Institutions, 380, 457
HST: Hubble Space Telescope, 5, 165, 301, 312, 352, 388

IAU: International Astronomical Union, 214
IDEAS: Initiative to Develop Education through Astronomy and Space Science, 77, 196, 257, 281
IDL: Interactive Data Language, 312

IITA: NASA Information Infrastructure Technology and Applications, 374
IKI: Space Research Institute, Moscow, 397
IMAGE: Imager for Magnetopause-to-Aurora Global Exploration, 278, 374
INSAP: Inspirations From Astronomical Phenomena, 139
INSPIRE: Interactive NASA Space Physics Ionosphere Radio Experiments, 397
IPS: International Planetarium Society, 139, 147, 182, 196, 260, 348
IRAF: Image Reduction and Analysis Facility, 312
IRTF: NASA Infrared Telescope Facility, 309
ISBE: Illinois State Board of Education, 109
ISS-AT: International Space Station Amateur Telescope, 254
ISS: International Space Station, 194, 254
ISTE: International Society for Technology in Education, 377
ISTP: International Solar-Terrestrial Physics Science Initiative, 178, 278

JPL: Jet Propulsion Laboratory, 5, 88, 102, 115, 178, 182, 239, 249, 267, 275, 312, 321, 323, 326, 333, 337, 355, 390
JSC: Johnson Space Center, 175, 249

KAO: Kuiper Airborne Observatory, 77
KBO: Kuiper Belt Object, 443
KIPR: KISS Institute for Practical Robotics, 252
KTI: Kindergarten Through Infinity Systemic Initiative, 111

LaSIP: Louisiana Systemic Initiatives Program, 120
LCER: Lewis Center for Educational Research, 321

Acronym Index

LCET: Louisiana Center Education Technology, 120
LEP: Limited English Proficiency, 17
LHEA: Lab for High Energy Astrophysics, 345
LHS: Lawrence Hall of Science, 95, 207, 348, 441, 463
LPI: Lunar and Planetary Institute, 85, 130, 318, 455
LSU: Louisiana State University, 120

MARSSB: Mid-Atlantic Region Space Science Broker, 450
McREL: Mid-continent Research for Education and Learning, 377
MECA: Mars Environmental Compatibility Assessment, 333
MESSENGER: Mercury Surface, Space Environment, Geochemistry, and Ranging, 241, 443
MFACM: Mexican Fine Arts Center Museum (Chicago), 191
MGS: Mars Global Surveyor, 402
MHD: Magneto-Hydrodynamics, 207
MIT: Massachusetts Institute of Technology, 365
MMSD: Madison Metropolitan School District, 341
MOC: Mars Orbital Camera, 402
MOLA: Mars Orbital Laser Altimeter, 402
MOSIE: Menu of Opportunities for Scientists In Education, 355, 459
MPI: Minority-Predominant Institution, 380
MPSP: Museum Partners Science Program (Chicago), 109
MSERC: Mathematics and Science Education Resources Center (Univ. of Delaware), 88
MSET: Math Science Engineering and Technology, 312
MSFC: NASA Marshall Space Flight Center, 235, 237, 397
MSIP: Mars Student Imaging Project, 323
MSS: Multi-Sensory Space Science, 423, 457

MSSE: Montana State University Master of Science – Science Education, 368, 386
MU-SPIN: Minority University-Space Interdisciplinary Network, 187, 257, 312
MUCERPI: NASA Minority University and College Education and Research Partnership Initiative, 380
MUI: Minority University Initiative, 207, 423

NAEP: National Assessment of Educational Progress, 17, 95, 107
NAI: National Association for Interpretation, 390
NCES: National Center for Education Statistics, 120, 463
NCLB: No Child Left Behind, 95
NCRR: National Center for Research Resources, 307
NCTM: National Council of Teachers of Mathematics, 17, 326, 377
NEA: National Education Association, 463
NEAR: Near-Earth Asteroid Rendezvous, 241, 257
NELS: National Education Longitudinal Study, 463
NESSIE: New England Space Science Initiative in Education, 286, 453
NEW: NASA Educational Workshops, 115
NEWEST: NASA Educational Workshops for Elementary School Teachers, 77
NEWMAST: NASA Educational Workshops for Mathematics, Science and Technology Teachers, 77
NFSI: Near and Far Sciences for Illinois, 109
NIH: National Institutes of Health, 17, 307

NLIST: Networking for Leadership, Inquiry and Systemic Thinking, 133
NOAA: National Oceanic and Atmospheric Administration, 374
NOAO: National Optical Astronomy Observatory, 264, 405, 406
NOBCChE: National Organization for the Professional Advancement of Black Chemists and Chemical Engineers, 85, 270, 380, 441
NOMISS: New Opportunities through Minority Initiatives in Space Science, 73, 102, 267
NPS: National Park Service, 284, 390
NRA: NASA Research Announcement, 147, 355
NRC: National Research Council, 17, 35, 44, 88, 95, 207, 315, 368, 463
NRL: US Naval Research Laboratory, 257
NRTS: Network Resources & Training Sites (MU-SPIN), 187
NSBP: National Society of Black Physicists, 187, 380
NSCAT: NASA Scatterometer, 329
NSES: National Science Education Standards, 55, 221, 326, 377, 386
NSF: National Science Foundation, 17, 109, 111, 147, 178, 207, 214, 221, 246, 275, 289, 463
NSIP: NASA Student Involvement Program, 87
NSSDC: National Space Science Data Center, 312
NSTA: National Science Teachers Association, 44, 55, 178, 326, 348, 441, 463
NSWP: National Space Weather Program, 278

OPO: Office of Public Outreach (STScI), 352
OSSE: Office of Space Science Education (Univ. of Wisconsin), 341

OSTP: Office of Science Technology Policy, 5

PACE: Physics and Chemistry Experiments in Space, 312
PDS: Planetary Data System, 337
PEDR: Precision Experiment Data Record (MOLA), 402
PERG: Program Evaluation Research Group (Lesley University), 260, 423
PIAFATI: Planetary Imaging and Analysis Facility and Advanced Training Institute, 323
PKAL: Project Kaleidoscope, 348
PLATO: Physics Learning and Astronomy Training Outreach, 120
PLATO: Planetarium Learning and Teaching Opportunities, 196
POLAR: Polar Plasma Laboratory, 178
Project ARIES: Astronomy Resources for Intercurricular Elementary Science, 35
Project DESIGNS: Doable Engineering Science Investigations Geared for Non-science Students, 35
Project LINK: Live and Interactive Network of Knowledge (Exploratorium), 77
Project STAR: Support and Training for Assessing Results, 35

RAC: Robotic Arm Camera, 333
RBSE: Research-Based Science Education, 405
REP: Robotics Education Project, 252
RESNA: Rehabilitation Engineering & Assistive Technology Society of North America, 301
ROBIE: Robots for Internet Experiences, 120
ROSIE: Regional Opportunities for Scientists in Education, 459
RPIF: Regional Planetary Image Facility, 337

RXTE: Rossi X-Ray Timing Explorer, 365
RYSS: Raul Yzaguirre School for Success, 175

S.M.A.R.T.: Science Mathematics Aerospace Research and Technology, 187, 257, 450
S.O.N.: Student Observation Network, 446
S2N2: Space Science Network Northwest, 460
SACNAS: Society for the Advancement of Chicanos and Native Americans in Science, 207, 307
SBN: Small Bodies Node (PDS), 337
SCORE: South Central Organization of Researchers and Educators (LPI), 455
SCSU: South Carolina State University, 312, 457
SCT: Schmidt-Cassegrain telescope, 254
SDP: Small Digital Planetarium, 165, 169
SDSS: Sloan Digital Sky Survey, 394
SEAP: Science and Engineering Apprentice Program (Department of Defense), 257
SEC: Sun-Earth Connection, 95, 187, 278, 423, 446
SECEF: Sun-Earth Connection Education Forum, 95, 187, 278, 423, 446
SEGway: Science Education Gateway, 262, 263
SEI: Space Education Initiatives, 115
SEMAA: Science Engineering and Mathematics Aeronautics Academy, 312
SEPA: Science Education Partnership Award, 307
SEPAC: Space Experiments with Particle ACcelerators, 397
SERCH: Southeast Regional Clearinghouse, 139, 147, 257, 423, 455, 457

SESTAT: Scientists and Engineers Statistical Data System, 463
SETI: Search for Extraterrestrial Intelligence, 286, 292
SEU: Structure & Evolution of the Universe, 260, 345, 445
SEUEF: Structure and Evolution of the Universe Education Forum, 260
SFUSD: San Francisco Unified School District, 263
SIGCSE: ACM Special Interest Group on Computer Science Education, 295
SIRTF: Space InfraRed Telescope Facility, 5, 77, 315
SLAC: Stanford Linear Accelerator Center, 260
SNG: http://science.nasa.gov/, 235
SNOOPY: Student Nanoexperiments for Outreach and Observational Planetary Inquiry, 333
SOFIA: Stratospheric Observatory for Infrared Astronomy, 5, 77, 109, 286, 309, 315
SOHO: The Solar and Heliospheric Observatory, 278
SPARC: Space Physics and Aeronomy Research Collaboratory, 178
SQUID: Superconducting QUantum Interference Device, 372
SSA: Solar System Ambassadors, 249
SScAC: NASA's Space Science Advisory Committee, 139, 423, 463
SSE: Solar System Exploration, 249, 348, 355, 443
SSEP: Solar System Educators Program, 115, 244
SSERD: Space Science Education Resource Directory, 73, 95, 400, 441, 463
SSI: Space Science Institute, 275, 278, 455, 459
SSIT: Space Science for Illinois Teachers, 109
SSU: Sonoma State University, 260

STARBASE: Students Training for Achievement in Research Based on Analytical Space-Science Experiences, 298
STEREO: Solar Terrestrial Relations Observatory, 262
STScI: Space Telescope Science Institute, 281, 301, 352, 441
STSDAS: Space Telescope Science Data Analysis System, 312
SWICS: Solar Wind Ion Composition Spectrometer (Ulysses), 178

TAEVIS: Tactile Access to Education for Visually Impaired Students, 301
TCU: Tribal Colleges and Universities, 380
TEA: Teachers Experiencing the Antarctic and Arctic program, 111
TERC: Now used as a the proper name; formerly referred to as Technical Education Research Center, 87, 107, 109, 275, 301, 365, 370, 463
TEXES: Texas Echelon Cross Echelle Spectrograph, 309
THEMIS: Thermal Emission Imaging System, 323
TIGR: Hawai'i Business Magazine's Targeted Industries Growth Report, 267
TIMSS: Third International Mathematics and Science Study, 95, 107
TLC: Technology Learning Center (S.M.A.R.T., 187
TLRBSE: Teacher Leaders in Research-Based Science Education, 405
TOPEX: TOPography EXperiment for Ocean Circulation, 329
TPS: The Planetary Society, 333
TRA: Teacher Resource Agent (HOU), 109
TSGC: Texas Space Grant Consortium, 244
TSU: Texas Southern University, 175

UHD: University of Houston Downtown, 175
UHH: University of Hawai'i at Hilo, 267
USGS: United States Geological Survey, 284, 337

VERITAS: Very Energetic Radiation Imaging Telescope Array System, 221
VLF: Very Low Frequency, 237, 397
VNR: Video News Release, 191
VPI: Visionary Products, Inc., 333

WIND: Interplanetary Physics Laboratory, 178

YMCA: Young Men's Christian Association, 257

THE FOLLOWING IS A LISTING OF THE VOLUMES

Published
by

ASTRONOMICAL SOCIETY OF THE PACIFIC
(ASP)

An international, nonprofit, scientific and educational organization
founded in 1889

All book orders or inquiries concerning

ASTRONOMICAL SOCIETY OF THE PACIFIC
CONFERENCE SERIES
(ASP - CS)

and

INTERNATIONAL ASTRONOMICAL UNION VOLUMES
(IAU)

should be directed to the:

Astronomical Society of the Pacific Conference Series
390 Ashton Avenue
San Francisco   CA   94112-1722   USA

Phone:	800-335-2624	(within USA)
Phone:	415-337-2126	
Fax:	415-337-5205	

E-mail: service@astrosociety.org
Web Site: http://www.astrosociety.org

Complete lists of proceedings of past IAU Meetings are maintained at the IAU Web site at the URL: http://www.iau.org/publicat.html

Volumes 32 - 189 in the IAU Symposia Series may be ordered from:

Kluwer Academic Publishers
P. O. Box 117
NL 3300 AA Dordrecht
The Netherlands

Kluwer@wKap.com

ASP CONFERENCE SERIES VOLUMES
Published by the Astronomical Society of the Pacific

PUBLISHED: 1988 (* asterisk means OUT OF PRINT)

Vol. CS-1 PROGRESS AND OPPORTUNITIES IN SOUTHERN HEMISPHERE
OPTICAL ASTRONOMY: CTIO 25TH Anniversary Symposium
eds. V. M. Blanco and M. M. Phillips
ISBN 0-937707-18-X

Vol. CS-2 PROCEEDINGS OF A WORKSHOP ON OPTICAL SURVEYS FOR QUASARS
eds. Patrick S. Osmer, Alain C. Porter, Richard F. Green, and Craig B. Foltz
ISBN 0-937707-19-8

Vol. CS-3 FIBER OPTICS IN ASTRONOMY
ed. Samuel C. Barden
ISBN 0-937707-20-1

Vol. CS-4 THE EXTRAGALACTIC DISTANCE SCALE:
Proceedings of the ASP 100th Anniversary Symposium
eds. Sidney van den Bergh and Christopher J. Pritchet
ISBN 0-937707-21-X

Vol. CS-5 THE MINNESOTA LECTURES ON CLUSTERS OF GALAXIES AND LARGE-SCALE STRUCTURE
ed. John M. Dickey
ISBN 0-937707-22-8

PUBLISHED: 1989

Vol. CS-6 * SYNTHESIS IMAGING IN RADIO ASTRONOMY: A Collection of Lectures from the Third NRAO Synthesis Imaging Summer School
eds. Richard A. Perley, Frederic R. Schwab, and Alan H. Bridle
ISBN 0-937707-23-6

PUBLISHED: 1990

Vol. CS-7 PROPERTIES OF HOT LUMINOUS STARS: Boulder-Munich Workshop
ed. Catharine D. Garmany
ISBN 0-937707-24-4

Vol. CS-8 * CCDs IN ASTRONOMY
ed. George H. Jacoby
ISBN 0-937707-25-2

Vol. CS-9 COOL STARS, STELLAR SYSTEMS, AND THE SUN:
Sixth Cambridge Workshop
ed. George Wallerstein
ISBN 0-937707-27-9

Vol. CS-10 EVOLUTION OF THE UNIVERSE OF GALAXIES:
Edwin Hubble Centennial Symposium
ed. Richard G. Kron
ISBN 0-937707-28-7

Vol. CS-11 CONFRONTATION BETWEEN STELLAR PULSATION AND EVOLUTION
eds. Carla Cacciari and Gisella Clementini
ISBN 0-937707-30-9

ASP CONFERENCE SERIES VOLUMES
Published by the Astronomical Society of the Pacific

PUBLISHED: 1991 (* asterisk means OUT OF PRINT)

Vol. CS-12   THE EVOLUTION OF THE INTERSTELLAR MEDIUM
ed. Leo Blitz
ISBN 0-937707-31-7

Vol. CS-13   THE FORMATION AND EVOLUTION OF STAR CLUSTERS
ed. Kenneth Janes
ISBN 0-937707-32-5

Vol. CS-14   ASTROPHYSICS WITH INFRARED ARRAYS
ed. Richard Elston
ISBN 0-937707-33-3

Vol. CS-15   LARGE-SCALE STRUCTURES AND PECULIAR MOTIONS IN THE UNIVERSE
eds. David W. Latham and L. A. Nicolaci da Costa
ISBN 0-937707-34-1

Vol. CS-16   Proceedings of the 3rd Haystack Observatory Conference on ATOMS, IONS, AND MOLECULES: NEW RESULTS IN SPECTRAL LINE ASTROPHYSICS
eds. Aubrey D. Haschick and Paul T. P. Ho
ISBN 0-937707-35-X

Vol. CS-17   LIGHT POLLUTION, RADIO INTERFERENCE, AND SPACE DEBRIS
ed. David L. Crawford
ISBN 0-937707-36-8

Vol. CS-18   THE INTERPRETATION OF MODERN SYNTHESIS OBSERVATIONS OF SPIRAL GALAXIES
eds. Nebojsa Duric and Patrick C. Crane
ISBN 0-937707-37-6

Vol. CS-19   RADIO INTERFEROMETRY: THEORY, TECHNIQUES, AND APPLICATIONS,
IAU Colloquium 131
eds. T. J. Cornwell and R. A. Perley
ISBN 0-937707-38-4

Vol. CS-20   FRONTIERS OF STELLAR EVOLUTION:
50th Anniversary McDonald Observatory (1939-1989)
ed. David L. Lambert
ISBN 0-937707-39-2

Vol. CS-21   THE SPACE DISTRIBUTION OF QUASARS
ed. David Crampton
ISBN 0-937707-40-6

PUBLISHED: 1992

Vol. CS-22   NONISOTROPIC AND VARIABLE OUTFLOWS FROM STARS
eds. Laurent Drissen, Claus Leitherer, and Antonella Nota
ISBN 0-937707-41-4

Vol CS-23 * ASTRONOMICAL CCD OBSERVING AND REDUCTION TECHNIQUES
ed. Steve B. Howell
ISBN 0-937707-42-4

Vol. CS-24   COSMOLOGY AND LARGE-SCALE STRUCTURE IN THE UNIVERSE
ed. Reinaldo R. de Carvalho
ISBN 0-937707-43-0

Vol. CS-25   ASTRONOMICAL DATA ANALYSIS, SOFTWARE AND SYSTEMS (ADASS) I
eds. Diana M. Worrall, Chris Biemesderfer, and Jeannette Barnes
ISBN 0-937707-44-9

ASP CONFERENCE SERIES VOLUMES
Published by the Astronomical Society of the Pacific

PUBLISHED: 1992 (asterisk (*) means OUT OF PRINT)

Vol. CS-26   COOL STARS, STELLAR SYSTEMS, AND THE SUN:
Seventh Cambridge Workshop
eds. Mark S. Giampapa and Jay A. Bookbinder
ISBN 0-937707-45-7

Vol. CS-27   THE SOLAR CYCLE: Proceedings of the National Solar
Observatory/Sacramento Peak 12th Summer Workshop
ed. Karen L. Harvey
ISBN 0-937707-46-5

Vol. CS-28   AUTOMATED TELESCOPES FOR PHOTOMETRY AND IMAGING
eds. Saul J. Adelman, Robert J. Dukes, Jr., and Carol J. Adelman
ISBN 0-937707-47-3

Vol. CS-29   Viña del Mar Workshop on CATACLYSMIC VARIABLE STARS
ed. Nikolaus Vogt
ISBN 0-937707-48-1

Vol. CS-30   VARIABLE STARS AND GALAXIES
ed. Brian Warner
ISBN 0-937707-49-X

Vol. CS-31   RELATIONSHIPS BETWEEN ACTIVE GALACTIC NUCLEI AND STARBURST
GALAXIES
ed. Alexei V. Filippenko
ISBN 0-937707-50-3

Vol. CS-32   COMPLEMENTARY APPROACHES TO DOUBLE AND MULTIPLE STAR
RESEARCH, IAU Colloquium 135
eds. Harold A. McAlister and William I. Hartkopf
ISBN 0-937707-51-1

Vol. CS-33   RESEARCH AMATEUR ASTRONOMY
ed. Stephen J. Edberg
ISBN 0-937707-52-X

Vol. CS-34   ROBOTIC TELESCOPES IN THE 1990's
ed. Alexei V. Filippenko
ISBN 0-937707-53-8

PUBLISHED: 1993

Vol. CS-35 *   MASSIVE STARS: THEIR LIVES IN THE INTERSTELLAR MEDIUM
eds. Joseph P. Cassinelli and Edward B. Churchwell
ISBN 0-937707-54-6

Vol. CS-36   PLANETS AROUND PULSARS
ed. J. A. Phillips, S. E. Thorsett, and S. R. Kulkarni
ISBN 0-937707-55-4

Vol. CS-37   FIBER OPTICS IN ASTRONOMY II
ed. Peter M. Gray
ISBN 0-937707-56-2

Vol. CS-38   NEW FRONTIERS IN BINARY STAR RESEARCH: Pacific Rim Colloquium
eds. K. C. Leung and I.-S. Nha
ISBN 0-937707-57-0

ASP CONFERENCE SERIES VOLUMES
Published by the Astronomical Society of the Pacific

PUBLISHED: 1993 (* asterisk means OUT OF PRINT)

Vol. CS-39 THE MINNESOTA LECTURES ON THE STRUCTURE AND DYNAMICS OF THE MILKY WAY
ed. Roberta M. Humphreys
ISBN 0-937707-58-9

Vol. CS-40 INSIDE THE STARS, IAU Colloquium 137
eds. Werner W. Weiss and Annie Baglin
ISBN 0-937707-59-7

Vol. CS-41 ASTRONOMICAL INFRARED SPECTROSCOPY: FUTURE OBSERVATIONAL DIRECTIONS
ed. Sun Kwok
ISBN 0-937707-60-0

Vol. CS-42 GONG 1992: SEISMIC INVESTIGATION OF THE SUN AND STARS
ed. Timothy M. Brown
ISBN 0-937707-61-9

Vol. CS-43 SKY SURVEYS: PROTOSTARS TO PROTOGALAXIES
ed. B. T. Soifer
ISBN 0-937707-62-7

Vol. CS-44 PECULIAR VERSUS NORMAL PHENOMENA IN A-TYPE AND RELATED STARS, IAU Colloquium 138
eds. M. M. Dworetsky, F. Castelli, and R. Faraggiana
ISBN 0-937707-63-5

Vol. CS-45 LUMINOUS HIGH-LATITUDE STARS
ed. Dimitar D. Sasselov
ISBN 0-937707-64-3

Vol. CS-46 THE MAGNETIC AND VELOCITY FIELDS OF SOLAR ACTIVE REGIONS, IAU Colloquium 141
eds. Harold Zirin, Guoxiang Ai, and Haimin Wang
ISBN 0-937707-65-1

Vol. CS-47 THIRD DECENNIAL US-USSR CONFERENCE ON SETI --
Santa Cruz, California, USA
ed. G. Seth Shostak
ISBN 0-937707-66-X

Vol. CS-48 THE GLOBULAR CLUSTER-GALAXY CONNECTION
eds. Graeme H. Smith and Jean P. Brodie
ISBN 0-937707-67-8

Vol. CS-49 GALAXY EVOLUTION: THE MILKY WAY PERSPECTIVE
ed. Steven R. Majewski
ISBN 0-937707-68-6

Vol. CS-50 STRUCTURE AND DYNAMICS OF GLOBULAR CLUSTERS
eds. S. G. Djorgovski and G. Meylan
ISBN 0-937707-69-4

Vol. CS-51 OBSERVATIONAL COSMOLOGY
eds. Guido Chincarini, Angela Iovino, Tommaso Maccacaro, and Dario Maccagni
ISBN 0-937707-70-8

ASP CONFERENCE SERIES VOLUMES
Published by the Astronomical Society of the Pacific

PUBLISHED: 1994 (* asterisk means OUT OF PRINT)

Vol. CS-52   ASTRONOMICAL DATA ANALYSIS SOFTWARE AND SYSTEMS (ADASS) II
eds. R. J. Hanisch, R. J. V. Brissenden, and Jeannette Barnes
ISBN 0-937707-71-6

Vol. CS-53   BLUE STRAGGLERS
ed. Rex A. Saffer
ISBN 0-937707-72-4

Vol. CS-54   THE FIRST STROMLO SYMPOSIUM: THE PHYSICS OF ACTIVE GALAXIES
eds. Geoffrey V. Bicknell, Michael A. Dopita, and Peter J. Quinn
ISBN 0-937707-73-2

Vol. CS-55   OPTICAL ASTRONOMY FROM THE EARTH AND MOON
eds. Diane M. Pyper and Ronald J. Angione
ISBN 0-937707-74-0

Vol. CS-56 *   INTERACTING BINARY STARS
ed. Allen W. Shafter
ISBN 0-937707-75-9

Vol. CS-57   STELLAR AND CIRCUMSTELLAR ASTROPHYSICS
eds. George Wallerstein and Alberto Noriega-Crespo
ISBN 0-937707-76-7

Vol. CS-58   THE FIRST SYMPOSIUM ON THE INFRARED CIRRUS AND DIFFUSE INTERSTELLAR CLOUDS
eds. Roc M. Cutri and William B. Latter
ISBN 0-937707-77-5

Vol. CS-59   ASTRONOMY WITH MILLIMETER AND SUBMILLIMETER WAVE INTERFEROMETRY, IAU Colloquium 140
eds. M. Ishiguro and Wm. J. Welch
ISBN 0-937707-78-3

Vol. CS-60   THE MK PROCESS AT 50 YEARS: A POWERFUL TOOL FOR ASTRO-PHYSICAL INSIGHT, A Workshop of the Vatican Observatory -- Tucson, Arizona, USA
eds. C. J. Corbally, R. O. Gray, and R. F. Garrison
ISBN 0-937707-79-1

Vol. CS-61   ASTRONOMICAL DATA ANALYSIS SOFTWARE AND SYSTEMS (ADASS) III
eds. Dennis R. Crabtree, R. J. Hanisch, and Jeannette Barnes
ISBN 0-937707-80-5

Vol. CS-62   THE NATURE AND EVOLUTIONARY STATUS OF HERBIG Ae/Be STARS
eds. Pik Sin Thé, Mario R. Pérez, and Ed P. J. van den Heuvel
ISBN 0-9837707-81-3

Vol. CS-63   SEVENTY-FIVE YEARS OF HIRAYAMA ASTEROID FAMILIES: THE ROLE OF COLLISIONS IN THE SOLAR SYSTEM HISTORY
eds. Yoshihide Kozai, Richard P. Binzel, and Tomohiro Hirayama
ISBN 0-937707-82-1

Vol. CS-64 *   COOL STARS, STELLAR SYSTEMS, AND THE SUN:
Eighth Cambridge Workshop
ed. Jean-Pierre Caillault
ISBN 0-937707-83-X

ASP CONFERENCE SERIES VOLUMES
Published by the Astronomical Society of the Pacific

PUBLISHED: 1994 (* asterisk means OUT OF PRINT)

Vol. CS-65 *   CLOUDS, CORES, AND LOW MASS STARS:
                The Fourth Haystack Observatory Conference
                eds. Dan P. Clemens and Richard Barvainis
                ISBN 0-937707-84-8

Vol. CS-66 *   PHYSICS OF THE GASEOUS AND STELLAR DISKS OF THE GALAXY
                ed. Ivan R. King
                ISBN 0-937707-85-6

Vol. CS-67   UNVEILING LARGE-SCALE STRUCTURES BEHIND THE MILKY WAY
                eds. C. Balkowski and R. C. Kraan-Korteweg
                ISBN 0-937707-86-4

Vol. CS-68   SOLAR ACTIVE REGION EVOLUTION: COMPARING MODELS WITH
                OBSERVATIONS
                eds. K. S. Balasubramaniam and George W. Simon
                ISBN 0-937707-87-2

Vol. CS-69   REVERBERATION MAPPING OF THE BROAD-LINE REGION IN ACTIVE
                GALACTIC NUCLEI
                eds. P. M. Gondhalekar, K. Horne, and B. M. Peterson
                ISBN 0-937707-88-0

Vol. CS-70 *   GROUPS OF GALAXIES
                eds. Otto-G. Richter and Kirk Borne
                ISBN 0-937707-89-9

PUBLISHED: 1995

Vol. CS-71   TRIDIMENSIONAL OPTICAL SPECTROSCOPIC METHODS IN
                ASTROPHYSICS, IAU Colloquium 149
                eds. Georges Comte and Michel Marcelin
                ISBN 0-937707-90-2

Vol. CS-72   MILLISECOND PULSARS: A DECADE OF SURPRISE
                eds. A. S Fruchter, M. Tavani, and D. C. Backer
                ISBN 0-937707-91-0

Vol. CS-73   AIRBORNE ASTRONOMY SYMPOSIUM ON THE GALACTIC ECOSYSTEM:
                FROM GAS TO STARS TO DUST
                eds. Michael R. Haas, Jacqueline A. Davidson, and Edwin F. Erickson
                ISBN 0-937707-92-9

Vol. CS-74   PROGRESS IN THE SEARCH FOR EXTRATERRESTRIAL LIFE:
                1993 Bioastronomy Symposium
                ed. G. Seth Shostak
                ISBN 0-937707-93-7

Vol. CS-75   MULTI-FEED SYSTEMS FOR RADIO TELESCOPES
                eds. Darrel T. Emerson and John M. Payne
                ISBN 0-937707-94-5

Vol. CS-76   GONG '94: HELIO- AND ASTERO-SEISMOLOGY FROM THE EARTH AND
                SPACE
                eds. Roger K. Ulrich, Edward J. Rhodes, Jr., and Werner Däppen
                ISBN 0-937707-95-3

ASP CONFERENCE SERIES VOLUMES
Published by the Astronomical Society of the Pacific

PUBLISHED: 1995 (* asterisk means OUT OF PRINT)

Vol. CS-77   ASTRONOMICAL DATA ANALYSIS SOFTWARE AND SYSTEMS (ADASS) IV
eds. R. A. Shaw, H. E. Payne, and J. J. E. Hayes
ISBN 0-937707-96-1

Vol. CS-78   ASTROPHYSICAL APPLICATIONS OF POWERFUL NEW DATABASES:
Joint Discussion No. 16 of the 22nd General Assembly of the IAU
eds. S. J. Adelman and W. L. Wiese
ISBN 0-937707-97-X

Vol. CS-79   ROBOTIC TELESCOPES: CURRENT CAPABILITIES, PRESENT DEVELOPMENTS, AND FUTURE PROSPECTS FOR AUTOMATED ASTRONOMY
eds. Gregory W. Henry and Joel A. Eaton
ISBN 0-937707-98-8

Vol. CS-80 *   THE PHYSICS OF THE INTERSTELLAR MEDIUM AND INTERGALACTIC MEDIUM
eds. A. Ferrara, C. F. McKee, C. Heiles, and P. R. Shapiro
ISBN 0-937707-99-6

Vol. CS-81   LABORATORY AND ASTRONOMICAL HIGH RESOLUTION SPECTRA
eds. A. J. Sauval, R. Blomme, and N. Grevesse
ISBN 1-886733-01-5

Vol. CS-82 *   VERY LONG BASELINE INTERFEROMETRY AND THE VLBA
eds. J. A. Zensus, P. J. Diamond, and P. J. Napier
ISBN 1-886733-02-3

Vol. CS-83 *   ASTROPHYSICAL APPLICATIONS OF STELLAR PULSATION,
IAU Colloquium 155
eds. R. S. Stobie and P. A. Whitelock
ISBN 1-886733-03-1

ATLAS   INFRARED ATLAS OF THE ARCTURUS SPECTRUM, 0.9–5.3 μm
eds. Kenneth Hinkle, Lloyd Wallace, and William Livingston
ISBN: 1-886733-04-X

Vol. CS-84   THE FUTURE UTILIZATION OF SCHMIDT TELESCOPES, IAU Colloquium 148
eds. Jessica Chapman, Russell Cannon, Sandra Harrison, and Bambang Hidayat
ISBN 1-886733-05-8

Vol. CS-85   CAPE WORKSHOP ON MAGNETIC CATACLYSMIC VARIABLES
eds. D. A. H. Buckley and B. Warner
ISBN 1-886733-06-6

Vol. CS-86   FRESH VIEWS OF ELLIPTICAL GALAXIES
eds. Alberto Buzzoni, Alvio Renzini, and Alfonso Serrano
ISBN 1-886733-07-4

PUBLISHED: 1996

Vol. CS-87   NEW OBSERVING MODES FOR THE NEXT CENTURY
eds. Todd Boroson, John Davies, and Ian Robson
ISBN 1-886733-08-2

ASP CONFERENCE SERIES VOLUMES
Published by the Astronomical Society of the Pacific

PUBLISHED: 1996 (* asterisk means OUT OF PRINT)

Vol. CS-88 * CLUSTERS, LENSING, AND THE FUTURE OF THE UNIVERSE
eds. Virginia Trimble and Andreas Reisenegger
ISBN 1-886733-09-0

Vol. CS-89 ASTRONOMY EDUCATION: CURRENT DEVELOPMENTS, FUTURE COORDINATION
ed. John R. Percy
ISBN 1-886733-10-4

Vol. CS-90 THE ORIGINS, EVOLUTION, AND DESTINIES OF BINARY STARS IN CLUSTERS
eds. E. F. Milone and J.-C. Mermilliod
ISBN 1-886733-11-2

Vol. CS-91 BARRED GALAXIES, IAU Colloquium 157
eds. R. Buta, D. A. Crocker, and B. G. Elmegreen
ISBN 1-886733-12-0

Vol. CS-92 * FORMATION OF THE GALACTIC HALO .... INSIDE AND OUT
eds. Heather L. Morrison and Ata Sarajedini
ISBN 1-886733-13-9

Vol. CS-93 RADIO EMISSION FROM THE STARS AND THE SUN
eds. A. R. Taylor and J. M. Paredes
ISBN 1-886733-14-7

Vol. CS-94 MAPPING, MEASURING, AND MODELING THE UNIVERSE
eds. Peter Coles, Vicent J. Martinez, and Maria-Jesus Pons-Borderia
ISBN 1-886733-15-5

Vol. CS-95 SOLAR DRIVERS OF INTERPLANETARY AND TERRESTRIAL DISTURBANCES: Proceedings of 16th International Workshop National Solar Observatory/Sacramento Peak
eds. K. S. Balasubramaniam, Stephen L. Keil, and Raymond N. Smartt
ISBN 1-886733-16-3

Vol. CS-96 HYDROGEN-DEFICIENT STARS
eds. C. S. Jeffery and U. Heber
ISBN 1-886733-17-1

Vol. CS-97 POLARIMETRY OF THE INTERSTELLAR MEDIUM
eds. W. G. Roberge and D. C. B. Whittet
ISBN 1-886733-18-X

Vol. CS-98 FROM STARS TO GALAXIES: THE IMPACT OF STELLAR PHYSICS ON GALAXY EVOLUTION
eds. Claus Leitherer, Uta Fritze-von Alvensleben, and John Huchra
ISBN 1-886733-19-8

Vol. CS-99 COSMIC ABUNDANCES:
Proceedings of the 6th Annual October Astrophysics Conference
eds. Stephen S. Holt and George Sonneborn
ISBN 1-886733-20-1

Vol. CS-100 ENERGY TRANSPORT IN RADIO GALAXIES AND QUASARS
eds. P. E. Hardee, A. H. Bridle, and J. A. Zensus
ISBN 1-886733-21-X

ASP CONFERENCE SERIES VOLUMES
Published by the Astronomical Society of the Pacific

PUBLISHED: 1996 (* asterisk means OUT OF PRINT)

Vol. CS-101 ASTRONOMICAL DATA ANALYSIS SOFTWARE AND SYSTEMS (ADASS) V
eds. George H. Jacoby and Jeannette Barnes
ISBN 1080-7926

Vol. CS-102 THE GALACTIC CENTER, 4th ESO/CTIO Workshop
ed. Roland Gredel
ISBN 1-886733-22-8

Vol. CS-103 THE PHYSICS OF LINERS IN VIEW OF RECENT OBSERVATIONS
eds. M. Eracleous, A. Koratkar, C. Leitherer, and L. Ho
ISBN 1-886733-23-6

Vol. CS-104* PHYSICS, CHEMISTRY, AND DYNAMICS OF INTERPLANETARY DUST,
IAU Colloquium 150
eds. Bo Å. S. Gustafson and Martha S. Hanner
ISBN 1-886733-24-4

Vol. CS-105 PULSARS: PROBLEMS AND PROGRESS, IAU Colloquium 160
ed. S. Johnston, M. A. Walker, and M. Bailes
ISBN 1-886733-25-2

Vol. CS-106 THE MINNESOTA LECTURES ON EXTRAGALACTIC NEUTRAL HYDROGEN
ed. Evan D. Skillman
ISBN 1-886733-26-0

Vol. CS-107 COMPLETING THE INVENTORY OF THE SOLAR SYSTEM:
A Symposium held in conjunction with the 106th Annual Meeting of the ASP
eds. Terrence W. Rettig and Joseph M. Hahn
ISBN 1-886733-27-9

Vol. CS-108 M.A.S.S. -- MODEL ATMOSPHERES AND SPECTRUM SYNTHESIS:
5th Vienna - Workshop
eds. Saul J. Adelman, Friedrich Kupka, and Werner W. Weiss
ISBN 1-886733-28-7

Vol. CS-109 COOL STARS, STELLAR SYSTEMS, AND THE SUN:
Ninth Cambridge Workshop
eds. Roberto Pallavicini and Andrea K. Dupree
ISBN 1-886733-29-5

Vol. CS-110 BLAZAR CONTINUUM VARIABILITY
eds. H. R. Miller, J. R. Webb, and J. C. Noble
ISBN 1-886733-30-9

Vol. CS-111 MAGNETIC RECONNECTION IN THE SOLAR ATMOSPHERE:
Proceedings of a Yohkoh Conference
eds. R. D. Bentley and J. T. Mariska
ISBN 1-886733-31-7

Vol. CS-112 THE HISTORY OF THE MILKY WAY AND ITS SATELLITE SYSTEM
eds. Andreas Burkert, Dieter H. Hartmann, and Steven R. Majewski
ISBN 1-886733-32-5

ASP CONFERENCE SERIES VOLUMES

Published by the Astronomical Society of the Pacific

PUBLISHED: 1997 (* asterisk means OUT OF PRINT)

Vol. CS-113 EMISSION LINES IN ACTIVE GALAXIES: NEW METHODS AND
TECHNIQUES, IAU Colloquium 159
eds. B. M. Peterson, F.-Z. Cheng, and A. S. Wilson
ISBN 1-886733-33-3

Vol. CS-114 YOUNG GALAXIES AND QSO ABSORPTION-LINE SYSTEMS
eds. Sueli M. Viegas, Ruth Gruenwald, and Reinaldo R. de Carvalho
ISBN 1-886733-34-1

Vol. CS-115 GALACTIC CLUSTER COOLING FLOWS
ed. Noam Soker
ISBN 1-886733-35-X

Vol. CS-116 THE SECOND STROMLO SYMPOSIUM: THE NATURE OF ELLIPTICAL
GALAXIES
eds. M. Arnaboldi, G. S. Da Costa, and P. Saha
ISBN 1-886733-36-8

Vol. CS-117 DARK AND VISIBLE MATTER IN GALAXIES
eds. Massimo Persic and Paolo Salucci
ISBN-1-886733-37-6

Vol. CS-118 FIRST ADVANCES IN SOLAR PHYSICS EUROCONFERENCE:
ADVANCES IN THE PHYSICS OF SUNSPOTS
eds. B. Schmieder. J. C. del Toro Iniesta, and M. Vázquez
ISBN 1-886733-38-4

Vol. CS-119 PLANETS BEYOND THE SOLAR SYSTEM AND THE NEXT GENERATION OF
SPACE MISSIONS
ed. David R. Soderblom
ISBN 1-886733-39-2

Vol. CS-120 LUMINOUS BLUE VARIABLES: MASSIVE STARS IN TRANSITION
eds. Antonella Nota and Henny J. G. L. M. Lamers
ISBN 1-886733-40-6

Vol. CS-121 ACCRETION PHENOMENA AND RELATED OUTFLOWS, IAU Colloquium 163
eds. D. T. Wickramasinghe, G. V. Bicknell, and L. Ferrario
ISBN 1-886733-41-4

Vol. CS-122 FROM STARDUST TO PLANETESIMALS:
Symposium held as part of the 108th Annual Meeting of the ASP
eds. Yvonne J. Pendleton and A. G. G. M. Tielens
ISBN 1-886733-42-2

Vol. CS-123 THE 12th 'KINGSTON MEETING': COMPUTATIONAL ASTROPHYSICS
eds. David A. Clarke and Michael J. West
ISBN 1-886733-43-0

Vol. CS-124 DIFFUSE INFRARED RADIATION AND THE IRTS
eds. Haruyuki Okuda, Toshio Matsumoto, and Thomas Roellig
ISBN 1-886733-44-9

Vol. CS-125 ASTRONOMICAL DATA ANALYSIS SOFTWARE AND SYSTEMS (ADASS) VI
eds. Gareth Hunt and H. E. Payne
ISBN 1-886733-45-7

ASP CONFERENCE SERIES VOLUMES

Published by the Astronomical Society of the Pacific

PUBLISHED: 1997 (* asterisk means OUT OF PRINT)

Vol. CS-126   FROM QUANTUM FLUCTUATIONS TO COSMOLOGICAL STRUCTURES
eds. David Valls-Gabaud, Martin A. Hendry, Paolo Molaro, and
Khalil Chamcham
ISBN 1-886733-46-5

Vol. CS-127   PROPER MOTIONS AND GALACTIC ASTRONOMY
ed. Roberta M. Humphreys
ISBN 1-886733-47-3

Vol. CS-128   MASS EJECTION FROM AGN (Active Galactic Nuclei)
eds. N. Arav, I. Shlosman, and R. J. Weymann
ISBN 1-886733-48-1

Vol. CS-129   THE GEORGE GAMOW SYMPOSIUM
eds. E. Harper, W. C. Parke, and G. D. Anderson
ISBN 1-886733-49-X

Vol. CS-130   THE THIRD PACIFIC RIM CONFERENCE ON RECENT DEVELOPMENT ON
BINARY STAR RESEARCH
eds. Kam-Ching Leung
ISBN 1-886733-50-3

PUBLISHED: 1998

Vol. CS-131   BOULDER-MUNICH II: PROPERTIES OF HOT, LUMINOUS STARS
ed. Ian D. Howarth
ISBN 1-886733-51-1

Vol. CS-132   STAR FORMATION WITH THE INFRARED SPACE OBSERVATORY (ISO)
eds. João L. Yun and René Liseau
ISBN 1-886733-52-X

Vol. CS-133   SCIENCE WITH THE NGST (Next Generation Space Telescope)
eds. Eric P. Smith and Anuradha Koratkar
ISBN 1-886733-53-8

Vol. CS-134 *   BROWN DWARFS AND EXTRASOLAR PLANETS
eds. Rafael Rebolo, Eduardo L. Martin, and Maria Rosa Zapatero Osorio
ISBN 1-886733-54-6

Vol. CS-135   A HALF CENTURY OF STELLAR PULSATION INTERPRETATIONS:
A TRIBUTE TO ARTHUR N. COX
eds. P. A. Bradley and J. A. Guzik
ISBN 1-886733-55-4

Vol. CS-136   GALACTIC HALOS: A UC SANTA CRUZ WORKSHOP
ed. Dennis Zaritsky
ISBN 1-886733-56-2

Vol. CS-137   WILD STARS IN THE OLD WEST: PROCEEDINGS OF THE 13th NORTH
AMERICAN WORKSHOP ON CATACLYSMIC VARIABLES AND RELATED
OBJECTS
eds. S. Howell, E. Kuulkers, and C. Woodward
ISBN 1-886733-57-0

Vol. CS-138   1997 PACIFIC RIM CONFERENCE ON STELLAR ASTROPHYSICS
eds. Kwing Lam Chan, K. S. Cheng, and H. P. Singh
ISBN 1-886733-58-9

ASP CONFERENCE SERIES VOLUMES
Published by the Astronomical Society of the Pacific

PUBLISHED: 1998 (* asterisk means OUT OF PRINT)

Vol. CS-139   PRESERVING THE ASTRONOMICAL WINDOWS:
Proceedings of Joint Discussion No. 5 of the 23rd General Assembly of the IAU
eds. Syuzo Isobe and Tomohiro Hirayama
ISBN 1-886733-59-7

Vol. CS-140   SYNOPTIC SOLAR PHYSICS --18th NSO/Sacramento Peak Summer Workshop
eds. K. S. Balasubramaniam, J. W. Harvey, and D. M. Rabin
ISBN 1-886733-60-0

Vol. CS-141   ASTROPHYSICS FROM ANTARCTICA:
A Symposium held as a part of the 109th Annual Meeting of the ASP
eds. Giles Novak and Randall H. Landsberg
ISBN 1-886733-61-9

Vol. CS-142   THE STELLAR INITIAL MASS FUNCTION: 38th Herstmonceux Conference
eds. Gerry Gilmore and Debbie Howell
ISBN 1-886733-62-7

Vol. CS-143 *   THE SCIENTIFIC IMPACT OF THE GODDARD HIGH RESOLUTION SPECTROGRAPH (GHRS)
eds. John C. Brandt, Thomas B. Ake III, and Carolyn Collins Petersen
ISBN 1-886733-63-5

Vol. CS-144   RADIO EMISSION FROM GALACTIC AND EXTRAGALACTIC COMPACT SOURCES, IAU Colloquium 164
eds. J. Anton Zensus, G. B. Taylor, and J. M. Wrobel
ISBN 1-886733-64-3

Vol. CS-145   ASTRONOMICAL DATA ANALYSIS SOFTWARE AND SYSTEMS (ADASS) VII
eds. Rudolf Albrecht, Richard N. Hook, and Howard A. Bushouse
ISBN 1-886733-65-1

Vol. CS-146   THE YOUNG UNIVERSE GALAXY FORMATION AND EVOLUTION AT INTERMEDIATE AND HIGH REDSHIFT
eds. S. D'Odorico, A. Fontana, and E. Giallongo
ISBN 1-886733-66-X

Vol. CS-147   ABUNDANCE PROFILES: DIAGNOSTIC TOOLS FOR GALAXY HISTORY
eds. Daniel Friedli, Mike Edmunds, Carmelle Robert, and Laurent Drissen
ISBN 1-886733-67-8

Vol. CS-148   ORIGINS
eds. Charles E. Woodward, J. Michael Shull, and Harley A. Thronson, Jr.
ISBN 1-886733-68-6

Vol. CS-149   SOLAR SYSTEM FORMATION AND EVOLUTION
eds. D. Lazzaro, R. Vieira Martins, S. Ferraz-Mello, J. Fernández, and C. Beaugé
ISBN 1-886733-69-4

Vol. CS-150   NEW PERSPECTIVES ON SOLAR PROMINENCES, IAU Colloquium 167
eds. David Webb, David Rust, and Brigitte Schmieder
ISBN 1-886733-70-8

ASP CONFERENCE SERIES VOLUMES
Published by the Astronomical Society of the Pacific

PUBLISHED: 1998 (* asterisk means OUT OF PRINT)

Vol. CS-151   COSMIC MICROWAVE BACKGROUND AND LARGE SCALE STRUCTURES
OF THE UNIVERSE
eds. Yong-Ik Byun and Kin-Wang Ng
ISBN 1-886733-71-6

Vol. CS-152   FIBER OPTICS IN ASTRONOMY III
eds. S. Arribas, E. Mediavilla, and F. Watson
ISBN 1-886733-72-4

Vol. CS-153   LIBRARY AND INFORMATION SERVICES IN ASTRONOMY III -- (LISA III)
eds. Uta Grothkopf, Heinz Andernach, Sarah Stevens-Rayburn,
and Monique Gomez
ISBN 1-886733-73-2

PUBLISHED: 1999

Vol. CS-154   COOL STARS, STELLAR SYSTEMS AND THE SUN:
Tenth Cambridge Workshop
eds. Robert A. Donahue and Jay A. Bookbinder
ISBN 1-886733-74-0

Vol. CS-155   SECOND ADVANCES IN SOLAR PHYSICS EUROCONFERENCE:
THREE-DIMENSIONAL STRUCTURE OF SOLAR ACTIVE REGIONS
eds. Costas E. Alissandrakis and Brigitte Schmieder
ISBN 1-886733-75-9

Vol. CS-156   HIGHLY REDSHIFTED RADIO LINES
eds. C. L. Carilli, S. J. E. Radford, K. M. Menten, and G. I. Langston
ISBN 1-886733-76-7

Vol. CS-157   ANNAPOLIS WORKSHOP ON MAGNETIC CATACLYSMIC VARIABLES
eds. Coel Hellier and Koji Mukai
ISBN 1-886733-77-5

Vol. CS-158   SOLAR AND STELLAR ACTIVITY: SIMILARITIES AND DIFFERENCES
eds. C. J. Butler and J. G. Doyle
ISBN 1-886733-78-3

Vol. CS-159   BL LAC PHENOMENON
eds. Leo O. Takalo and Aimo Sillanpää
ISBN 1-886733-79-1

Vol. CS-160   ASTROPHYSICAL DISCS: An EC Summer School
eds. J. A. Sellwood and Jeremy Goodman
ISBN 1-886733-80-5

Vol. CS-161   HIGH ENERGY PROCESSES IN ACCRETING BLACK HOLES
eds. Juri Poutanen and Roland Svensson
ISBN 1-886733-81-3

Vol. CS-162   QUASARS AND COSMOLOGY
eds. Gary Ferland and Jack Baldwin
ISBN 1-886733-83-X

Vol. CS-163   STAR FORMATION IN EARLY-TYPE GALAXIES
eds. Jordi Cepa and Patricia Carral
ISBN 1-886733-84-8

ASP CONFERENCE SERIES VOLUMES
Published by the Astronomical Society of the Pacific

PUBLISHED: 1999 (* asterisk means OUT OF PRINT)

Vol. CS-164   ULTRAVIOLET–OPTICAL SPACE ASTRONOMY BEYOND HST
eds. Jon A. Morse, J. Michael Shull, and Anne L. Kinney
ISBN 1-886733-85-6

Vol. CS-165   THE THIRD STROMLO SYMPOSIUM: THE GALACTIC HALO
eds. Brad K. Gibson, Tim S. Axelrod, and Mary E. Putman
ISBN 1-886733-86-4

Vol. CS-166   STROMLO WORKSHOP ON HIGH-VELOCITY CLOUDS
eds. Brad K. Gibson and Mary E. Putman
ISBN 1-886733-87-2

Vol. CS-167   HARMONIZING COSMIC DISTANCE SCALES IN A POST-HIPPARCOS ERA
eds. Daniel Egret and André Heck
ISBN 1-886733-88-0

Vol. CS-168   NEW PERSPECTIVES ON THE INTERSTELLAR MEDIUM
eds. A. R. Taylor, T. L. Landecker, and G. Joncas
ISBN 1-886733-89-9

Vol. CS-169   11th EUROPEAN WORKSHOP ON WHITE DWARFS
eds. J.-E. Solheim and E. G. Meištas
ISBN 1-886733-91-0

Vol. CS-170   THE LOW SURFACE BRIGHTNESS UNIVERSE, IAU Colloquium 171
eds. J. I. Davies, C. Impey, and S. Phillipps
ISBN 1-886733-92-9

Vol. CS-171   LiBeB, COSMIC RAYS, AND RELATED X- AND GAMMA RAYS
eds. Reuven Ramaty, Elisabeth Vangioni-Flam, Michel Cassé, and Keith Olive
ISBN 1-886733-93-7

Vol. CS-172   ASTRONOMICAL DATA ANALYSIS SOFTWARE AND SYSTEMS (ADASS) VIII
eds. David M. Mehringer, Raymond L. Plante, and Douglas A. Roberts
ISBN 1-886733-94-5

Vol. CS-173   THEORY AND TESTS OF CONVECTION IN STELLAR STRUCTURE:
First Granada Workshop
ed. Álvaro Giménez, Edward F. Guinan, and Benjamín Montesinos
ISBN 1-886733-95-3

Vol. CS-174   CATCHING THE PERFECT WAVE: ADAPTIVE OPTICS AND
INTERFEROMETRY IN THE 21st CENTURY,
A Symposium held as a part of the 110th Annual Meeting of the ASP
eds. Sergio R. Restaino, William Junor, and Nebojsa Duric
ISBN 1-886733-96-1

Vol. CS-175   STRUCTURE AND KINEMATICS OF QUASAR BROAD LINE REGIONS
eds. C. M. Gaskell, W. N. Brandt, M. Dietrich, D. Dultzin-Hacyan,
and M. Eracleous
ISBN 1-886733-97-X

Vol. CS-176   OBSERVATIONAL COSMOLOGY: THE DEVELOPMENT OF GALAXY
SYSTEMS
eds. Giuliano Giuricin, Marino Mezzetti, and Paolo Salucci
ISBN 1-58381-000-5

ASP CONFERENCE SERIES VOLUMES
Published by the Astronomical Society of the Pacific

PUBLISHED: 1999 (* asterisk means OUT OF PRINT)

Vol. CS-177   ASTROPHYSICS WITH INFRARED SURVEYS: A Prelude to SIRTF
eds. Michael D. Bicay, Chas A. Beichman, Roc M. Cutri, and Barry F. Madore
ISBN 1-58381-001-3

Vol. CS-178   STELLAR DYNAMOS: NONLINEARITY AND CHAOTIC FLOWS
eds. Manuel Núñez and Antonio Ferriz-Mas
ISBN 1-58381-002-1

Vol. CS-179   ETA CARINAE AT THE MILLENNIUM
eds. Jon A. Morse, Roberta M. Humphreys, and Augusto Damineli
ISBN 1-58381-003-X

Vol. CS-180   SYNTHESIS IMAGING IN RADIO ASTRONOMY II
eds. G. B. Taylor, C. L. Carilli, and R. A. Perley
ISBN 1-58381-005-6

Vol. CS-181   MICROWAVE FOREGROUNDS
eds. Angelica de Oliveira-Costa and Max Tegmark
ISBN 1-58381-006-4

Vol. CS-182   GALAXY DYNAMICS: A Rutgers Symposium
eds. David Merritt, J. A. Sellwood, and Monica Valluri
ISBN 1-58381-007-2

Vol. CS-183   HIGH RESOLUTION SOLAR PHYSICS: THEORY, OBSERVATIONS, AND TECHNIQUES
eds. T. R. Rimmele, K. S. Balasubramaniam, and R. R. Radick
ISBN 1-58381-009-9

Vol. CS-184   THIRD ADVANCES IN SOLAR PHYSICS EUROCONFERENCE: MAGNETIC FIELDS AND OSCILLATIONS
eds. B. Schmieder, A. Hofmann, and J. Staude
ISBN 1-58381-010-2

Vol. CS-185   PRECISE STELLAR RADIAL VELOCITIES, IAU Colloquium 170
eds. J. B. Hearnshaw and C. D. Scarfe
ISBN 1-58381-011-0

Vol. CS-186   THE CENTRAL PARSECS OF THE GALAXY
eds. Heino Falcke, Angela Cotera, Wolfgang J. Duschl, Fulvio Melia, and Marcia J. Rieke
ISBN 1-58381-012-9

Vol. CS-187   THE EVOLUTION OF GALAXIES ON COSMOLOGICAL TIMESCALES
eds. J. E. Beckman and T. J. Mahoney
ISBN 1-58381-013-7

Vol. CS-188   OPTICAL AND INFRARED SPECTROSCOPY OF CIRCUMSTELLAR MATTER
eds. Eike W. Guenther, Bringfried Stecklum, and Sylvio Klose
ISBN 1-58381-014-5

Vol. CS-189   CCD PRECISION PHOTOMETRY WORKSHOP
eds. Eric R. Craine, Roy A. Tucker, and Jeannette Barnes
ISBN 1-58381-015-3

ASP CONFERENCE SERIES VOLUMES
Published by the Astronomical Society of the Pacific

PUBLISHED: 1999 (* asterisk means OUT OF PRINT)

Vol. CS-190   GAMMA-RAY BURSTS: THE FIRST THREE MINUTES
eds. Juri Poutanen and Roland Svensson
ISBN 1-58381-016-1

Vol. CS-191   PHOTOMETRIC REDSHIFTS AND HIGH REDSHIFT GALAXIES
eds. Ray J. Weymann, Lisa J. Storrie-Lombardi, Marcin Sawicki, and Robert J. Brunner
ISBN 1-58381-017-X

Vol. CS-192   SPECTROPHOTOMETRIC DATING OF STARS AND GALAXIES
eds. I. Hubeny, S. R. Heap, and R. H. Cornett
ISBN 1-58381-018-8

Vol. CS-193   THE HY-REDSHIFT UNIVERSE:
GALAXY FORMATION AND EVOLUTION AT HIGH REDSHIFT
eds. Andrew J. Bunker and Wil J. M. van Breugel
ISBN 1-58381-019-6

Vol. CS-194   WORKING ON THE FRINGE:
OPTICAL AND IR INTERFEROMETRY FROM GROUND AND SPACE
eds. Stephen Unwin and Robert Stachnik
ISBN 1-58381-020-X

PUBLISHED: 2000

Vol. CS-195   IMAGING THE UNIVERSE IN THREE DIMENSIONS:
Astrophysics with Advanced Multi-Wavelength Imaging Devices
eds. W. van Breugel and J. Bland-Hawthorn
ISBN 1-58381-022-6

Vol. CS-196   THERMAL EMISSION SPECTROSCOPY AND ANALYSIS OF DUST, DISKS, AND REGOLITHS
eds. Michael L. Sitko, Ann L. Sprague, and David K. Lynch
ISBN: 1-58381-023-4

Vol. CS-197   XVth IAP MEETING DYNAMICS OF GALAXIES:
FROM THE EARLY UNIVERSE TO THE PRESENT
eds. F. Combes, G. A. Mamon, and V. Charmandaris
ISBN: 1-58381-24-2

Vol. CS-198   EUROCONFERENCE ON "STELLAR CLUSTERS AND ASSOCIATIONS: CONVECTION, ROTATION, AND DYNAMOS"
eds. R. Pallavicini, G. Micela, and S. Sciortino
ISBN: 1-58381-25-0

Vol. CS-199   ASYMMETRICAL PLANETARY NEBULAE II:
FROM ORIGINS TO MICROSTRUCTURES
eds. J. H. Kastner, N. Soker, and S. Rappaport
ISBN: 1-58381-026-9

Vol. CS-200   CLUSTERING AT HIGH REDSHIFT
eds. A. Mazure, O. Le Fèvre, and V. Le Brun
ISBN: 1-58381-027-7

Vol. CS-201   COSMIC FLOWS 1999: TOWARDS AN UNDERSTANDING OF LARGE-SCALE STRUCTURES
eds. Stéphane Courteau, Michael A. Strauss, and Jeffrey A. Willick
ISBN: 1-58381-028-5

ASP CONFERENCE SERIES VOLUMES
Published by the Astronomical Society of the Pacific

PUBLISHED: 2000 (* asterisk means OUT OF PRINT)

Vol. CS-202 *   PULSAR ASTRONOMY – 2000 AND BEYOND, IAU Colloquium 177
                eds. M. Kramer, N. Wex, and R. Wielebinski
                ISBN: 1-58381-029-3

Vol. CS-203   THE IMPACT OF LARGE-SCALE SURVEYS ON PULSATING STAR
                RESEARCH, IAU Colloquium 176
                eds. L. Szabados and D. W. Kurtz
                ISBN: 1-58381-030-7

Vol. CS-204   THERMAL AND IONIZATION ASPECTS OF FLOWS FROM HOT STARS:
                OBSERVATIONS AND THEORY
                eds. Henny J. G. L. M. Lamers and Arved Sapar
                ISBN: 1-58381-031-5

Vol. CS-205   THE LAST TOTAL SOLAR ECLIPSE OF THE MILLENNIUM IN TURKEY
                eds. W. C. Livingston and A. Özgüç
                ISBN: 1-58381-032-3

Vol. CS-206   HIGH ENERGY SOLAR PHYSICS – *ANTICIPATING HESSI*
                eds Reuven Ramaty and Natalie Mandzhavidze
                ISBN: 1-58381-033-1

Vol. CS-207   NGST SCIENCE AND TECHNOLOGY EXPOSITION
                eds. Eric P. Smith and Knox S. Long
                ISBN: 1-58381-036-6

ATLAS   VISIBLE AND NEAR INFRARED ATLAS OF THE ARCTURUS SPECTRUM
                3727–9300 Å
                eds. Kenneth Hinkle, Lloyd Wallace, Jeff Valenti, and Dianne Harmer
                ISBN: 1-58381-037-4

Vol. CS-208   POLAR MOTION: HISTORICAL AND SCIENTIFIC PROBLEMS,
                IAU Colloquium 178
                eds. Steven Dick, Dennis McCarthy, and Brian Luzum
                ISBN: 1-58381-039-0

Vol. CS-209   SMALL GALAXY GROUPS, IAU Colloquium 174
                eds. Mauri J. Valtonen and Chris Flynn
                ISBN: 1-58381-040-4

Vol. CS-210   DELTA SCUTI AND RELATED STARS: Reference Handbook and Proceedings
                of the 6th Vienna Workshop in Astrophysics
                eds. Michel Breger and Michael Houston Montgomery
                ISBN: 1-58381-043-9

Vol. CS-211   MASSIVE STELLAR CLUSTERS
                eds. Ariane Lançon and Christian M. Boily
                ISBN: 1-58381-042-0

Vol. CS-212   FROM GIANT PLANETS TO COOL STARS
                eds. Caitlin A. Griffith and Mark S. Marley
                ISBN: 1-58381-041-2

Vol. CS-213   BIOASTRONOMY '99: A NEW ERA IN BIOASTRONOMY
                eds. Guillermo A. Lemarchand and Karen J. Meech
                ISBN: 1-58381-044-7

ASP CONFERENCE SERIES VOLUMES
Published by the Astronomical Society of the Pacific

PUBLISHED: 2000 (* asterisk means OUT OF PRINT)

Vol. CS-214   THE Be PHENOMENON IN EARLY-TYPE STARS, IAU Colloquium 175
               eds. Myron A. Smith, Huib F. Henrichs and Juan Fabregat
               ISBN: 1-58381-045-5

Vol. CS-215   COSMIC EVOLUTION AND GALAXY FORMATION:
               STRUCTURE, INTERACTIONS AND FEEDBACK
               The 3rd Guillermo Haro Astrophysics Conference
               eds. José Franco, Elena Terlevich, Omar López-Cruz, and Itziar Aretxaga
               ISBN: 1-58381-046-3

Vol. CS-216   ASTRONOMICAL DATA ANALYSIS SOFTWARE AND SYSTEMS (ADASS) IX
               eds. Nadine Manset, Christian Veillet, and Dennis Crabtree
               ISBN: 1-58381-047-1   ISSN: 1080-7926

Vol. CS-217   IMAGING AT RADIO THROUGH SUBMILLIMETER WAVELENGTHS
               eds. Jeffrey G. Mangum and Simon J. E. Radford
               ISBN: 1-58381-049-8

Vol. CS-218   MAPPING THE HIDDEN UNIVERSE: THE UNIVERSE BEHIND THE MILKY
               WAY – THE UNIVERSE IN HI
               eds. Renée C. Kraan-Korteweg, Patricia A. Henning, and Heinz Andernach
               ISBN: 1-58381-050-1

Vol. CS-219   DISKS, PLANETESIMALS, AND PLANETS
               eds. F. Garzón, C. Eiroa, D. de Winter, and T. J. Mahoney
               ISBN: 1-58381-051-X

Vol. CS-220   AMATEUR – PROFESSIONAL PARTNERSHIPS IN ASTRONOMY:
               The 111th Annual Meeting of the ASP
               eds. John R. Percy and Joseph B. Wilson
               ISBN: 1-58381-052-8

Vol. CS-221   STARS, GAS AND DUST IN GALAXIES: EXPLORING THE LINKS
               eds. Danielle Alloin, Knut Olsen, and Gaspar Galaz
               ISBN: 1-58381-053-6

PUBLISHED: 2001

Vol. CS-222   THE PHYSICS OF GALAXY FORMATION
               eds. M. Umemura and H. Susa
               ISBN: 1-58381-054-4

Vol. CS-223   COOL STARS, STELLAR SYSTEMS AND THE SUN:
               Eleventh Cambridge Workshop
               eds. Ramón J. García López, Rafael Rebolo, and María Zapatero Osorio
               ISBN: 1-58381-056-0

Vol. CS-224   PROBING THE PHYSICS OF ACTIVE GALACTIC NUCLEI BY
               MULTIWAVELENGTH MONITORING
               eds. Bradley M. Peterson, Ronald S. Polidan, and Richard W. Pogge
               ISBN: 1-58381-055-2

Vol. CS-225   VIRTUAL OBSERVATORIES OF THE FUTURE
               eds. Robert J. Brunner, S. George Djorgovski, and Alex S. Szalay
               ISBN: 1-58381-057-9

ASP CONFERENCE SERIES VOLUMES
Published by the Astronomical Society of the Pacific

PUBLISHED: 2001 (* asterisk means OUT OF PRINT)

Vol. CS-226　12th EUROPEAN CONFERENCE ON WHITE DWARFS
eds. J. L. Provencal, H. L. Shipman, J. MacDonald, and S. Goodchild
ISBN: 1-58381-058-7

Vol. CS-227　BLAZAR DEMOGRAPHICS AND PHYSICS
eds. Paolo Padovani and C. Megan Urry
ISBN: 1-58381-059-5

Vol. CS-228　DYNAMICS OF STAR CLUSTERS AND THE MILKY WAY
eds. S. Deiters, B. Fuchs, A. Just, R. Spurzem, and R. Wielen
ISBN: 1-58381-060-9

Vol. CS-229　EVOLUTION OF BINARY AND MULTIPLE STAR SYSTEMS
A Meeting in Celebration of Peter Eggleton's 60th Birthday
eds. Ph. Podsiadlowski, S. Rappaport, A. R. King, F. D'Antona, and L. Burderi
IBSN: 1-58381-061-7

Vol. CS-230　GALAXY DISKS AND DISK GALAXIES
eds. Jose G. Funes, S. J. and Enrico Maria Corsini
ISBN: 1 58381-063-3

Vol. CS-231　TETONS 4: GALACTIC STRUCTURE, STARS, AND THE INTERSTELLAR MEDIUM
eds. Charles E. Woodward, Michael D. Bicay, and J. Michael Shull
ISBN: 1-58381-064-1

Vol. CS-232　THE NEW ERA OF WIDE FIELD ASTRONOMY
eds. Roger Clowes, Andrew Adamson, and Gordon Bromage
ISBN: 1-58381-065-X

Vol. CS-233　P CYGNI 2000: 400 YEARS OF PROGRESS
eds. Mart de Groot and Christiaan Sterken
ISBN: 1-58381-070-6

Vol. CS-234　X-RAY ASTRONOMY 2000
eds. R. Giacconi, S. Serio, and L. Stella
ISBN: 1-58381-071-4

Vol. CS-235　SCIENCE WITH THE ATACAMA LARGE MILLIMETER ARRAY (ALMA)
ed. Alwyn Wootten
ISBN: 1-58381-072-2

Vol. CS-236　ADVANCED SOLAR POLARIMETRY: THEORY, OBSERVATION, AND INSTRUMENTATION, The 20th Sacramento Peak Summer Workshop
ed. M. Sigwarth
ISBN: 1-58381-073-0

Vol. CS-237　GRAVITATIONAL LENSING: RECENT PROGRESS AND FUTURE GOALS
eds. Tereasa G. Brainerd and Christopher S. Kochanek
ISBN: 1-58381-074-9

Vol. CS-238　ASTRONOMICAL DATA ANALYSIS SOFTWARE AND SYSTEMS (ADASS) X
eds. F. R. Harnden, Jr., Francis A. Primini, and Harry E. Payne
ISBN: 1-58381-075-7

ASP CONFERENCE SERIES VOLUMES
Published by the Astronomical Society of the Pacific

PUBLISHED: 2001 (* asterisk means OUT OF PRINT)

Vol. CS-239   MICROLENSING 2000: A NEW ERA OF MICROLENSING ASTROPHYSICS
eds. John Menzies and Penny D. Sackett
ISBN: 1-58381-076-5

Vol. CS-240   GAS AND GALAXY EVOLUTION,
A Conference in Honor of the 20th Anniversary of the VLA
eds. J. E. Hibbard, M. P. Rupen, and J. H. van Gorkom
ISBN: 1-58381-077-3

Vol. CS-241   THE 7TH TAIPEI ASTROPHYSICS WORKSHOP ON
COSMIC RAYS IN THE UNIVERSE
ed. Chung-Ming Ko
ISBN: 1-58381-079-X

Vol. CS-242   ETA CARINAE AND OTHER MYSTERIOUS STARS:
THE HIDDEN OPPORTUNITIES OF EMISSION SPECTROSCOPY
eds. Theodore R. Gull, Sveneric Johannson, and Kris Davidson
ISBN: 1-58381-080-3

Vol. CS-243   FROM DARKNESS TO LIGHT:
ORIGIN AND EVOLUTION OF YOUNG STELLAR CLUSTERS
eds. Thierry Montmerle and Philippe André
ISBN: 1-58381-081-1

Vol. CS-244   YOUNG STARS NEAR EARTH: PROGRESS AND PROSPECTS
eds. Ray Jayawardhana and Thomas P. Greene
ISBN: 1-58381-082-X

Vol. CS-245   ASTROPHYSICAL AGES AND TIME SCALES
eds. Ted von Hippel, Chris Simpson, and Nadine Manset
ISBN: 1-58381-083-8

Vol. CS-246   SMALL TELESCOPE ASTRONOMY ON GLOBAL SCALES,
IAU Colloquium 183
eds. Wen-Ping Chen, Claudia Lemme, and Bohdan Paczyński
ISBN: 1-58381-084-6

Vol. CS-247   SPECTROSCOPIC CHALLENGES OF PHOTOIONIZED PLASMAS
eds. Gary Ferland and Daniel Wolf Savin
ISBN: 1-58381-085-4

Vol. CS-248   MAGNETIC FIELDS ACROSS THE HERTZSPRUNG-RUSSELL DIAGRAM
eds. G. Mathys, S. K. Solanki, and D. T. Wickramasinghe
ISBN: 1-58381-088-9

Vol. CS-249   THE CENTRAL KILOPARSEC OF STARBURSTS AND AGN:
THE LA PALMA CONNECTION
eds. J. H. Knapen, J. E. Beckman, I. Shlosman, and T. J. Mahoney
ISBN: 1-58381-089-7

Vol. CS-250   PARTICLES AND FIELDS IN RADIO GALAXIES CONFERENCE
eds. Robert A. Laing and Katherine M. Blundell
ISBN: 1-58381-090-0

Vol. CS-251   NEW CENTURY OF X-RAY ASTRONOMY
eds. H. Inoue and H. Kunieda
ISBN: 1-58381-091-9

ASP CONFERENCE SERIES VOLUMES
Published by the Astronomical Society of the Pacific

PUBLISHED: 2001 (* asterisk means OUT OF PRINT)

Vol. CS-252   HISTORICAL DEVELOPMENT OF MODERN COSMOLOGY
eds. Vicent J. Martínez, Virginia Trimble, and María Jesús Pons-Bordería
ISBN: 1-58381-092-7

PUBLISHED: 2002

Vol. CS-253   CHEMICAL ENRICHMENT OF INTRACLUSTER AND INTERGALACTIC MEDIUM
eds. Roberto Fusco-Femiano and Francesca Matteucci
ISBN: 1-58381-093-5

Vol. CS-254   EXTRAGALACTIC GAS AT LOW REDSHIFT
eds. John S. Mulchaey and John T. Stocke
ISBN: 1-58381-094-3

Vol. CS-255   MASS OUTFLOW IN ACTIVE GALACTIC NUCLEI: NEW PERSPECTIVES
eds. D. M. Crenshaw, S. B. Kraemer, and I. M. George
ISBN: 1-58381-095-1

Vol. CS-256   OBSERVATIONAL ASPECTS OF PULSATING B AND A STARS
eds. Christiaan Sterken and Donald W. Kurtz
ISBN: 1-58381-096-X

Vol. CS-257   AMiBA 2001: HIGH-Z CLUSTERS, MISSING BARYONS, AND CMB POLARIZATION
eds. Lin-Wen Chen, Chung-Pei Ma, Kin-Wang Ng, and Ue-Li Pen
ISBN: 1-58381-097-8

Vol. CS-258   ISSUES IN UNIFICATION OF ACTIVE GALACTIC NUCLEI
eds. Roberto Maiolino, Alessandro Marconi, and Neil Nagar
ISBN: 1-58381-098-6

Vol. CS-259   RADIAL AND NONRADIAL PULSATIONS AS PROBES OF STELLAR PHYSICS, IAU Colloquium 185
eds. Conny Aerts, Timothy R. Bedding, and Jørgen Christensen-Dalsgaard
ISBN: 1-58381-099-4

Vol. CS-260   INTERACTING WINDS FROM MASSIVE STARS
eds. Anthony F. J. Moffat and Nicole St-Louis
ISBN: 1-58381-100-1

Vol. CS-261   THE PHYSICS OF CATACLYSMIC VARIABLES AND RELATED OBJECTS
eds. B. T. Gänsicke, K. Beuermann, and K. Reinsch
ISBN: 1-58381-101-X

Vol. CS-262   THE HIGH ENERGY UNIVERSE AT SHARP FOCUS: CHANDRA SCIENCE, held in conjunction with the 113th Annual Meeting of the ASP
eds. Eric M. Schlegel and Saeqa Dil Vrtilek
ISBN: 1-58381-102-8

Vol. CS-263   STELLAR COLLISIONS, MERGERS AND THEIR CONSEQUENCES
ed. Michael M. Shara
ISBN: 1-58381-103-6

ASP CONFERENCE SERIES VOLUMES
Published by the Astronomical Society of the Pacific

PUBLISHED: 2002 (* asterisk means OUT OF PRINT)

Vol. CS-264　CONTINUING THE CHALLENGE OF EUV ASTRONOMY: CURRENT ANALYSIS AND PROSPECTS FOR THE FUTURE
eds. Steve B. Howell, Jean Dupuis, Daniel Golombek, Frederick M. Walter, and Jennifer Cullison
ISBN: 1-58381-104-4

Vol. CS-265　ω CENTAURI, A UNIQUE WINDOW INTO ASTROPHYSICS
eds. Floor van Leeuwen, Joanne D. Hughes, and Giampaolo Piotto
ISBN: 1-58381-105-2

Vol. CS-266　ASTRONOMICAL SITE EVALUATION IN THE VISIBLE AND RADIO RANGE, IAU Technical Workshop
eds. J. Vernin, Z. Benkhaldoun, and C. Muñoz-Tuñón
ISBN: 1-58381-106-0

Vol. CS-267*　HOT STAR WORKSHOP III: THE EARLIEST STAGES OF MASSIVE STAR BIRTH
ed. Paul A. Crowther
ISBN: 1-58381-107-9

Vol. CS-268　TRACING COSMIC EVOLUTION WITH GALAXY CLUSTERS
eds. Stefano Borgani, Marino Mezzetti, and Riccardo Valdarnini
ISBN: 1-58381-108-7

Vol. CS-269　THE EVOLVING SUN AND ITS INFLUENCE ON PLANETARY ENVIRONMENTS
eds. Benjamín Montesinos, Álvaro Giménez, and Edward F. Guinan
ISBN: 1-58381-109-5

Vol. CS-270　ASTRONOMICAL INSTRUMENTATION AND THE BIRTH AND GROWTH OF ASTROPHYSICS: A Symposium held in honor of Robert G. Tull
eds. Frank N. Bash and Christopher Sneden
ISBN: 1-58381-110-9

Vol. CS-271　NEUTRON STARS IN SUPERNOVA REMNANTS
eds. Patrick O. Slane and Bryan M. Gaensler
ISBN: 1-58381-111-7

Vol. CS-272　THE FUTURE OF SOLAR SYSTEM EXPLORATION, 2003–2013 Community Contributions to the NRC Solar System Exploration Decadal Survey
ed. Mark V. Sykes
ISBN: 1-58381-113-3

Vol. CS-273　THE DYNAMICS, STRUCTURE AND HISTORY OF GALAXIES
eds. G. S. Da Costa and H. Jerjen
ISBN: 1-58381-114-1

Vol. CS-274　OBSERVED HR DIAGRAMS AND STELLAR EVOLUTION
eds. Thibault Lejeune and João Fernandes
ISBN: 1-58381-116-8

Vol. CS-275　DISKS OF GALAXIES: KINEMATICS, DYNAMICS AND PERTURBATIONS
eds. E. Athanassoula, A. Bosma, and R. Mujica
ISBN: 1-58381-117-6

ASP CONFERENCE SERIES VOLUMES
Published by the Astronomical Society of the Pacific

PUBLISHED: 2002 (* asterisk means OUT OF PRINT)

Vol. CS-276   SEEING THROUGH THE DUST:
THE DETECTION OF HI AND THE EXPLORATION OF THE ISM IN GALAXIES
eds. A. R. Taylor, T. L. Landecker, and A. G. Willis
ISBN: 1-58381-118-4

Vol. CS 277   STELLAR CORONAE IN THE CHANDRA AND XMM-NEWTON ERA
eds. Fabio Favata and Jeremy J. Drake
ISBN: 1-58381-119-2

Vol. CS 278   NAIC–NRAO SCHOOL ON SINGLE-DISH ASTRONOMY:
TECHNIQUES AND APPLICATIONS
eds. Snezana Stanimirovic, Daniel Altschuler, Paul Goldsmith, and Chris Salter
ISBN: 1-58381-120-6

Vol. CS 279   EXOTIC STARS AS CHALLENGES TO EVOLUTION, IAU Colloquium 187
eds. Christopher A. Tout and Walter Van Hamme
ISBN: 1-58381-122-2

Vol. CS 280   NEXT GENERATION WIDE-FIELD MULTI-OBJECT SPECTROSCOPY
eds. Michael J. I. Brown and Arjun Dey
ISBN: 1-58381-123-0

Vol. CS 281   ASTRONOMICAL DATA ANALYSIS SOFTWARE AND SYSTEM (ADASS) XI
eds. David A. Bohlender, Daniel Durand, and Thomas H. Handley
ISBN: 1-58381-124-9   ISSN: 1080-7926

Vol. CS 282   GALAXIES: THE THIRD DIMENSION
eds. Margarita Rosado, Luc Binette, and Lorena Arias
ISBN: 1-58381-125-7

Vol. CS 283   A NEW ERA IN COSMOLOGY
eds. Nigel Metcalfe and Tom Shanks
ISBN: 1-58381-126-5

Vol. CS 284   AGN SURVEYS
eds. R. F. Green, E. Ye. Khachikian, and D. B. Sanders
ISBN: 1-58381-127-3

Vol. CS 285   MODES OF STAR FORMATION AND THE ORIGIN OF FIELD POPULATIONS
eds. Eva K. Grebel and Walfgang Brandner
ISBN: 1-58381-128-1

PUBLISHED: 2003

Vol. CS 286   CURRENT THEORETICAL MODESL AND HIGH RESOLUTION SOLAR
OBSERVATIONS: PREPARING FOR ATST
eds. Alexei A. Pevtsov and Han Uitenbroek
ISBN: 1-58381-129-X

Vol. CS 287   GALACTIC STAR FORMATION ACROSS THE STELLAR MASS SPECTRUM
eds. J. M. De Buizer and N. S. van der Bliek
ISBN:1-58381-130-3

Vol. CS 288   STELLAR ATMOSPHERE MODELING
eds. I. Hubeny, D. Mihalas and K. Werner
ISBN: 1-58381-131-1

ASP CONFERENCE SERIES VOLUMES
Published by the Astronomical Society of the Pacific

PUBLISHED: 2003 (* asterisk means OUT OF PRINT)

Vol. CS 289   THE PROCEEDINGS OF THE IAU 8TH ASIAN-PACIFIC REGIONAL MEETING, VOLUME 1
eds. Satoru Ikeuchi, John Hearnshaw and Tomoyuki Hanawa
ISBN: 1-58381-134-6

Vol. CS 290   ACTIVE GALACTIC NUCLEI: FROM CENTRAL ENGINE TO HOST GALAXY
eds. S. Collin, F. Combes and I. Shlosman
ISBN: 1-58381-135-4

Vol. CS-291   HUBBLE'S SCIENCE LEGACY:
FUTURE OPTICAL/ULTRAVIOLET ASTRONOMY FROM SPACE
eds. Kenneth R. Sembach, J. Chris Blades, Garth D. Illingworth and Robert C. Kennicutt, Jr.
ISBN: 1-58381-136-2

Vol. CS-292   INTERPLAY OF PERIODIC, CYCLIC AND STOCHASTIC VARIABILITY IN SELECTED AREAS OF THE H-R DIAGRAM
ed. Christiaan Sterken
ISBN: 1-58381-138-9

Vol. CS-293   3D STELLAR EVOLUTION
eds. S. Turcotte, S. C. Keller and R. M. Cavallo
ISBN: 1-58381-140-0

Vol. CS-294   SCIENTIFIC FRONTIERS IN RESEARCH ON EXTRASOLAR PLANETS
eds. Drake Deming and Sara Seager
ISBN: 1-58381-141-9

Vol. CS-295   ASTRONOMICAL DATA ANALYSIS SOFTWARE AND SYSTEMS (ADASS) XII
eds. Harry E. Payne, Robert I. Jedrzejewski and Richard N. Hook
ISBN: 1-58381-142-7

Vol. CS-296   NEW HORIZONS IN GLOBULAR CLUSTER ASTRONOMY
eds. Giampaolo Piotto, Georges Meylan, S. George Djorgovski and Marco Riello
ISBN: 1-58381-143-5

Vol. CS-297   STAR FORMATION THROUGH TIME, A Conference to Honour Robert J. Terlevich
eds. Enrique Pérez, Rosa M. González Delgado and Guillermo Tenorio-Tagle
ISBN: 1-58381-144-3

Vol. CS-298   GAIA SPECTROSCOPY: SCIENCE AND TECHNOLOGY
ed. Ulisse Munari
ISBN: 1-58381-145-1

Vol. CS-299   HIGH ENERGY BLAZAR ASTRONOMY, An International Conference held to Celebrate the 50th Anniversary of Tuorla Observatory
eds. Leo O. Takalo and Esko Valtaoja
ISBN: 1-58381-146-X

Vol. CS-300   RADIO ASTRONOMY AT THE FRINGE, A Conference held in honor of Kenneth I. Kellermann, on the occasion of his 65th Birthday
eds. J. Anton Zensus, Marshall H. Cohen and Eduardo Ros
ISBN: 1-58381-147-8

ASP CONFERENCE SERIES VOLUMES
Published by the Astronomical Society of the Pacific

PUBLISHED: 2003 (* asterisk means OUT OF PRINT)

Vol. CS-301 MATTER AND ENERGY IN CLUSTERS OF GALAXIES
eds. Stuart Bowyer and Chorng-Yuan Hwang
ISBN: 1-58381-149-4

Vol. CS-302 RADIO PULSARS, In celebration of the contributions of Andrew Lyne, Dick Manchester and Joe Taylor – A Festschrift honoring their 60th Birthdays
eds. Matthew Bailes, David J. Nice and Stephen E. Thorsett
ISBN: 1-58381-151-6

Vol. CS-303 SYMBIOTIC STARS PROBING STELLAR EVOLUTION
eds. R. L. M. Corradi, J. Mikołajewska and T. J. Mahoney
ISBN: 1-58381-152-4

Vol. CS-304 CNO IN THE UNIVERSE
eds. Corinne Charbonnel, Daniel Schaerer and Georges Meynet
ISBN: 1-58381-153-2

Vol. CS-305 International Conference on MAGNETIC FIELDS IN O, B AND A STARS: ORIGIN AND CONNECTION TO PULSATION, ROTATION AND MASS LOSS
eds. Luis A. Balona, Huib F. Henrichs and Rodney Medupe
ISBN: 1-58381-154-0

Vol. CS-306 NEW TECHNOLOGIES IN VLBI
ed. Y. C. Minh
ISBN: 1-58381-155-9

Vol. CS-307 SOLAR POLARIZATION 3
eds. Javier Trujillo Bueno and Jorge Sanchez Almeida
ISBN: 1-58381-156-7

Vol. CS-308 FROM X-RAY BINARIES TO GAMMA-RAY BURSTS
eds. Edward P. J. van den Heuvel, Lex Kaper, Evert Rol and Ralph A. M. J. Wijers
ISBN: 1-58381-158-3

PUBLISHED: 2004

Vol. CS-309 ASTROPHYSICS OF DUST
eds. Adolf N. Witt, Geoffrey C. Clayton and Bruce T. Draine
ISBN: 1-58381-159-1

Vol. CS-310 VARIABLE STARS IN THE LOCAL GROUP, IAU Colloquium 193
eds. Donald W. Kurtz and Karen R. Pollard
ISBN: 1-58381-162-1

Vol. CS-311 AGN PHYSICS WITH THE SLOAN DIGITAL SKY SURVEY
eds. Gordon T. Richards and Patrick B. Hall
ISBN: 1-58381-164-8

Vol. CS-312 Third Rome Workshop on GAMMA-RAY BURSTS IN THE AFTERGLOW ERA
eds. Marco Feroci, Filippo Frontera, Nicola Masetti and Luigi Piro
ISBN: 1-58381-165-6

Vol. CS-313 ASYMMETRICAL PLANETARY NEBULAE III: WINDS, STRUCTURE AND THE THUNDERBIRD
eds. Margaret Meixner, Joel H. Kastner, Bruce Balick and Noam Soker
ISBN: 1-58381-168-0

ASP CONFERENCE SERIES VOLUMES
Published by the Astronomical Society of the Pacific

PUBLISHED: 2004 (* asterisk means OUT OF PRINT)

Vol. CS 314   ASTRONOMICAL DATA ANALYSIS SOFTWARE AND SYSTEMS (ADASS) XIII
eds. Francois Ochsenbein, Mark G. Allen and Daniel Egret
ISBN: 1-58381-169-9   ISSN: 1080-7926

Vol. CS 315   MAGNETIC CATACLYSMIC VARIABLES, IAU Colloquium 190
eds. Sonja Vrielmann and Mark Cropper
ISBN: 1-58381-170-2

Vol. CS 316   ORDER AND CHAOS IN STELLAR AND PLANETARY SYSTEMS
eds. Gene G. Byrd, Konstantin V. Kholshevnikov, Aleksandr A. Mylläri, Igor' I. Nikiforov and Victor V. Orlov
ISBN: 1-58381-172-9

Vol. CS 317   MILKY WAY SURVEYS: THE STRUCTURE AND EVOLUTION OF OUR GALAXY, The 5th Boston University Astrophysics Conference
eds: Dan Clemens, Ronak Shah and Tereasa Brainerd
ISBN: 1-58381-177-X

Vol. CS 318   SPECTROSCOPICALLY AND SPATIALLY RESOLVING THE COMPONENTS OF CLOSE BINARY STARS
eds. Ronald W. Hilditch, Herman Hensberge and Krešimir Pavlovski
ISBN: 1-58381-179-6

Vol. CS-319   NASA OFFICE OF SPACE SCIENCE EDUCATION AND PUBLIC OUTREACH CONFERENCE
eds. Carolyn Narasimhan, Bernhard Beck-Winchatz, Isabel Hawkins and Cassandra Runyon
ISBN: 1-58381-181-8

A listing of the IAU Volumes published by the ASP follows on the next page.

INTERNATIONAL ASTRONOMICAL UNION (IAU) VOLUMES
Published by the Astronomical Society of the Pacific

PUBLISHED: 1999 (* asterisk means OUT OF STOCK)

Vol. No. 190　　NEW VIEWS OF THE MAGELLANIC CLOUDS
　　　　　　　　eds. You-Hua Chu, Nicholas B. Suntzeff, James E. Hesser, and
　　　　　　　　David A. Bohlender
　　　　　　　　ISBN: 1-58381-021-8

Vol. No. 191　　ASYMPTOTIC GIANT BRANCH STARS
　　　　　　　　eds. T. Le Bertre, A. Lèbre, and C. Waelkens
　　　　　　　　ISBN: 1-886733-90-2

Vol. No. 192　　THE STELLAR CONTENT OF LOCAL GROUP GALAXIES
　　　　　　　　eds. Patricia Whitelock and Russell Cannon
　　　　　　　　ISBN: 1-886733-82-1

Vol. No. 193　　WOLF-RAYET PHENOMENA IN MASSIVE STARS AND STARBURST
　　　　　　　　GALAXIES
　　　　　　　　eds. Karel A. van der Hucht, Gloria Koenigsberger, and Philippe R. J. Eenens
　　　　　　　　ISBN: 1-58381-004-8

Vol. No. 194　　ACTIVE GALACTIC NUCLEI AND RELATED PHENOMENA
　　　　　　　　eds. Yervant Terzian, Daniel Weedman, and Edward Khachikian
　　　　　　　　ISBN: 1-58381-008-0

PUBLISHED: 2000

Vol. XXIVA　　TRANSACTIONS OF THE INTERNATIONAL ASTRONOMICAL UNION
　　　　　　　　REPORTS ON ASTRONOMY 1996–1999
　　　　　　　　ed. Johannes Andersen
　　　　　　　　ISBN: 1-58381-035-8

Vol. No. 195　　HIGHLY ENERGETIC PHYSICAL PROCESSES AND MECHANISMS FOR
　　　　　　　　EMISSION FROM ASTROPHYSICAL PLASMAS
　　　　　　　　eds. P. C. H. Martens, S. Tsuruta, and M. A. Weber
　　　　　　　　ISBN: 1-58381-038-2

Vol. No. 197 *　ASTROCHEMISTRY: FROM MOLECULAR CLOUDS TO PLANETARY
　　　　　　　　SYSTEMS
　　　　　　　　eds. Y. C. Minh and E. F. van Dishoeck
　　　　　　　　ISBN: 1-58381-034-X

Vol. No. 198　　THE LIGHT ELEMENTS AND THEIR EVOLUTION
　　　　　　　　eds. L. da Silva, M. Spite, and J. R. de Medeiros
　　　　　　　　ISBN: 1-58381-048-X

PUBLISHED: 2001

IAU SPS　　　　ASTRONOMY FOR DEVELOPING COUNTRIES
　　　　　　　　Special Session of the XXIV General Assembly of the IAU
　　　　　　　　ed. Alan H. Batten
　　　　　　　　ISBN: 1-58381-067-6

Vol. No. 196　　PRESERVING THE ASTRONOMICAL SKY
　　　　　　　　eds. R. J. Cohen and W. T. Sullivan, III
　　　　　　　　ISBN: 1-58381-078-1

Vol. No. 200 *　THE FORMATION OF BINARY STARS
　　　　　　　　eds. Hans Zinnecker and Robert D. Mathieu
　　　　　　　　ISBN: 1-58381-068-4

INTERNATIONAL ASTRONOMICAL UNION (IAU) VOLUMES
Published by the Astronomical Society of the Pacific

PUBLISHED: 2001 (* asterisk means OUT OF STOCK)

Vol. No. 203   RECENT INSIGHTS INTO THE PHYSICS OF THE SUN AND HELIOSPHERE: HIGHLIGHTS FROM SOHO AND OTHER SPACE MISSIONS
eds. Pål Brekke, Bernhard Fleck, and Joseph B. Gurman
ISBN: 1-58381-069-2

Vol. No. 204   THE EXTRAGALACTIC INFRARED BACKGROUND AND ITS COSMOLOGICAL IMPLICATIONS
eds. Martin Harwit and Michael G. Hauser
ISBN: 1-58381-062-5

Vol. No. 205   GALAXIES AND THEIR CONSTITUENTS AT THE HIGHEST ANGULAR RESOLUTIONS
eds. Richard T. Schilizzi, Stuart N. Vogel, Francesco Paresce, and Martin S. Elvis
ISBN: 1-58381-066-8

Vol. XXIVB   TRANSACTIONS OF THE INTERNATIONAL ASTRONOMICAL UNION REPORTS ON ASTRONOMY
ed. Hans Rickman
ISBN: 1-58381-087-0

PUBLISHED: 2002

Vol. No. 12   HIGHLIGHTS OF ASTRONOMY
ed. Hans Rickman
ISBN: 1-58381-086-2

Vol. No. 199   THE UNIVERSE AT LOW RADIO FREQUENCIES
eds. A. Pramesh Rao, G. Swarup, and Gopal-Krishna
ISBN: 00001-121-4

Vol. No. 206   COSMIC MASERS: FROM PROTOSTARS TO BLACKHOLES
eds. Victor Migenes and Mark J. Reid
ISBN: 1-58381-112-5

Vol. No. 207   EXTRAGALACTIC STAR CLUSTERS
eds. Doug Geisler, Eva K. Grebel, and Dante Minniti
ISBN: 1-58381-115-X

PUBLISHED: 2003

Vol. XXVA   TRANSACTIONS OF THE INTERNATIONAL ASTRONOMICAL UNION REPORTS ON ASTRONOMY 1999–2002
ed. Hans Rickman
ISBN: 1-58381-137-0

Vol. No. 208   ASTROPHYSICAL SUPERCOMPUTING USING PARTICLE SIMULATIONS
eds. Junichiro Makino and Piet Hut
ISBN: 1-58381-139-7

Vol. No. 209   PLANETARY NEBULAE: THEIR EVOLUTION AND ROLE IN THE UNIVERSE
eds. Sun Kwok, Michael Dopita and Ralph Sutherland
ISBN: 1-58381-148-6

Vol. No. 210   MODELLING OF STELLAR ATMOSPHERES
eds. N. Piskunov, W. W. Weiss and D. F. Gray
ISBN: 1-58381-160-5

INTERNATIONAL ASTRONOMICAL UNION (IAU) VOLUMES
Published by the Astronomical Society of the Pacific

PUBLISHED: 2003 (* asterisk means OUT OF STOCK)

Vol. No. 211　　BROWN DWARFS
　　　　　　　　ed. Eduardo Martín
　　　　　　　　ISBN: 1-58381-132-X

Vol. No. 212　　A MASSIVE STAR ODYSSEY: FROM MAIN SEQUENCE TO SUPERNOVA
　　　　　　　　eds. Karel A. van der Hucht, Artemio Herrero and César Esteban
　　　　　　　　ISBN: 1-58381-133-8

Vol. No. 214　　HIGH ENERGY PROCESSES AND PHENOMENA IN ASTROPHYSICS
　　　　　　　　eds. X. D. Li, V. Trimble and Z. R. Wang
　　　　　　　　ISBN: 1-58381-157-5

PUBLISHED: 2004

Vol. No. 202　　PLANETARY SYSTEMS IN THE UNIVERSE: OBSERVATION, FORMATION AND EVOLUTION
　　　　　　　　eds. Alan Penny, Pawel Artymowicz, Anne-Marie LaGrange and Sara Russell
　　　　　　　　ISBN: 1-58381-176-1

Vol. No. 213　　BIOASTRONOMY 2002: LIFE AMONG THE STARS
　　　　　　　　eds. Ray P. Norris and Frank H. Stootman
　　　　　　　　ISBN: 1-58381-171-0

Vol. Nol 215　　STELLAR ROTATION
　　　　　　　　eds. André Maeder and Philippe Eenens
　　　　　　　　ISBN: 1-58381-180-X

Vol. No. 217　　RECYCLING INTERGALACTIC AND INTERSTELLAR MATTER
　　　　　　　　eds. Pierre-Alain Duc, Jonathan Braine and Elias Brinks
　　　　　　　　ISBN: 1-58381-166-4

Vol. No. 218　　YOUNG NEUTRON STARS AND THEIR ENVIRONMENTS
　　　　　　　　eds. Fernando Camilo and Bryan M. Gaensler
　　　　　　　　ISBN: 1-58381-178-8

Vol. No. 219　　STARS AS SUNS: ACTIVITY, EVOLUTION AND PLANETS
　　　　　　　　eds. A. K. Dupree and A. O. Benz
　　　　　　　　ISBN: 1-58381-163-X

Vol. No. 220　　DARK MATTER IN GALAXIES
　　　　　　　　eds. S. D. Ryder, D. J. Pisano, M. A. Walker and K. C. Freeman
　　　　　　　　ISBN: 1-58381-167-2

Vol. No. 221　　STAR FORMATION AT HIGH ANGULAR RESOLUTION
　　　　　　　　eds. Michael Burton, Ray Jayawardhana and Tyler Bourke
　　　　　　　　ISBN: 1-58381-161-3

Vol. No. 13　　 HIGHLIGHTS OF ASTRONOMY
　　　　　　　　ed. Oddbjorn Engvold
　　　　　　　　ISBN: 1-58381-189-3

Ordering information is available at the beginning of the listing